LOOKING UP

LOOKING UP
A History of the Royal Astronomical Society of Canada

R. PETER BROUGHTON

DUNDURN PRESS
Toronto & Oxford

Copyright © The Royal Astronomical Society of Canada, 1994

All rights reserved. No part of this publication may be reproduced, stored in a retrieval system, or transmitted in any form or by any means, electronic, mechanical, photocopying, recording, or otherwise (except brief passages for purposes of review) without the prior permission of Dundurn Press Limited. Permission to photocopy should be requested from the Canadian Reprography Collective.

Editor: Leslie Derbecker
Printed and bound in Canada by Friesen Printers

The publisher wishes to acknowledge the generous assistance and ongoing support of the **Canada Council**, the **Book Publishing Industry Development Program** of the **Department of Canadian Heritage**, the **Ontario Arts Council**, the **Ontario Publishing Centre** of the **Ministry of Culture, Tourism and Recreation**, and the **Ontario Heritage Foundation**.
  Care has been taken to trace the ownership of copyright material used in the text (including the illustrations). The author and publisher welcome any information enabling them to rectify any reference or credit in subsequent editions.

J. Kirk Howard, Publisher

Canadian Cataloguing in Publication Data

Broughton, R. Peter, 1940–
    Looking up : a history of the Royal Astronomical Society of Canada

Includes bibliographical references and index.
ISBN 1-55002-208-3

1. Royal Astronomical Society of Canada – History.
2. Astronomy – Canada – Societies, etc. – History.
I. Title.

QB1.B7 1994   520'.6'071   C94-930819-6

Dundurn Press Limited	Dundurn Distribution	Dundurn Press Limited
2181 Queen Street East	73 Lime Walk	1823 Maryland Avenue
Suite 301	Headington, Oxford	P.O. Box 1000
Toronto, Canada	England	Niagara Falls, N.Y.
M4E 1E5	0X3 7AD	U.S.A. 14302-1000

TABLE OF CONTENTS

	PREFACE AND ACRONYMS		vii
1.	HOW I WONDER WHAT YOU ARE	The Society, its Name, Objectives, By-laws, Characteristics and Governance	1
2.	THE CENTRE OF THE UNIVERSE	The origins of the Society in Toronto, and the headquarters of the RASC	19
3.	GIANTS AND DWARFS	Classes of membership, Women's Issues, and Professional/Amateur mix, Award winners	35
4.	BRIGHT LIGHTS	Officers	63
5.	KEEPING A BALANCE	Income and expenditure, Grants, Funds and Donations	73
6.	OF VOLUMES AND SPACE	Library	83
7.	THE THREE-BODY PROBLEM	The Journal, Observer's Handbook and Bulletin	91
8.	ENCOUNTERS OF ALL KINDS	Public Lectures, Shows and Displays, Star Nights, Astronomy Day, Youth Programs, Media	117
9.	THE SCOPE OF OBSERVING	Star Parties, Light Pollution, Solar System, Stars and Nebulae, Occultations and Eclipses	131
10.	CLUSTERS	Members' Meetings at Centres, the Annual Meetings and General Assemblies	163
11.	SATELLITES AND THE FIRST STEPS BEYOND	Early Expansion in Ontario, Ottawa, Peterborough, Hamilton and Guelph	177
12.	NEW FRONTIERS	Winnipeg, Regina, Edmonton, Calgary, Saskatoon, Victoria and Vancouver	197
13.	FROM WEST TO EAST	Montreal, Quebec, Halifax and St John's	227
14.	THE ENTERPRISE RETURNS	The Other Ontario Centres and back to Toronto	251
	FINAL THOUGHTS		277
	GENERAL BIBLIOGRAPHY		278
	NOTES		279
	INDEX		283

Preface

The motivation for writing this book was the centenary of the Royal Astronomical Society of Canada, an event which was celebrated in 1990. As this significant date approached, the ruling body, or Council, of the RASC set up a special fund and appointed a Centennial Committee to look into appropriate ways of marking the occasion. The most enduring outcome was the inauguration of a new publication, *The Beginner's Observing Guide*, aimed primarily at young people, and thus looking toward the society's future. In observance of the year itself, a public symposium entitled "You and the Universe" was held on June 30, and the five outstanding speeches delivered that day were published in the October, 1990 issue of the Society's *Journal*. Astronomy Day, an annual event intended to focus public attention on the universe beyond the Earth, was extended to an ambitious Astronomy Week. Many of the local chapters, or Centres of the Society, put on programs which attracted strong public interest, the Toronto Centre winning an award from the well-known American astronomy magazine, *Sky and Telescope*.

The committee, myself included, did consider the propriety of arranging for a history of the Society to be written and published, and decided against it! We knew that some articles had been written on this subject over the years and felt that they were still available for anyone interested enough to retrieve them from back issues of the Society's *Journal*. It was not until the year 1990 was almost upon us that a gnawing feeling began to emerge that many members, particularly the newer ones, had some interest in the development of the Society but did not have easy access to those older papers. The more recent advancement of the Society was also due for some attention. So, in a moment of self-delusion, I offered to write a book about the RASC, a proposal which the Centennial Committee and the Council very generously supported.

There was also a growing realization that the earlier histories generally had little to say about the individuals who were prominent in the growth of the RASC. The strength of any organization is largely that of its leaders and most active members, and so brief vignettes of all former presidents and winners of the Service Award or Chant Medal will be found scattered at what I hope are appropriate places throughout the book. These are the people who have made an extensive commitment to the Society or who, as amateurs, have advanced astronomy through their investigations over a period of years. I am pleased that Nominations and Awards Committees of the past have made these selections and that I can hide behind their shield if anyone wants to shoot arrows of outrage at my oversights and immodesty. I also made the decision not to include biographies of award winners after 1990, since their contributions are fresh in our minds and properly belong in a history of the Society's second century. There are, of course, hundreds of other members who have made heavy commitments of time and talent. They should not be forgotten, though their names may not appear here. My only defence is to say that I have tried to write a book, not a directory. I would be very pleased if this history prompted the nomination of worthy candidates not already recognized through awards.

The Royal Astronomical Society of Canada has had a remarkable career, unique in many ways. Its story deserves to be known beyond the ranks of its own members. Canadians should be better aware of its wonderful contribution to the cultural and scientific life of the nation. I hope that organizations of all kinds will find inspiration and practical ideas from its progress and the experiences of some of its members, and that readers, whatever their interests, will see parallel opportunities for enriching their own lives.

In stating these aspirations, I am acutely aware of the honour and responsibility I have accepted by undertaking to write this book. The task has been made both easier and more difficult by the great volume of source material which the Society retains — over one hundred years of publications, minutes, correspondence and other files. One area where the RASC Archives were lacking was in photographs, particularly pictures of illustrious members of the past. So I am especially pleased to have gathered hundreds of photographs; many of them are published here but all will enrich the Society's Archives and, I hope, bring pleasure to others exploring the Society's history. A number of photographs in this book originally appeared in the Society's publication. Those which are specifically noted as having been previously published in the *Journal* are reproduced here from original prints in the David Dunlap Observatory Archive with the co-operation of the Director. Negatives made from these prints are now kept in the RASC Archives, with the permission of the Observatory Director.

The author is indebted to a great many people for help of one sort or another. All those members featured on special pages who are still living have been given the opportunity to read and correct their write-ups. To them I am very grateful for assistance, photographs and interesting letters. Though I will not list their names here, many of these "special people" willingly read portions of the manuscript in addition to their own biographies and made constructive suggestions. Many other present members of the Society, including George Ascroft, Louie Bernstein, Bob Bishop, Eric Clinton, Terence Dickinson, David Dodge, Otmar Eigler, Mary Anne Harrington, John Howell, Bert Huneault, Pat Kelly, Ray Koenig, Réal Manseau, Robert May, Roger Nelson, Zdenko Saroch, Ernest Seaquist, Steven Spinney, Walter Stilwell, Paul Sykes, Joady Ulrich, Jean-Pierre Urbain, Mary Lou Whitehorne, Bert Widdop, Garry Woodcock and Scott Young have cheerfully supplied photographs, information and encouragement. Their kindness has made this project a wonderful experience. In addition there are relatives of former members whose cooperation I wish to acknowledge — Mrs C.E. (Jackson) Cansfield, Mlle Nolita Coallier, Prof. Ian Dalton, Mrs Frances (Asbury) Jamieson, Mrs Gwen (Clark) Newton, Martine Simard Normandin and Clifford Oliver. Many non-members have been very generous with their time: Rollande Chassé, Dave Duncan, Rosemary Freeman, Louise Herzberg, Karl Kamper, Huberte Palardy, William Peters, Fernand Richard, Guy St-Denis, Harry Turner, Tom Williams and many librarians, archivists and secretaries who have had had to tolerate numerous inquiries. The Ontario Heritage Foundation, an agency of the Ministry of Culture and Communications, provided generous financial assistance. Finally, I thank my family whose tolerance allowed me to pursue this absorbing project.

While pondering a century or more of history, one inevitably thinks of what is to come. I hope, as the title suggests, that research into the past will be a guide and source of optimism for the future.

ACRONYMS

Many other scientific and astronomical institutions and societies are mentioned in this book. Their names in abbreviated form are listed here for convenience, and also as a means of indicating some of the background and other information relevant to the history of the RASC. Members of the Society have played important leadership roles in many of these organizations.

A&P: The Astronomical and Physical Society of Toronto was a fore-runner of the RASC.

AAAS: The American Association for the Advancement of Science met in Montreal in 1857 and 1882, Toronto in 1889, 1921, and 1981, and in Ottawa in 1938. At least in 1921, the RASC helped with local arrangements and refreshments.

AAS: The American Astronomical Society traces its origins to a meeting of astronomers at the dedication of the Yerkes Observatory in 1897. This society of mainly professional astronomers held its first Canadian meeting in Ottawa in 1911, including an outdoor paper session in the Gatineau Hills. The AAS next met in Ottawa in 1929 when there was a joint session with the Ottawa Centre of the RASC and yet again in 1949. Many RASC members attended AAS meetings in Toronto in 1935, shortly after the opening of the DDO. The AAS meeting in Toronto in 1959 was their largest up to that time, with 200 astronomers and their families in attendance and the RASC assisted with refreshments for the delegates. In 1968, the fiftieth anniversary of the DAO brought the AAS to Victoria. Reports and group photos of many AAS meetings appeared in the RASC *Journal*. The AAS now comprises about 5,300 members.

AAVSO: The American Association of Variable Star Observers was begun in 1911 as a means of co-ordinating observations made by amateur astronomers. The AAVSO met in Toronto in 1940 and 1965, Montreal in 1957, Ottawa in 1961, Quebec in 1967 and 1983, and Winnipeg in 1974. Their annual reports and column, "Variable Star Notes," were published for many years in the *Journal*. Six AAVSO presidents were Canadians: F. DeKinder, G. Fortier, C. Good, H.S. Hogg, J.R. Percy, D.W. Rosebrugh. In recent years, a couple of dozen Canadians typically contribute about 10,000 variable star observations annually, about 10 percent of the United States total and about 5 percent of the worldwide total.

AGAA: L'Association des Groupes d'Astronomes Amateurs incorporated in Quebec in 1976 as an association of amateur astronomy clubs in the province. Their publication, *Magnitude Zéro*, came out quarterly until an agreement was reached with SAM to produce *Le Québec Astronomique* on a monthly basis in 1980. By 1993, AGAA was replaced by La Fédération des Astronomes Amateurs du Québec (FAAQ) comprising 32 clubs and 650 members.

ASP: The Astronomical Society of the Pacific, one of the largest astronomy groups in the world today with about 6,000 members, was established in California in 1889. Its founder and first director of the Lick Observatory, Edward S. Holden, sent best wishes and publications of the ASP to the A&P Society on the occasion of its incorporation in 1890. Following its Annual Meeting in Seattle in 1936, the ASP held an extra paper session in Victoria which was enjoyed by many RASC members. The ASP returned to Victoria in 1962 and in 1988 for its hundredth annual meeting. In addition to its technical *Publications*, its popular periodical, *Mercury*, and a quarterly newsletter aimed at

Photograph of Section D of the AAAS on the steps of the DO, 29 June, 1938. Included are three gold medal winners of the RASC: Millman, Harper and McDiarmid; seven RASC Presidents: Stewart, Millman, Harper, Northcott, Collins, R DeLury and Jackson; Reinhardt who was a benefactor, and Burland, a winner of the Service Award.

KEY TO PHOTOGRAPH

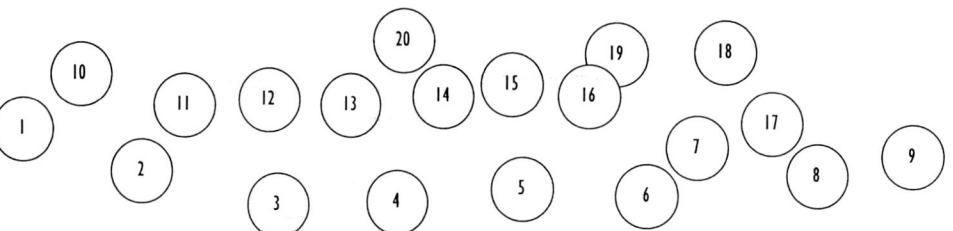

1. G.H. Ling, Saskatoon
2. J.P. Henderson, Ottawa
3. D.B. Nugent, Ottawa
4. J. Pawling, Washington
5. R.M. Stewart, Ottawa
6. H.T. Stetson, Cambridge
7. P.M. Millman, Toronto
8. W.E. Harper, Victoria
9. R.J. McDiarmid, Ottawa
10. J.B. Irwin, Washington
11. C. Reinhardt, Perth
12. R.J. Northcott, Toronto
13. M.S. Burland, Ottawa
14. J.R. Collins, Toronto
15. R.E. DeLury, Ottawa
16. J.L. O'Connor, Ottawa
17. W.E.W. Jackson, Ottawa
18. E.C. Arbogast, Ottawa
19. C.A. French, Ottawa
20. W.S. McClenahan, Ottawa

Biographical sketches like the following one will accompany many of the photographs throughout the book.

WALTER ERASTUS WEDDEL JACKSON (1880-1962) was born in Newmarket, Ontario. His mother's family name was Weddel; his father was Editor of the *Newmarket Era*. After attending local schools, he went to the University of Toronto where he received his BA in 1905 and MA in 1906. Even before graduating he joined the Meteorological Service in 1904 and the RASC in 1905. Jackson worked for a time at the Toronto Observatory, and made measurements of double stars and other observations with the 15-cm refractor. He then served as meteorologist with Captain Bernier in the first official Canadian explorations of the Arctic in 1908-09. This led to some papers in the *Proceedings* of the Royal Society of Canada and an address to the RASC. His special area of expertise was in terrestrial magnetism, at that time a vital part of the Meteorological Service and in the course of his duties he travelled in Hudson Bay and Strait, worked at the Magnetic Observatory at Agincourt, Ontario, and established the Magnetic Observatory at Meanook, Alberta, in 1916. Jackson became Assistant Director of the Meteorological Service in 1929, and in 1936 was transferred to the Dominion Observatory in Ottawa where he was in charge of all the magnetic work done at Canadian observatories.

Within the RASC, he was especially active in the 1920s when he spoke at a number of meetings, published several papers and reports in the *Journal* and served as Vice-President and President. Other professional connections included membership in the International Meteorological Organization and the Geodetic and Geophysical Union.

BAA: The British Astronomical Association was founded in 1890 primarily for amateur astronomers as the RAS had become largely a professional Society. Co-operation between observers in both the BAA and RASC has occurred intermittently during their century of co-existence. The two Societies have about the same number of members.

BAAS: (BA for short): The British Association for the Advancement of Science began in the 1820s. Partly as a result of a BA report in 1838, magnetic observatories were established in four British colonies, including one at Toronto in 1840. The Association had a number of meetings in Canada which brought renowned scientists like Kelvin and J.J. Thomson, to our young country. Montreal was the site in 1884, Toronto in 1897, Winnipeg in 1909 and Toronto again in 1924.

CASCA (CAS originally): Canadian Astronomical Society/Société Canadienne d'Astronomie. The idea for a professional association of Canadian Astronomers was first voiced publicly by R.M. Petrie in 1964. A Sub-Committee of the National Committee for Canada of the IAU was formed in 1969 but there was no strong support for the idea of a professional society, and the Sub-Committee recommended no action be taken. In the meanwhile, an umbrella association of Canadian scientific societies (SCITEC) was being planned, and the professional astronomers realized that they had better form a society if they wished to have an adequate voice in that organization. Ultimately CAS was established and held its inaugural meeting in Victoria in 1971. On that occasion, a message of best wishes and congratulations was sent by the RASC and a spirit of co-operation has been maintained between the two societies in a number of areas including the Plaskett medal, the Hogg lecture and the publication of CASCA Abstracts in the *Journal*. There are currently over 400 members of CASCA, of whom about one-fifth belong to the RASC.

DAO: Dominion Astrophysical Observatory, near Victoria, BC, was opened by the federal government in 1918.

DDO: David Dunlap Observatory, part of the University of Toronto, was opened in 1935.

DO: The Dominion Observatory in Ottawa was the first real centre of astrophysical research in Canada when it opened in 1905.

GA: General Assembly of the RASC, an annual convention.

IAU: The International Astronomical Union now comprising about 7,000 members began as the International Union for Co-operation in Solar Research in 1904. The RASC applied for admission in 1908, and sent representatives to the 1910 meetings at Mt Wilson. The IAU formed in 1919, and held the first of its triennial international assemblies in 1922. Canada, like all adhering countries, had a National Committee for the IAU to decide who its delegates would be and to discuss matters requiring international action. The RASC appointed five representatives to the National Committee until 1952, and only one afterwards. That policy was continued, even after the duties of the National Committee were subsumed by the NRC's Associate Committee on Astronomy, until a reorganization by NRC in 1989 terminated any direct voice in national or international astronomical affairs by the RASC. The 1979 IAU General Assembly was

teachers, the Society provides a very useful service by selling slides, posters and other educational materials.

held in Montreal and a number of special symposia were held in London, Ottawa, Toronto and Victoria. The RASC mounted a special display of its activities for the Montreal meeting and made a $500 grant (later returned and put into the Beals fund to help astronomers travel to international meetings). SAM organized a special evening of public talks and observing. In 1988, the Society contributed $500 to enable a delegate from Thailand to attend the IAU Colloquium in Williamstown, Massachusetts on on the teaching of Astronomy.

IUAA: International Union of Amateur Astronomers. The idea of such an organization was first suggested by the well-known British amateur, Patrick Moore, at the IAU meetings in Prague in 1967. RASC member, Ken Chilton, was their first secretary. Their third congress, in 1975, was held at his invitation in Hamilton. Dr Kennedy O'Brien of Newfoundland was later a president. The Union failed after a few years largely because of financial and language problems.

NFCAAA: Niagara Frontier Council of Amateur Astronomical Associations was established in New York State in 1968. Many RASC Centres in southern Ontario have held joint meetings with this group.

NRC: National Research Council. Upon the closing of the DO, NRC assumed responsibility for government astronomy in 1970 and the NRC Associate Committee for Astronomy subsumed the responsibilities previously exercised by the National Committee for the IAU.

RAS: Royal Astronomical Society (of London, England) was established in 1820. It began to exchange publications with the RASC in 1892.

KENNETH E. CHILTON (1939-1976) was a teacher by profession. He joined the Hamilton Centre in 1966 and during his short life, served as Director of Observations, Treasurer, Editor of *Orbit*, Vice-President, and President. He received the national Service Award in 1976.

His expertise as an educator came to the fore in a series of Cable TV programs which he started in 1971, called *The Sky Tonight*. Thirty-eight programs were aired in the next year.

As an observer, he made magnitude estimates for AAVSO from 1966-73. He took part in visual meteor programs, searched for transient lunar phenomena, timed a transit of Mercury, sketched Venus and observed its polarization, wrote articles on observing Mars and observing asteroids, found a correlation between the intensity of Jupiter's South equatorial belt and the position of the satellite Callisto, and studied Saturn's belt zones and rings with his 32 cm Gregorian reflector. Chilton was the obvious person to serve as the national Co-ordinator of Planetary Observations in the late 60's and early 70's when the Society attempted to systematize observations from active members across the country. Under his direction, the Planetary Section was one of the most active in the program. He was also the Chairman of the Co-ordinating Committee for the 1972 Solar Eclipse, visible in many parts of Canada.

Ken Chilton took a great interest in the organization of amateur astronomers. He was very eager to promote liaisons between RASC Centres in Ontario and clubs in Western New York State, and served a term as chairman of NFCAAA. But it was as President of the International Union of Amateur Astronomers that he received the recognition which he undoubtedly relished most. The IUAA was founded in 1969 as a sort of counterpart of the professional IAU, and indeed was recognized by that body and affiliated with it.

Although he endured several years of illness, his energy and enthusiasm were an inspiration to everyone. The Kenneth E. Chilton Prize was established as a tribute to his accomplishments and his spirit.

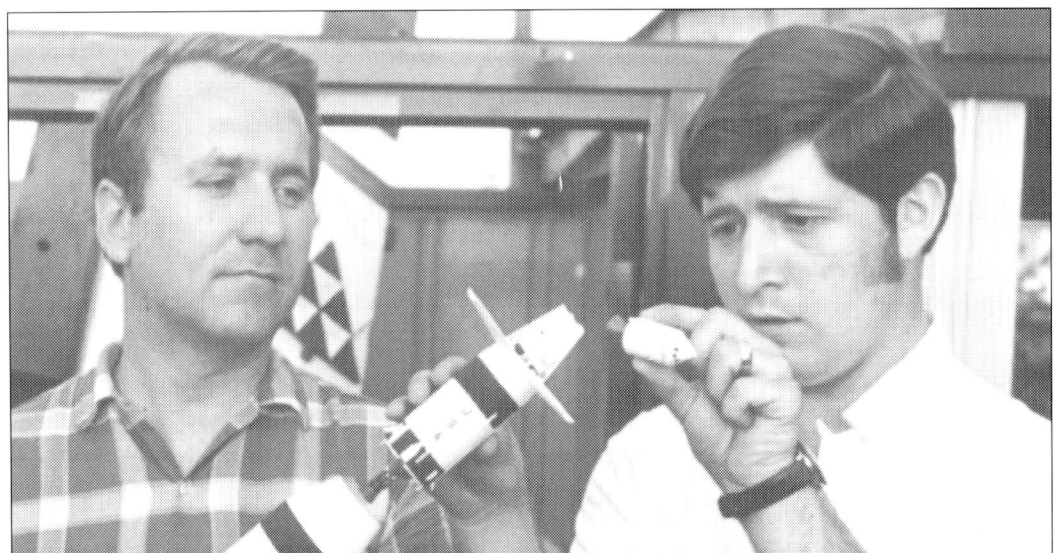

Ken Chilton (l) and Robert Speck simulate the Apollo-Saturn 11 separation at the time of the first lunar landing, July, 1969.

RASC: The Royal Astronomical Society of Canada.

RCI: Royal Canadian Institute was originally established in Toronto in 1849 as the Canadian Institute and their *Canadian Journal* occasionally contained some astronomical papers. Andrew Elvins joined in 1860, and over the years many who were prominent in the RCI were also leaders in the RASC. For nearly fifty years, the Society was a tenant of the Institute. Today, as in the past, the RASC Toronto Centre and the RCI jointly sponsor a lecture by some outstanding speaker nearly every year.

RSC: Royal Society of Canada, founded by the Governor General, the Marquess of Lorne, in 1882, admits only a limited number of the country's most distinguished scholars. The A&P Society, and later the RASC, sent representatives to meetings of the RSC until the 1930s. In 1931, there was a joint session of Section III of the RSC with the RASC in Toronto. Papers presented at a symposium sponsored by the RSC in 1973, "Chemical Evolution of the Universe," were published in the *Journal*, and the centennial of the RSC was marked by an article by Helen Hogg.

SAF: La Société Astronomique de France was established in 1887 by Camille Flammarion. In a historical account which appeared on its sixtieth anniversary, it was stated that the SAF had a distinct influence on the founders of our Society. SAF's publication *L'Astronomie* (started in 1882) was exchanged for our *Journal* from 1900 on. It was recently reported that SAF has a membership of about 5,000 including 100 professional astronomers.

SAM: La Société d'Astronomie de Montréal was legally incorporated in 1968. It grew out of the RASC's Centre français de Montréal.

SRAC: La Société Royale d'Astronomie du Canada became the French half of the official name of the Society when it was federally incorporated in 1968.

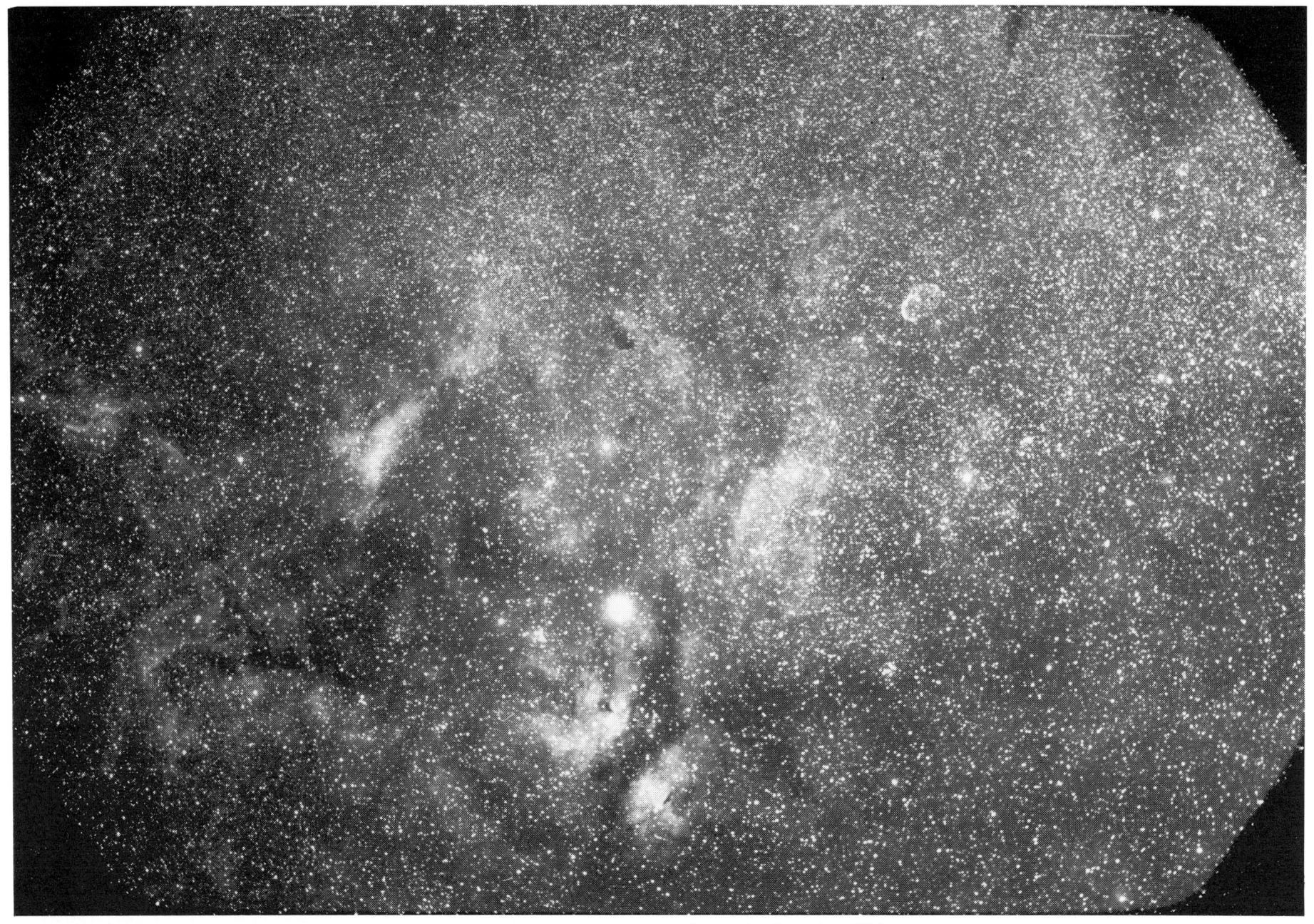

The nebulous region of the Milky Way near the bright star Gamma Cygni. Photographed by Damien Lemay, 5–6 September, 1991. North is at the left and east is at the bottom.

CHAPTER 1

How I Wonder What You Are

The RASC is an association of over 3,000 members. Relative to national population, there are few, if any, other astronomical societies in the world with as large a following. Those who belong to the RASC have diverse interests and levels of commitment which vary from participant to professional. In a sense, all members are amateurs since they love astronomy. The fact that some are paid to do research or to teach does not diminish their underlying devotion to their calling. Nonetheless, there is a recognized division between the small group who have the perseverance, specialized education and success in the scientific world to be called professionals and the great majority of members who call themselves amateurs. The linkage which the Society fosters between these two perceived groups is one of its great strengths. The Society also excels at affording diverse opportunities for members to serve the public. As an organization, the RASC thus exemplifies what Robert Stebbins, a sociologist at the University of Calgary, calls a professional-amateur-public (P-A-P) system of interrelationships.

Stebbins has made a study of the role of amateurs in fields as disparate as art, science and sport and finds many parallels in the interaction of amateurs with their professional counterparts, and in their relationships with their families and occupations. He has examined the amateurs' attitudes, what they see as the rewards and costs of their avocation, and the contributions they make to individuals, the profession and to society. For instance, an actor in a local troupe may have first-hand knowledge of a play being staged professionally and may be able to appreciate the nuances of pacing and stage direction just as an amateur astronomer may be able to relate to certain technical problems underlying scientific research. An amateur archaeologist unearthing a prehistoric relic experiences the same thrill as an astrophotographer who captures an elusive comet on film. A visit to the Hockey Hall of Fame or to the Royal Greenwich Observatory makes a devotee feel an association with the immortals. In all areas of strong amateur involvement, marital and family life may flourish or suffer. Some spouses share in their mate's enthusiasm, some find it best to cultivate independent interests, and some resent the intrusion of another passion into their lives.

But for many amateur astronomers, there is a unique dimension to what they are doing. There is something fundamentally uplifting and meaningful in probing the universe. It's hard to say from a child's point of view whether a first look at Saturn through a telescope exceeds the thrill of a first solid hit with a baseball bat. Both experiences can also be pretty satisfying to the adult who has helped make them happen. But we do like to think that something more broadening, or civilizing, or permanent will result from looking beyond ourselves, from encouraging a curiosity about the larger universe.

The reader will find in the brief biographies in this book that an interest in astronomy can be awakened at any age and in many ways. For those with mechanical or technical skills, satisfaction may come

CHARLES CARPMAEL (1846-94) was born in England and graduated in Mathematics from Cambridge. One of a large family, he and two brothers were fellows of the RAS. He came to Canada, commencing work with the Meteorological Service as G T Kingston's Assistant in 1872 and eventually succeeding him as Director in 1880. One of his first duties was to investigate a violent storm which caused much destruction along the southeast shore of Nova Scotia in 1873.

Carpmael had some astronomical background. He was a pupil of Charles Pritchard, later Savilian Professor of Astronomy at Oxford, and he participated in a British eclipse expedition to Spain in 1870 where he made spectroscopic observations of the corona. In Canada, astronomy was not often part of Carpmael's official duties but the Dominion Government did place him in charge of the transit of Venus observations in 1882 and he did determine the longitude of the Toronto Observatory in 1888. Within the Society he was always willing to act as an advisor, and welcomed members to the Observatory on a few occasions. But he was not in good health and only chaired 11 of the 125 meetings which were held during his four-year Presidency. He returned to England where he died at the age of 48.

LARRATT W. SMITH (1820-1905), though born in England, was educated in Toronto at Upper Canada College and at the University. His father had been with the Royal Artillery in Canada during the War of 1812 and he himself served with the militia during the rebellion of 1837.

Larratt W Smith had a distinguished career as a lawyer, heading his Toronto firm and serving as the Clerk of the Court of Appeals. He chaired a Royal Commission, was Vice-Chancellor of the U of T twice, was a Director of a number of companies and President of Consumers Gas. He had a large property and house named "Summerhill" near Yonge Street, between Bloor and St Clair. Astronomy was not a big part of his life, though he did write on the daily headings of his diary, notes on the phenomena of Jupiter's satellites, the minima of Algol and so on. Infirmities of old age prevented him from chairing more than a couple of meetings during his Presidency in the Society.

JOHN A. PATERSON (1846-1930) was born in Stornoway, Scotland, but emigrated to Canada as a young boy when his father accepted a post as pioneer Presbyterian missionary at Bobcaygeon, Ontario. Graduating from Upper Canada College with a double scholarship in classics and mathematics, he proceeded to get his MA from the University of Toronto. After a period of teaching mathematics at UCC, he turned to law in 1873, a profession which engaged him for the rest of his life. He was solicitor for the U of T, North American Life and the Presbyterian Church in Canada. One of his most memorable cases was a rigorous and successful defence of the Lord's Day Act which severely limited activities on the Sabbath. He was also an influential promoter of the establishment of the United Church in Canada.

He joined the Society in 1890 and contributed several papers and addresses, some of a biographical nature and some dealing with astronomical allusions found in poetry. During his Presidency he travelled to Ayr and Elora to deliver popular lectures.

The first six presidents of the Society, before it became the RASC, are featured on these pages. They were not professional astronomers but all were men of importance in the community and lent a measure of prestige to the fledgling Society, especially in a time when social status was definitely something to be cultivated.

ARTHUR HARVEY (1834-1905) was a polymath, publishing papers on the grain trade, the reciprocity treaty, the Canadian Census of 1871, metrication, botany, geology, pathology, anthropology, philosophy, physics and, of course, many on astronomy.

He came to Canada in 1856 after education in England, France, Holland and at Trinity College, Dublin. After working for a number of newspapers in Hamilton, Montreal and Quebec (where he married), he began a career as a statistician with the government of Canada 1862-70. During his time in Ottawa, Harvey also founded the *Year Book and Almanac of British North America* in 1867 was active in the Natural History Society, serving as its Secretary in 1867-68. He then moved to Toronto to become manager of an insurance company and president of a loan and land company.

His astronomical interests were very broad also. He saw and wrote about Donati's Comet of 1858 and the Comet 1861 II. He wrote on the parallax of the aurora and was invited to contribute his views to *Nature*. He spoke at meetings on several occasions on topics including eclipses, the telescope, observatories, Roentgen's discoveries, the synchronism of northern and southern auroras, meteors and meteorites and contributed 25 papers to the Society's publications on an equally dazzling array of subjects. His investigations into solar-terrestrial relationships led to his election as Honorary President and Director of the Institutio Solar Internacional Montevideo, Uruguay. He claimed to have discovered the emission in solar radiation of negatively charged particles and to have been the first to announce a 27.5 day periodicity in magnetic disturbances on earth.

GEORGE E. LUMSDEN (1847-1903) was the first Canadian-born President of the Society. His father was an Anglican minister, and it was from him that he acquired his love of astronomy. After a career in journalism, George Lumsden entered the Ontario Civil Service in 1880, rising to become Assistant Provincial Secretary, an office comparable to Deputy Minister. In spite of the heavy demands of this position he was devoted to the Astronomical Society, serving as Secretary, Editor and President. He had papers published by the Royal Society of Canada, in *Popular Astronomy* and in *Scientific American*. Fifteen of his contributions appear in the *Transactions* of the Society. A neighbour and friend said, "He was always ready to point out the great stars and constellations and to aid the visitor in using the telescope on his premises. His gentle enthusiasm, as he dwelt on the beauty and magnificence of the firmament, induced many to pursue such charming studies."

ROBERT FREDERIC STUPART (1857-1940) joined the permanent staff of the Canadian Meteorological Service as a map-drawer when he was only 15. Who would have thought that he would someday have a bay named for him on Hudson Strait, become a Fellow, and later President of Section III of the RSC, be President of the RCI, an honorary life member of the Royal Meteorological Society of London and be knighted Sir Frederic?

As Carpmael's successor as Director of the Meteorological Service, he graciously agreed to stand as Vice-President of the Society in 1896, becoming President in 1902. He contributed dozens of articles and served the RASC as Associate Editor. Sir Frederic spoke to Toronto (Centre) a number of times in the 1920s and served as their first Honorary President from 1930 to 1934.

(Photo credits: for Smith: Metropolitan Toronto Reference Library, T13792; for Paterson: Archives of the United Church of Canada, Victoria University, P 5040; for Stupart: City of Toronto Archives, J 2340.)

Ray Thompson shows his telescope to his seven-year-old son, David, in 1963, in his first observatory at Maple, just north of Toronto. David is now a laser physicist working on fusion at the Los Alamos Laboratory in New Mexico.

RAYMOND R. THOMPSON (1923–) remembers the marvellous views of the night sky he had from the darkened deck of a ship in mid-Atlantic when he was just fourteen. His father, with the British Admiralty, had been transferred to Bermuda, and the binoculars he had been given as a parting gift by friends in England gave young Ray his first taste of what lay beyond. Eventually Ray came to Toronto and took up his career as teacher and musician. Contact with the RASC and Jesse Ketchum rekindled his interest, and he was soon active in the Society. President of the Toronto Centre in 1963–64 and director of observations in 1965, Ray Thompson won the Chant Medal in 1967 for outstanding work both in instrumentation and observational astronomy, particularly solar, lunar and planetary sketching.

Thompson started his second observatory as a Centennial project in 1967 and equipped it with twin refractors – a 15-cm f/10 built by himself and a 20-cm f/15. These he found ideal not only for his solar system observations but also for visual photometry of variable stars. The smaller instrument was useful as a finder and for the brighter variables, while the larger instrument enabled him to reach 13th magnitude until his neighbourhood was drowned by streetlighting. He built a new two-storey observatory in 1971, housing the refractors under a 3.6-m dome and a 10-cm polar-axis telescope which was especially useful for the daily sunspot plot. A 25-cm Schmidt-Cassegrain 'scope replaced the refractors in 1991.

Over a period of thirty years Thompson has reported nearly 10,000 visual estimates of variable stars to the AAVSO. In recent years, he has used a solid-state photometer coupled to a computer which does all the reductions and statistical analysis, and his work has been especially useful in a program to detect and confirm small variations in red giants.

from building their own telescopes, or working with computers. Others may take a more passive approach, but get just as much pleasure from reading about quarks and quasars, and how both the very small and the very large are interrelated in an understanding of the early moments of the universe. Some, like hikers, will simply enjoy the exhilaration of following a trail that few others know, and will revel in the view of the night sky, and a very few will discover objects never seen before. Those with a historical or philosophical bent may find the growth of ideas and the concepts of scientific revolution and evolution fascinating. Travellers may find the incentive of a solar eclipse gives a purpose to their journeys, or visits to observatories a focus to their understanding.

A strong astronomical society has the resources and good management to stimulate and encourage all of these activities and more. Its strength is surely measured by the growth and happiness of its membership and on both counts the RASC seems to be doing well.

NAME

Social attitudes and political realities have influenced the Society over the years – even its name. Beginning as a group of friends in 1868, the Toronto Astronomical Club became a Society the following year. The style which was adopted about 1884, namely "The Astronomical and Physical Society of Toronto" continued when the Society became incorporated in Ontario in 1890. George Lumsden, one of the charter members and the corresponding secretary in the 1890s, tried on more than one occasion to have the name changed. In the Victorian era, refined people were certain, at least in public, that they did not belong to the animal kingdom and took pains never to discuss or display anything to do with the body or its functions. The story was that Mrs Lumsden objected to the use of the word "Physical" on these grounds and her husband did what he could to eliminate the embarrassment. In May, 1900, the year in which he became president, Lumsden got his way, and the official title became The Toronto Astronomical Society as it had originally been in 1869. The name only lasted for three years. It was really too restrictive as there were by this time many members outside Toronto, and three or four astronomical clubs in Ontario towns had affiliated with the Society.

In the meanwhile, in September and October, 1901, the Duke and Duchess of York made a coast-to-coast tour of Canada. Everywhere they went, the future King George V and Queen Mary were greeted with waves of pro-imperial sentiment, and members of the Society, feeling very much a part of this tide of support, decided it would be highly desirable to become a "Royal" Society. In the words of the minutes for 1902:

> After considerable discussion concerning the expediency of making application to the Crown for the privilege of styling the Society "Royal" and in the event of doing so whether "of Toronto," "of Ontario" or "of Canada" should be asked for, a committee of council was appointed to consider the whole question.

A petition was drawn up, dated January 7, 1903, signed by President R.F. Stupart, and Secretary J.R. Collins, and sent to His Excellency the Governor General of Canada soliciting from His Majesty the King (Edward VII) the privilege of prefixing the word Royal to the name Astronomical Society of Canada. The reply came in a letter dated February 27, 1903, from Joseph Pope, Under-Secretary of State, and himself a member of the Society. The letter stated that the Governor General had received a despatch from the Secretary of State for the Colonies [sic] acquainting His Excellency that His Majesty the King had been graciously pleased to grant permission to the Toronto Astronomical Society to adopt the title of the Royal Astronomical Society of Canada. Application was thereupon made to the Honorable Chief Justice of the Common Pleas Division of the High Court of Justice for Ontario, Sir William Meredith, to change the corporate name of the Society. This he granted on March 3, 1903.

The Society has gratefully retained the name with minor variants ever since. A change was necessitated in 1968 when the organization sought to become incorporated federally. At that time, the "new" Society had to have a different title from the "old" one so that the assets and members of the provincially incorporated body could be legally

The Duke and Duchess of York arrive at Toronto's City Hall, October 1901.

taken over by the federally incorporated society. For that reason the new name adopted was "The Royal Astronomical Society of Canada - 1968." With two French-speaking Centres in the organization, the title "La Société Royale d'Astronomie du Canada – 1968" was also officially recognized. The change was also appropriate as bilingualism became an official policy of the federal government in 1969. Finally, after a respectable waiting period, the "1968" was dropped in 1973, resulting in the familiar name of former years, but now in its bilingual form.

To outsiders, The Royal Astronomical Society of Canada may sound as remote and aloof as the distant stars. Perhaps images of the elite Royal Society of London come to mind, associated with intellectual icons like Newton and Halley. Though such impressions of the RASC are utterly wrong, the name and the regrettable reputation that science is too difficult for the average person have undoubtedly alienated some. Others may have felt reassured by the regal cachet or even been attracted by a perceived sense of prestige. Nonetheless, the whole thrust of the Society is and always was, to popularize astronomy and to make it accessible to all.

OBJECTIVES

The formal goals of the Society have been spelled out in the various versions of the constitution. At the time of incorporation in Ontario (1890), the objects of the Society were declared to be as follows:

(a) To encourage, advance and popularize the study of Astronomy and Physics, and to diffuse as widely as is practicable, information in those branches of Science.

(b) To publish, from time to time, the results of the work of the Society in the form of Transactions, which shall contain such Notes and Papers as shall have been approved of for publication, and

(c) To acquire and maintain a library and such apparatus and other property, both real and personal, as may be necessary and convenient for the due carrying out of the said objects of the Society or any of them.

When the by-Laws were thoroughly revised in connection with the name change in 1900, the objects were altered, the first one now reading:

(a) To study astronomical and astrophysical subjects, and such cognate subjects as the Society shall approve of and shall, in its opinion, tend to the better consideration and elucidation of astronomical and astro-physical problems, and to diffuse theoretical and practical information with respect to such subjects.

Clauses (b) and (c) were pretty much the same as before.

The change in wording was significant, signalling an alteration of the course to be taken. The emphasis was changed in 1900 from popularizing to studying and was broadened to include cognate subjects, with specific mention being made of theoretical information. These goals survived almost verbatim the major by-law revision of 1908 which was precipitated by the formal recognition of Centres.

A further overhaul in 1944 maintained the objects but streamlined the wording:

(a) To study Astronomy, including Astrophysics, Geophysics and such cognate subjects as shall be approved by the Society; and to diffuse theoretical and practical information with respect to such subjects.

(b) To publish from time to time the results of the work of the Society; and
(c) To acquire and maintain a Library and other property which may be desired for carrying into effect the objects of the Society.

With national incorporation in 1968 came the newly stated objectives which still guide the course of the Society. They are:

(a) to stimulate interest and to promote and increase knowledge in astronomy and related sciences;
(b) to acquire and maintain equipment, libraries and other property necessary for the pursuit of its aims;
(c) to publish journals, books and other material containing information on the progress of astronomy and the work of the Society;
(d) to receive and administer gifts, donations and bequests from members of the Society and others;
(e) to make contributions and render assistance to individuals and institutions engaged in the study and advancement of astronomy.

Notice that the first objective, "to stimulate interest and to promote and increase knowledge in astronomy and related sciences." is more in line with the original statement of 1890 than those of intervening years.

Though a cynic might doubt that the formal wording of objectives would truly reflect the actual operation of the organization, there is in these statements some real correspondence with the direction the Society took. Whether the stated goals were causes or effects is debatable, but the periods when popularizing and stimulating of interests was emphasized, that is the early years and more recent

Malcolm Thomson in the Time Room, National Research Council, 17 October, 1972.

MALCOLM M. THOMSON (1908–), as the son of a minister, moved quite a lot. He was born in Nelson, BC, and grew up in Edmonton and Winnipeg. After earning his BA at the University of Manitoba in 1929, he joined the staff of the DO in Ottawa as a member of the Time Service. There he made his career except for war-time service and a leave of absence to get his MSc degree at Yale University in 1954. Thomson was promoted to head of the Positional Astronomy Division in 1957, chief of the Astronomy Division in 1963, and head of the Time and Frequency Section of the Physics Division at NRC after astronomical work ended at the DO in 1970. Among his responsibilities was the installation of the Photographic Zenith Telescope at Pridis, Alberta, inaugurated at the time of the 1968 General Assembly. His book, *The Beginning of the Long Dash*, published in 1978, gives the whole story of timekeeping in Canada, including many developments in which he participated.

Malcolm Thomson joined the Ottawa Centre shortly after taking up his duties at the DO and held office in the Centre continuously from 1934 to 1948 as Secretary, Vice-President and President. From 1963–72, he moved through the sequence of national presidential offices. During his active years in the Ottawa Centre, Thomson was very helpful in directing teams of meteor observers and in working with groups of young people. He chaired the organizing committee for the 1973 General Assembly. During his national service, he worked tactfully but decisively to smooth out internal problems in the Quebec Centre, and put in a great deal of work preparing a new RASC Constitution prior to incorporation under Federal charter. Malcolm Thomson's substantial contribution to the Society was recognized with a Centennial Medal in 1967 and by the presentation to the RASC national office of an historic railway station pendulum clock in his honour.

Lloyd A. Higgs (1937–) is presently director of the Dominion Radio Astrophysical Observatory in Penticton, BC, where he has worked on such specialized interests as planetary nebulae and supernova remnants, recombination lines in the radio spectrum, and the development of software for data reduction. He was born in Moncton, NB, attended the University of New Brunswick and went to Oxford University as a Rhodes Scholar, receiving his D.Phil. degree in 1961. He then began his career in research with the Radio and Electrical Engineering Division of NRC in Ottawa where he worked until moving to Penticton.

Dr Higgs joined the RASC shortly after moving to Ottawa and took an active part in a number of capacities, including a term as Centre president 1971–72. Always an enthusiastic supporter of the RASC, he contributed many scientific papers and reviews to the Society's publications and edited the *Journal* with care and skill from 1976 to 1980. His election as the Society's Second Vice-President in 1984 led to a term as President in 1988–90, during which time he spoke at meetings of nearly all the twenty-two Centres.

As the citation for Higgs' Service Award stated in 1983, few professional astronomers have been willing or able to commit themselves so whole-heartedly to the Society, yet he always gave the impression that he was the beneficiary. On stepping down from his editorial role, Higgs wrote, "the frequent contacts with enthusiastic amateur astronomers were refreshing experiences which never failed to re-kindle my own flagging spirits." There are undoubtedly many amateur members who feel the same way about their encounters with him.

times, were marked by the greatest involvement of the amateur members as officers, speakers and writers. The time in between was marked by the greatest professional involvement. Not that the Society ever diminished its efforts to encourage public interest in astronomy, but professional concerns were quite naturally more evident during the middle period. Those were the years between the opening of the country's first major observatory in 1905 and the establishment of CASCA in 1971 which subsumed most of the professional astronomers' objectives.

Some further insight into the changing attitudes can be gained from speeches made by presidents and other officers from time to time. Andrew Elvins made it clear in 1891 that the aims of the Society were "to invite into the field of practical amateur work anyone interested in astronomy, and to assist students in their studies of the science." Five years later, President J.A. Paterson said, "It must not be supposed that what we do here is to attract the attention of astronomers but is rather to instruct each other." Lumsden, in 1901, looked for more lasting effects: "May the President of the Society who stands in my place one hundred years from tonight be able to speak to his audience of the usefulness [of the Society to Science] and may he be able to point to some beautiful arch, the keystone of which was put in place by a member of this Society."

The professional astronomers, even in presidential addresses, had very little to say about their perceptions of the Society's importance in their work. One who did was W.F. King, who as Canada's Chief Astronomer, was the first director of the Dominion Observatory. He concluded his 1907 inaugural address as the first president of the Ottawa Section of the Society by saying, "the meeting together of those engaged in study and investigation ... is the end which societies like this should strive to attain." Though the *Journal* was used as a medium for publishing papers, observatory reports and news, RASC meetings outside Ottawa never developed into scientific sessions as King seemed to have in mind. Besides there was really no basis for a national organization of professional astronomers as long as there were only three centres of research in the country – Ottawa, Victoria and Toronto. The need only developed in the 1960s with the burgeoning of universities, the opening of two radio observatories, and the beginning of space-related agencies, and by that time there were other structures in place for meetings of professional astronomers in Canada.

The role of the Society came under close scrutiny in 1976, with the formation of a committee to consider the future and finances of the RASC. Under Dr John Percy's leadership, a very thorough report comprising forty-nine recommendations was prepared. Fifteen years later these proposals seem as sensible as ever. The majority have been acted upon with beneficial results while a few still remain as good ideas. The preamble to the report captures the essence of the RASC just as accurately today as it did then:

> [The membership is] scattered clear across Canada (and elsewhere), comprising professional astronomers and elementary school students, anglophones and francophones, active observers and armchair astronomers, beginners and 50-year veterans. This diversity is both the greatest asset and greatest problem in the Society.

Further brainstorming took place at the 1987 General Assembly when delegates were encouraged to bring innovative ideas concerning future directions and projects to consider. As a result of this work by Dr Lloyd Higgs and his committee, attractive brochures were produced which have proved beneficial in fundraising. The general nature of the Society was summed up as "an organization that is

This 1989 photo shows Toronto Centre President Ralph Chou (l) presenting the Bert Winnearls Award to John Percy for his outstanding contribution to the operation of the Centre.

JOHN R. PERCY (1941–) combines research, education and Society affairs in one very busy life. His stature as a professional astronomer is evident from the dozens of research papers on variable stars he has published, his election as president of the AAVSO in 1989 and his presidency of the Commission on Variable Stars of the IAU.

His contributions to the promotion and communication of science to students and the public are second to none. Dr Percy has written a number of papers on the teaching of Astronomy, and co-edited a book on the subject. He has given innumerable talks at schools and to teachers, to clubs and to the public generally. As RASC President in 1978–80, he spoke to all Centres from St John's to Victoria. Within the Society, he has chaired a committee on adult education, inaugurated "Education Notes" in the *Journal*, was the first editor of the *National Newsletter*, and editor of the *Observer's Handbook* for eleven years. He has been Toronto Centre's Second Vice-President a number of times (a position which he currently holds) and has served the Centre as president. His outstanding contribution to the Society was marked in the presentation of the RASC Service Award in 1977.

Percy's leadership has been vital to a number of other organizations including the RCI, the Editorial Board of *Science Affairs* (a magazine devoted to Canadian Science in the 1960s) and the Science Teachers Association of Ontario.

A University of Toronto man to the core, John Percy received the RASC Gold Medal on graduation in 1962 and subsequently earned his post-graduate degrees there. Since 1967 he has been a member of the faculty as a professor of astronomy based on the Erindale campus where he is now Associate Dean and Vice-Principal.

devoted to the advancement of public awareness and appreciation of all aspects of astronomy." Philosophically, such a statement seems to have brought the Society right back to its initial goal of 1890. Careful reading, however, might suggest a subtle difference. Both here and in the first formal objective stated in 1968, the role of study and work has been played down. Perhaps it is characteristic of our times to avoid reminders that advancement and appreciation come only at the expense of individual effort.

BY-LAWS

The by-laws under which the RASC presently functions were approved at the Annual Meeting of 1989 (with some amendments since) and anyone hoping to have a thorough understanding of the operation of the Society would have to consult those rules. All that can be done here is to give a feeling for the organization without belabouring all the details.

The Society's by-laws have evolved over the years with complete revisions being published in 1893, 1900, 1908, 1944, 1957, 1969 and 1989. Amendments which have come as a response to needs, as has generally been the case, usually have stood the test of time. Instances where some individual's bright idea seems to have won support have normally withered from lack of application and led to eventual removal.

Drafting the by-laws which eventually were published in 1908 was not an easy matter. As an outcome of the establishment of the Dominion Observatory, there was a desire to form an Astronomical Society in Ottawa. A group there met and passed a motion in 1906 that the new organization would be a section of the RASC, but their president, Dr W.F. King, had very definite ideas about the way the Society should govern new Centres such as this. There had been already "affiliated societies" in various Ontario towns but these groups had little input into the operation of the general Society. As King said, "If the Ottawa Section is to be in a position of inferiority, I, as Director of the Government Observatory cannot agree." He and his colleagues felt that the regular meetings of the Society held in Toronto should be no more or less important than those held in some other centre, and proposed that Toronto should be a Section on an equal footing with others like Ottawa. Each would have its own governing body, with an overall general Council elected by the Centres, rather than by the individual members, to regulate the affairs of the Society as a whole. The Toronto group, some of them charter members, felt equally strongly that the Society had originated with them and were opposed to any move which would diminish their status. Paterson especially was afraid that Ottawa's new-found prominence on the astronomical scene might eventually lead to a relocation of the Society's headquarters. After much correspondence, King and his assistant, Plaskett, were invited at the Society's expense to come to a meeting in Toronto to explain their point of view. An agreement was reached and an article was inserted in the new by-laws stipulating that "The Executive Headquarters, the Library, and other property of the Society shall be located in the City of Toronto." It was also agreed that the new constitution would not refer to Sections at all. The wording worked out for the Constitution of 1908 was as follows:

> When, at any Centre, a sufficient number of members of the Society desire to organize regular meetings, the Council may by resolution authorize such meetings. These shall be known as meetings of the Royal Astronomical Society of Canada, and shall be under the control of the Council, which shall make an annual appropriation to defray the cost thereof. Their organization and conduct shall be deputed to a Board of the Members comprising them, who shall be

elected annually at a special meeting called for the purpose.

The only direct representation the Centres got on the (national) Council was that the presiding officer of each Centre was now included along with the Officers, six additional elective members, and all the past-presidents. It was evidently enough to satisfy the Centres.

Over the years, the Centres have increased in number and in representation to the point where the majority of Council members are now Centre representatives. The increasing autonomy of the Centres has also made a uniform set of by-laws more and more difficult to maintain.

CHARACTERISTICS

This decentralized structure is a distinctive characteristic of the RASC. Across the country, spanning six time zones from St John's to Victoria, there are now twenty-two branches of the Society, many of them incorporated in their own province. As long as a Centre has objectives and by-laws which are in accord with those of the Society as a whole, it may take whatever initiatives are necessary to achieve those goals. The Centres are responsible for planning meetings and activities for their own members and in some cases for maintaining a library, telescopes or an observatory and issuing newsletters. Although they retain 40 percent of the fees of their attached members, the Centres in some cases find it necessary to levy a surcharge on their members, to cover the costs of their services. Each Centre is administered by a Council comprising at least a president, secretary, and treasurer, all of whom are elected by members of that Centre. The local by-laws determine the voting procedures, as well as the terms and duties of those elected.

Members do not always belong to a Centre. Some cannot attend any Centre meetings because of

JOHN STANLEY PLASKETT (1865–1941) spent a happy youth on the family farm near Woodstock, Ontario, and did mechanical and electrical work for a number of years before moving to Toronto. There he earned his BA in 1899 while employed as a mechanical assistant in the University of Toronto Physics Department. His name first appears in Society records as assisting Chant with lecture demonstrations. His astronomical career began in 1903 when he accepted a job in Ottawa superintending the installation of equipment at the Dominion Observatory then under construction. Plaskett went on to use the 38-cm refractor to obtain spectra of binaries. He soon faced the limitations of the equipment there and designed a new and faster spectrograph, but in the long run, he knew that a much larger telescope was needed to put Canada in the forefront of astrophysical research. His persistent lobbying of the federal authorities led eventually to the establishment of the DAO under his directorship.

Plaskett's discovery in 1922 of a binary system which had the greatest mass of any then known, attracted a lot of media attention but his pre-eminent work, carried out between 1928 and 1935 in collaboration with J.A. Pearce, confirmed the rotation rate of the Galaxy. For his outstanding work, Plaskett received some of the most prestigious international awards that any astronomer can ever earn.

Plaskett was a kind and constant supporter of the RASC. Between 1907 and 1918, he held many offices in the Ottawa Centre and on the national Council, including the presidency at both levels. He contributed numerous papers to the *Journal* of which he was Associate Editor for twenty-eight years, and throughout his career he frequently spoke at Centre meetings, especially in later years in Victoria and in Winnipeg where he was Honorary President. One of his sons, Harry H. Plaskett, became Savilian Professor of Astronomy at Oxford University.

In 1895, President Larratt W. Smith suggested that the Society, as a corporation, should have a seal to affix to its legal documents. Nothing was done for a couple of years until Dr Edmund Meredith, another lawyer, and later Vice-President of the Society, suggested a design. Again no action was taken, perhaps because of the impending name change. The idea was not dropped completely, however, and in 1905, Dr Chant was able to announce that a final design had been selected, and the work completed:

> The central portion of the design is the figure of Urania, the muse of Astronomy. This sketch is after a sculpture by Flaxman. Above her head, on a starry background, is the motto, "Quo Ducit Urania" (i.e. where Urania leads, we follow), suggested by Professor John Fletcher, of University College. Above this again is the royal crown, and surrounding all is the name of our Society. The sketches from which the seal was cut were made by Mr John Ellis.

The Society incurred significant expenses to have the seal made: $20 in 1905 and a further $34.75 in 1906. John Ellis (1837–1923) was a lithographer of considerable skill. He was an enthusiastic amateur astronomer who hosted open-air meetings at his home on several occasions. Ellis Avenue, just west of Toronto's High Park is named for his family who originally owned 250 acres extending from the lakefront to Bloor Street.

The idea of Urania as subject for the seal seems to have been in circulation for some time. The classics were still a part of a good education, and everyone would have known who Urania was. President Paterson frequently alluded to her in his speeches and W.B. Musson in 1899 described an engraving of Urania copied from one in the Vatican, and suggested a motto "Mens Agitat Molem" (the mind is stirred by thy greatness).

A poem of four verses entitled "An Astronomer's Apostrophe to Urania" by Frederick C. Leonard was published in the *Journal* for October, 1923. It can be given a distinctive Canadian flavour by singing it to the tune of the haunting Huron Indian carol, "Jesus Ahatonhia." A bit of care is needed in going from lines six to seven, and the first line must be repeated at the end. Here is the first verse:

Celestial Urania,
    Thy wonders we would know —
Those mysteries and marvels which
    Alone inspire us so
That nightly from this Earthly sphere
    We try to comprehend
Thy Universe as infinite,
    And ages without end.

The Seal of the Society in its present form

impractical distances; some simply prefer to support the Society at arm's length, receiving the national publications but none of the local services or social benefits. Such members are called "unattached," a label which might be interpreted for better or for worse depending on a member's marital aspirations. At the present time, of the 3,200 members, about 20 percent are unattached.

Speaking of being unattached, it should be noted that the ratio of men to women amongst regular members is about 12:1. This imbalance may not be quite as lopsided as it seems however, for there are many examples of married couples (for which the RASC should get at least some of the credit) and consequently there are an unknown number of wives who take an active interest in the Society, but who do not show up on the membership roll. There is provision for another class, called Associate Membership, which some Centres require spouses and other family members to belong to before they can participate in Centre activities. However, these associate members are not included in the national membership list and, as they pay only a nominal fee, do not receive any national membership privileges.

Regular members comprise 86 percent of the total membership. Youth Members (under twenty-one) make up 4 percent, and the remaining 10 percent are Life Members.

On a regional basis, Ontario claims 57 percent of the attached members, Quebec 8 percent, the Maritimes 14 percent, and the remaining 21 percent from the western provinces. The high proportion from Ontario is in part attributable to the Society's Toronto origin, but also is a reflection of the relatively low representation from francophone Quebec. Amongst the unattached members, about half are in the United States, 8 percent overseas, and the rest in Canada.

All members, whether attached or unattached, may participate in the Annual Meeting of the Society and the concurrent General Assembly. This three-day event is generally hosted by one of the Centres and is held on the Victoria Day or Canada Day weekend. It is a wonderful time for making and renewing friendships, sharing ideas and conducting the business of the Society.

All members receive the *Journal* and *Bulletin* which come out six times per year, and the annual *Observer's Handbook*. Another benefit of membership is the library maintained at the Society's national office in Toronto. There are well over a thousand books plus many periodicals, slides and some videotapes but the emphasis of the collection is on works relating to the historical development of astronomy.

GOVERNANCE

The national Society is governed by a Council of individuals who volunteer their time for the benefit of the membership as a whole.

A president and two vice-presidents are each elected by the membership for two-year terms, and normally the president stays on Council as past-president for two more terms. Thus a person elected as second vice-president usually has a ten-year span of service ahead. Moreover, these dedicated people have likely contributed a lot to the Society before undertaking presidential responsibilities. Such commitment and continuity assure strong and informed leadership. Of course, new ideas and fresh energies are needed to stimulate the Society, and so the by-laws prohibit anyone from holding any of these presidential offices for more than one term. Also elected to the National Council are a treasurer and secretary, each for three-year terms, with re-election possible once.

A number of other officers with somewhat more specialized roles are on the National Council, though they are appointed by the councillors, rather than being elected by the membership at large. These are the four editors of the *Journal*, *The Observer's Handbook*, the *Bulletin* (formerly the *National Newsletter*) and *The Beginner's Observing Guide*, a librarian, a recorder (responsible for preparing minutes) and an honorary president.

Completing the makeup of the National Council are representatives from the Centres — one for every 200 voting members or portion thereof. This formula means that most Centres have one representative, Calgary and Ottawa have two each, and Toronto has five. At the present time, then, a meeting of National Council could involve close to fifty people — twenty-eight members representing Centres, thirteen officers and some chairpersons of standing committees.

The Council generally meets four times per year, typically for an all-day meeting on a Saturday late in the fall and again in midwinter, and just before and just after the Annual Meeting in May or July. The fall and winter meetings are usually held in Toronto as a convenience to the majority of councillors, but no matter what site is chosen, some members end up travelling thousands of kilometres. Since the Annual Meeting moves from city to city, councillors at least sometimes get to meet on home ground.

While any member of Council is free to bring up any matter of concern, most of the meeting time is taken up with reports of officers and committees which naturally cover a wide range of topics. As important as all this administrative business is, the really vital work which keeps the Society flourishing is carried out most effectively at the Centre level.

Ever since 1890, the Council has overseen "the general management of the Society's affairs." At first they met irregularly as the need arose, the earliest occasion on record being January 13, 1891, when they had to deal with the treasurer's resignation. With a small number of members and nearly all the activities in Toronto, there was a tendency to conduct most of the business at the regular fortnightly meetings. This occasionally caused problems if controversial items led the discussion to encroach on the time needed by the speaker for the evening.

The photograph (by Edgar Fleming and published in the Journal **26**, 360) shows part of a group of Victoria members and distinguished visitors in 1932. Brydon is seen at the front left, Mrs Brydon is at the right (wearing a hat). Higher up on the steps are Dr and Mrs J.A. Pearce. The others from left to right are C. Hartley, J.P. Hibben, W. Goodacre and Miss K. Williams.

H. Boyd Brydon (?–1947) was born and educated in England. After a career as a steam engineer in the Chicago area, he retired to Victoria in 1925 and took up astronomy as a hobby. In 1931, he purchased A.F. Miller's historic 10-cm refractor and installed it in an observatory which he built at his home at 2390 Oak Bay Avenue. This was the focus of Centre observing activities, not only while Brydon was still active, but even after the Centre purchased the refractor from him in 1943.

A tireless promoter of astronomy for amateurs, he spoke at meetings and on radio, and published twenty-eight papers in the *Journal*, a record unmatched by any amateur member. Many of these articles were concerned with improvements in mountings and drives for small telescopes but others showed the range of his observing interests including auroras, occultations, variable stars, and sunspots. Occasionally he ventured into more speculative topics such as the origin of comets and extra-terrestrial life.

Brydon served the Centre as a councillor, Secretary, Treasurer, President and Honorary President and then, in 1938, went on the National Council. He was elected Second Vice-President of the Society in 1943, and First Vice-President in 1945, but had to decline the presidency because of ill-health. At the presentation to him of the Chant Medal for 1941, K.O. Wright noted that, "The great increase in activity among amateur astronomers in Victoria during the past ten years can be directly traced to the work of Mr Brydon."

So as the Society evolved, more and more of its administrative work was delegated to the Council. To counterbalance this centralization of authority the Council was gradually enlarged to democratize the decision making.

Starting in 1899, Council meetings were scheduled once a month, and continued at that frequency for over a decade. Since every bill over $5, every book ordered, and every letter received came to their attention, it is not hard to imagine that monthly meetings did not lack for busy-ness. The Society's only paid employee assumed most of the clerical duties in 1912, and from then on, Council has met only three or four times per year.

As a result of infrequent Council meetings and increasing business, the need arose for an Executive Committee to attend to routine matters or any emergency which came up between times, though it was always understood to be accountable to the Council for any action it took. The amount of work done by the Executive Committee has varied greatly. Some presidents have called no meetings during their tenure of office. Helen Hogg, on the other hand called the Executive to six meetings in 1957 and twelve in 1958.

While the Council certainly moved away from the old ways of making every decision no matter how petty, concerns were raised from time to time that too much time was spent on routine matters and not enough on items of genuine concern to Centres. Boyd Brydon of Victoria expressed his feelings in 1935 when he said, "The General Council could do much to help but except for its grant, each Centre is left to work out its problems by itself. Vision, enthusiasm, and a definite programme are needed ... Privates may win battles but they do not plan them; that is the duty of their officers." In time-honoured tradition, the officers soon responded by nominating Brydon for a seat on Council!

Whether Brydon or other councillors from afar could actually attend meetings was another matter. A round trip by train from the west coast would have taken a week, and even after plane travel became possible, the costs were prohibitive. The classic Canadian problem of vast distances made cohesiveness very difficult to attain. The only hope for representation from Centres far from Toronto was that members might occasionally combine a business trip with a Council meeting.

For most of the Society's history, the Council, General Council, or National Council as it became officially in 1949, met wherever the Society was headquartered, though for several years (roughly from 1911 to 1933) they met in the Meteorological Office on Bloor Street through the courtesy of Sir Frederic Stupart, president of the Society in 1902–03. The first time the Council met outside Toronto was in 1958, but starting in the early 1960s, one or two of the four meetings each year were always held in conjunction with the General Assemblies. There have been a few other occasions when the Council ventured outside Toronto. In 1979, they gathered in Winnipeg at the time of the total solar eclipse, and later the same year at Edmonton. In 1987, since the General Assembly was held in Toronto in May, the September Council meeting took place in Halifax. With the availability of funding for representatives to travel to meetings, attendance has increased noticeably, and the need to move around the country is less essential.

Most of the time, Council deals with routine business of the Society, but it occasionally takes steps having broader implications. For instance, it authorized the preparation and presentation of briefs to the Royal Commission on the Development of Arts and Sciences in 1949, to the Royal Commission on Bilingualism and Biculturalism in 1964, and to a Special Senate Committee on Science Policy in 1969. The first recommendation in the Society's brief in 1964 was that the teaching of astronomy leading to a professional career be included in French-Canadian centres of learning such as Laval University and l'Université de Montréal. Whether the RASC can take any credit is a moot point, but both institutions do offer graduate degrees in Astronomy now and the proportion of francophone astronomers in Canada has increased from 6 percent in 1964 to an estimated 14 percent presently.

Committees have always provided a convenient way for Council to deal with time-consuming problems or situations where advice or information was not at hand. In the early days of the Society, committees were entirely appointed as needed. Sometimes their duties were familiar ones like drafting by-laws or arranging dates and places of meetings. But some of the committees in the first decade of the Society had interesting and surprising mandates: to suggest revisions to Webb's *Celestial Objects*, to collect information as to the teaching of Astronomy in the city and Provincial schools and to suggest means by which to promote the study of that science, to frame a series of questions to be submitted to astronomers respecting the change in the the start of the astronomical day from noon to midnight, to consider and report upon the propriety of petitioning the Dominion Parliament to reduce or remove the duty upon astronomical apparatus not made in Canada, to enquire into the best means of promoting the investigations of terrestrial magnetism and earth currents, and to explain and support the claim of Messrs Collins to the invention by them of monoplane telescopes. In carrying out these projects, the Society gained recognition which in the long run was probably more significant than any of the specific results. One special committee whose work indirectly bore fruit was the Society's Planetarium Committee. Prompted by a provision in member Carl Reinhardt's will which set aside $10,000 to assist with the construction of a planetarium if a start were made by the Society within ten years, a revitalized Planetarium Committee began in 1963 to investigate and propose a planetarium for Toronto. Their hopes were given some media coverage in 1964 and as a result of this,

This historic photo shows the National Council at its meeting of March 28, 1958, at McMaster University. This was the first time that the Council had ever met outside Toronto.

Seated around the table are (l to r) J.B. Oke (Librarian), W. Wehlau (London), R.J. Lockhart (Winnipeg), J.F. Heard (Past-President), J.H. Horning (Treasurer), C. Reinhardt (Trustee), H.S. Hogg (President), J.E. Kennedy (Secretary), G.H. Hall (Montreal), R.J. Northcott (Editor), A.R. MacLennan (Montreal), M.W. Burke-Gaffney (Halifax), K.M. Heaton (Toronto). Standing (l to r) are J.L. Locke (Ottawa), W.H. Adamson (London), H.B. Fox (Hamilton), N. Green (Hamilton) and G.A. Cooper (Toronto).

HERBERT FOX (1910–81) worked as chief chemist for the Steel Company of Canada in Hamilton. For thirty-eight years, he was on the Council of the Hamilton Centre, and held the offices of Secretary, Treasurer, Vice-President and President, the latter position in 1942–43.

The Service Award was bestowed on him in 1968, "for his efficiency, for his guidance, and for the quiet way in which he ... influenced so many members and so many decisions." He is shown standing in the middle of this group photograph.

Colonel R.S. McLaughlin decided to donate a million dollars for the construction of the facility which now bears his name.

Gradually, the Society acquired standing committees. The first was one to deal with editing in 1908, then one for finance was formed in 1918. Programme, Membership and Building Committees came along in 1930. At present there are nine standing committees as follows: Awards, Constitution, Executive, Finance, Historical, Library, Nominating, Property, and Publications, with ten additional special committees. All are rather prosaic by the standards of the 1890s.

The usefulness or success of any committee always depends primarily on the enthusiasm and organization of the person chairing the group. In an organization of volunteers all of whom are very much occupied with their regular jobs or studies or families, nothing can be done to ensure prompt action except to try to appoint chairpersons who have the necessary drive to get the job done. In the absence of keen leadership, the rest of the committee is unlikely or unable to take any initiative. Our Canadian geography of course doesn't help matters. Committees of members from different parts of the country can only meet face to face on rare occasions. Fortunately, electronic mail, faxes, and long-distance phone calls, all make the consultative process easier than it used to be.

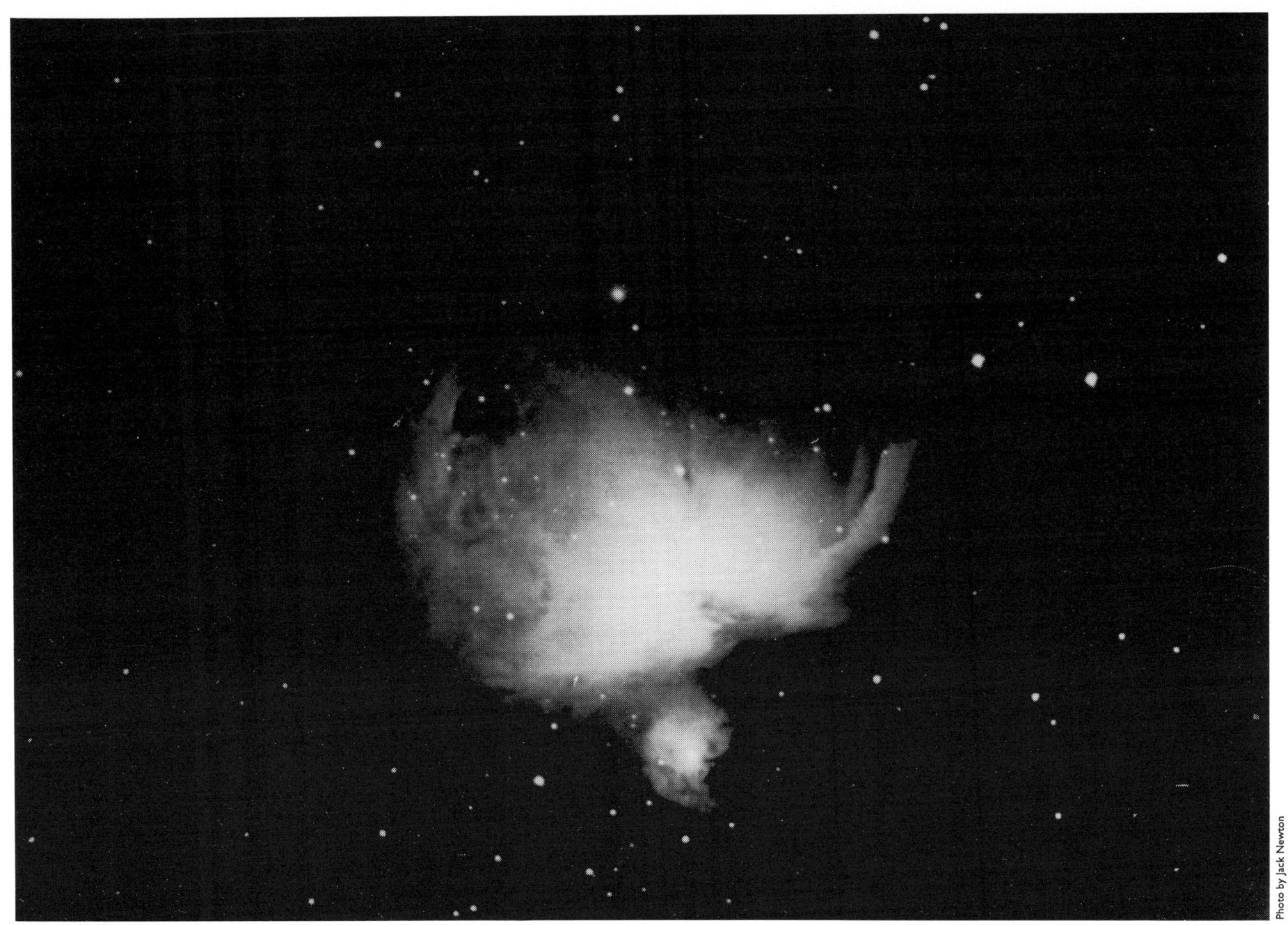
One example of the birthplace of stars is the emission nebula, M42, in the sword of the constellation Orion. Clouds of gas and dust, attracted by mutual gravitation, contract to form stars.

CHAPTER 2

The Centre of the Universe

Torontonians may not believe it, but the rest of the country claims that those who live in Canada's largest city think they are at the centre of the universe. Like Ptolemaists of old, they allegedly presume that everything revolves around them. A more enlightened point of view is that, at least as far as the RASC is concerned, there are many centres of attraction across the country. Historically, however, the Society did originate in Toronto and it has always been headquartered there.

ORIGINS

Though 1890 is considered the Society's birth date, 1868 must mark its conception, for on December 1 of that year, a meeting was held in the Mechanics' Institute of Toronto to consider "the propriety of forming a society for the prosecution of astronomical science." Without question, Andrew Elvins was the Society's father. The only clues to motherhood to be found are the social, economic, technical and other circumstances which provided the fertile ground for the embryo Society to develop through a long and difficult gestation period. As often happens in such circumstances, older and wiser voices opposed the very conception.

G.T. Kingston, the Director of the Toronto Meteorological Observatory, was not in favour of forming an astronomical society and the Reverend William Hincks, president of the Canadian Institute, (and brother of the last pre-Confederation prime minister), advised against it:

Considering what I know of the many difficulties attending the organization of societies, the cost of rooms, printing and various services and the interference of one society with another, I am compelled to conclude that, except where the votaries of science are very numerous and abundant in means, it is incomparably the best plan to have a Society [like the Canadian Institute] embracing all scientific and learned pursuits and pursuing its objects in common. ... Members specially interested in one branch may organise special meetings ... so as to unite any real advantages of a separate society with the solid benefits of union.

On the face of it, Hincks' opinion seemed sound enough; the Canadian Institute was by this time well-established and Toronto's population was but 54,000. (Subsequent history shows that RASC Centres which have survived, with the exception of Victoria, have all drawn on an initial population base exceeding 50,000.)

Hincks' letter, dated December 2, was a day too late for the meeting, though undoubtedly the eight men who gathered on December 1 already knew the alternatives. Undaunted, they unanimously adopted Elvins' motion:

That a society be formed under the name of "The Toronto Astronomical Club" having for its object the aiding of each other in the pursuit of astronomical knowledge; – in order to which it is proposed:

Photo courtesy of Clifford Oliver, great-grandson of Elvins.

ANDREW ELVINS (1823-1918) was a child of the industrial revolution; his life epitomized those qualities of optimism, faith and perseverance which characterized many of his generation who rose from humble beginnings to achieve something of lasting importance. He was born near the Cornish mining town of St Austell, and by the age of ten, his formal schooling evidently over, he faced a lifetime of drudgery in the tin mines. But a fascination with geology and a keen interest in the universe developed in his young mind. Apparently, he used to sketch the constellations as he saw them with his naked eye, and only later discovered, to his great delight, that they had names and were depicted in books, much the way he had drawn them. Elvins recalled that he and his school mates longed for a telescope, but the 25 pounds required for a 6 cm refractor was far beyond the resources of the school master or the boys. The bright side to the story is that Elvins' burning desire never faded, and through whatever opportunities he had for self-improvement, he was eventually able to realize his dream and to share his enthusiasm for astronomy with others.

His young life took a turn for the better when, at age 12, he was apprenticed to a tailor, a trade which would give him employment for all his working life. During his seven-year apprenticeship, he read assiduously and attended night-school. At age 21, he emigrated to Canada, settling and marrying in Cobourg. On excursions to nearby quarries, he found and collected many fossils which led him to start a museum and a Mechanics' Institute in the town. For two years, 1858-59, the Elvinses lived in nearby Port Hope where he and a group of kindred spirits formed a scientific society. Moving to Toronto in 1860, he joined a large clothing firm called The Mammoth House on King Street East. At the same time he became a member of the Canadian Institute, to which he contributed some of his best fossil specimens and before which he read several papers on archaeology, astronomy and meteorology.

With only the evening hours free from work, he turned more and more to astronomy, founding the Toronto Astronomical Club in 1868, and the Astronomical and Physical Society in 1884. He read widely in the library of the Canadian Institute, and corresponded with such well-known scientists as Cleveland Abbe and Norman Lockyer, and wrote articles for Toronto papers, *The Leader*, and *The Telegraph*, and an English journal, *The Astronomical Register*. He enjoyed viewing and sketching the Moon, planets and Sun and sometimes made observations of the aurora, solar halos and the zodiacal light. He was not content to stop there, however, but went on to concoct theories in which he took considerable pride. He claimed, for instance, that activity observed on Jupiter was evidence that material was still being ejected from the giant planet, and that satellites were even then being formed. When the fifth satellite was discovered in 1892, he took this as confirmation of his theory. He also believed that he had discovered evidence for two pulses of rainfall during each sunspot cycle, and was chagrined to see Lockyer subsequently publishing similar conclusions without crediting him.

Whatever we might think of his quaint ideas a hundred years later, there can be no doubt that Andrew Elvins was a driving force in the formation of our Society, and a very positive influence in encouraging others to pursue scientific studies.

I To meet monthly at such time and place as may be agreed upon

II To spend the evening somewhat as follows:

(a) Reading extracts from papers or publications, of anything new or otherwise interesting, bearing on the subject of Astronomy.
(b) Reading original papers connected with any department of Astronomy.
(c) Examining anything new in Astronomical Science.
(d) Observing celestial objects if circumstances should favor our doing so.
(e) Conversation &c.

Although Elvins was always credited with originating the Society by those who had been associated with it in its early days, he did not become its president. Seemingly, Elvins always deferred to others with more formal education or higher social status. So it was that Daniel K. Winder was elected president in 1868. He had taught astronomy at a college in Ohio, but came to Canada as a pacifist during the American Civil War. Here he earned his living as a printer and lay preacher. Mungo Turnbull, a cabinet maker with a good Scottish education, volunteered to provide the January meeting with "A brief notice of the past and present state of Optical Science, viewed chiefly in its bearing on Celestial discovery," and Winder agreed to speak at a subsequent meeting on "The spectroscope – its construction and application to Celestial Chemistry."

For most of the next year, members met on the first Tuesday of each month at 7 pm at one another's homes. Usually about half a dozen attended, but the meeting at Mr Winder's attracted an additional eight including his wife and daughter and the secretary's wife, Mrs Clare. During the year Turnbull completed building a speculum reflecting telescope, which drew praise for making the eighth magnitude companion to Polaris appear like a fourth magnitude star. The nights scheduled for group observing were cloudy but individual members did make their own observations and reported on them at meetings. Elvins studied lunar craters and found that the changes seen by others could be explained by the varying angles of illumination. Winder reported on the form of the aurora of April 15 and May 3 and noted the relationship to solar activity and magnetic disturbances. He even examined the aurora with his homemade spectroscope and polariscope, and concluded from the absence of polarization that the light was not reflected – a clear demonstration against the once-popular view that

An early meeting (about 1888) at D.J. Howell's home, 218 Bleeker Street, Toronto. From left to right are Barters, Elvins, A.F. Miller, G.G. Pursey and W.J. Moore.

ANDREW F. HUNTER (1863-1940) joined the Astronomical Society in the 1880s while he was a student in Mathematics and Physics at the University of Toronto. He had grown up near Barrie, about 80 km north of Toronto, and returned there as owner and editor of the local newspaper following graduation in 1889. He became a town Councillor in Barrie and was a talented amateur archaeologist and local historian. His *History of the County of Simcoe* is still a valuable reference work. He also had a lasting interest in geology stemming from his employment with the Canadian Geological Survey 1904-08. All this probably explains why he held no office in the Society for many years, but after he returned to Toronto in 1913 to become Librarian and Secretary of the Ontario Historical Society, he was able to take a more active part in the RASC. From 1918 until 1927 he was successively Recorder, Secretary, Vice-President and President. Most of the lectures he gave to the RASC and articles he wrote for the *Journal* relate to atmospheric phenomena, halos and the like, and the aurora.

This photo of a meeting, probably in Mr Roberts' house in 1888, shows (l to r) Pursey, Elvins, S. Roberts, Howell, Hunter and Miller.

the aurora was due to light reflected from the polar ice.

A highlight of the year was the solar eclipse of August 7, 1869. Turnbull had set the stage by delivering a long and detailed paper at the June meeting on eclipses in general and the forthcoming one in particular, for which he predicted the circumstances at Toronto. Subsequent observations showed that his largest error, the time of first contact, was off by nearly four minutes but his calculations were certainly good enough to ensure that the members knew the eclipse would not be total at Toronto. How strange then, that the list of proposed experiments drawn up on August 3, in anticipation of the great event, had members designated to look for Bailey's Beads, corona and visible stars. The enthusiasm of the little band clearly outreached their erudition. On the day itself, late on a Saturday afternoon, they gathered at the end of Nassau Street, where Turnbull lived, and assiduously timed and recorded contacts with the limb and with sunspots, all the while carefully noting the sharpness and jaggedness of the moon's limb as evidence of the lack of any lunar atmosphere. Temperatures in the sunlight and in the shade, as well as the barometric pressure were written down every five minutes. To a man, they compensated for a lack of formal knowledge by learning what they could from the phenomena they observed.

Minutes of the monthly meetings up to and including September 6, 1869, were recorded in beautiful script by Samuel Clare, a writing teacher at the Normal School. One more meeting with only four in attendance, that of December 7, 1869, was recorded in a different handwriting and there the minute book ended. Was this a case of an infant death, not so uncommon in those days? Were the voices of dissent able to say, "I told you so?"

The Toronto Astronomical Society as it became known in May, 1869, barely survived, but as Andrew Elvins recalled many years later, it did exist in a very precarious way. The precise sequence of events is unknown. Apparently in 1879, an organization called the Recreative Science Association attracted to its ranks some of those previously associated with the Astronomical Society. The Recreative Association or Club as it was sometimes called, may itself have been formed by Elvins; he did serve as its president. Those who had belonged later spoke of it as a predecessor of the RASC. In Elvins words:

> Mr Clare's death and Mr Winder's removal to the United States were deeply felt, but weekly meetings were held at my house by a few who felt interested in astronomy or scientific subjects. We embraced other subjects, and found it useful as far as attendance was concerned. ...
>
> Natural history found more enthusiasts than the astronomical part, and it was at last decided to join the Natural History Society of Toronto which had already obtained a charter. Several of our members became its members but yet kept up an interest in astronomy.

The Natural History Society rented space in the Canadian Institute's Richmond Street building in 1881 (and possibly earlier), and then in 1885 amalgamated with the Institute becoming its biological section. Other sections devoted to architecture, philology, photography, geology and mining formed in 1885 but not astronomy.

Early in the 1880s, Elvins, according to his recollection, had met George Lumsden and his friend Allan Miller, both of whom had an interest in astronomy and later became presidents of the Astronomical Society. Miller recalled that "a small party of amateurs met occasionally, from house to house, to discuss scientific matters and papers under Mr Elvins leadership." Sometimes as many as seven or eight people came, and as Miller said, "all sorts of theories were expounded, and subsequently pounded, complete freedom of speech being the rule. Among all the habitués, I was only impressed by Mr Elvins, whose remarkable fund of knowledge and originality of thought seemed to me far beyond the common." In 1884, they founded The Astronomical and Physical Society of Toronto. Six years later, Elvins, Lumsden and Miller were the three most prominent proponents of the decision to secure a legal charter as an astronomical society, with Lumsden, as Assistant Provincial Secretary, in a good position to see it through. And who would be president? Apparently there was some maneuvering behind the scenes. Elvins was certainly entitled to the honour. He had acted as virtual chairman for many years and had been steadfast in encouraging the development of the society since the very beginning. Miller, however, seems to have favoured Lumsden for president. Anyway, the three men agreed to visit Charles Carpmael, the director of the Meteorological Observatory, president of the Royal Canadian Institute, and a man of some stature in the world of science, to see if he would be patron of the infant Society. He said, "Why not let me join?" and Elvins naturally replied that Carpmael would be a most welcome member. Thus it was that the director of the Observatory became virtually the only member with any formal scientific credentials. It is easy to see that he was the natural choice for president if the fledgling Society was to become anything more than a group of well-meaning amateurs.

Charles Carpmael was very much like a godfather. His presence at the christening and from time to time in the formative years of the young Society gave a comforting assurance to those doing most of the work that a force beyond their own efforts would support and sustain them.

The happy event, the birth of the Society, took place on Tuesday, February 25, 1890, at Lumsden's home at 719 Ontario Street, with ten members present. The Constitution and by-laws had already been drafted and were approved with some amendments. Six officers were all elected by acclamation, and a committee consisting of Miller and Lumsden was appointed to secure the incorporation of the

CHARLES AUGUSTUS CHANT (1865–1956) grew up in a village near Toronto. A school book entitled *Geography Generalized*, by Robert Sullivan, first aroused his interest in astronomy. After high school, Chant taught for a couple of years as a means of earning some money, then entered the University of Toronto and graduated in mathematics and physics in 1890. After a short stint working for the Finance Department in Ottawa, he accepted a position as Lecturer in Physics at the University of Toronto. Soon, as he recalled, he joined the A&P Society where he "met a number of practical observing astronomers who further stimulated my interest in astronomy."

Though his PhD was in Physics (earned during a year's leave of absence at Harvard in 1901), Dr Chant became a highly respected and influential teacher of astronomy at the University of Toronto. For thirty years, nearly all Canadian astronomers were trained by him. Ironically, though, his heroic efforts to obtain an observatory for Toronto were only realized with the opening of the David Dunlap Observatory on the very day of his seventieth birthday and his official retirement from the university.

Chant's connection with the Society spanned sixty-four years, and for fifty of these he edited the *Journal* and the *Handbook*, building them into internationally recognized publications. He was national President for four years and Librarian for sixteen years. He contributed countless papers to the *Journal*, the last being published when he was ninety-one. He was a great popular educator and in the words of J.F. Heard, "lived for the public lecture platform." Chant's introductory book, *Our Wonderful Universe*, was translated into five languages. He co-authored physics texts which sold in the hundreds of thousands. Undoubtedly, royalties contributed to his large estate of $250,000 which he left to his daughters and to the David Dunlap Observatory.

Chant and one of his daughters in front of their summer home at Go-Home Bay, 1904. The designs in the railing illustrate geometric theorems. His daughter, Dr Elizabeth Robertson, became an outstanding pediatrician.

Chant, seated in the library at the DDO, about 1939. Notice the design of the andirons for which he was responsible. ▶

Society under Ontario law. By the time of the following meeting, just two weeks later, the committee reported that they had appeared before Judge MacDougall and that he had made the necessary declaration recognizing the Society as a legal entity.

The custom of meeting on Tuesday evenings at members' homes which had begun in 1868, continued even after incorporation. With only about ten to twenty people coming out regularly, this arrangement was satisfactory, and did a lot to build esprit de corps. The elected officers took most of the responsibility for hosting the meetings. In 1890, President Carpmael invited the members to the Observatory on three occasions for viewing through the six-inch (15-cm) Cooke refractor, and each of the other officers opened their homes three or four times. A special event took place a few days before Christmas when the group gathered at the Spadina Crescent home of a distinguished member, the former Toronto mayor, MP and Judge, Sir Adam Wilson, to admire his new six-inch speculum telescope. Following the regular meeting, he and Lady Wilson entertained their guests at an informal supper.

HEADQUARTERS

As numbers grew, and the Society began to receive donated instruments (including Sir Adam's telescope) and to acquire books as gifts and periodicals in exchange for its *Transactions*, a more permanent home became desirable. Members tried to help out. Dr Foster offered his rooms at Gerrard and Yonge Streets for the first meeting of every month, and Mr G.G. Pursey, the librarian, made room for the books in his house. Charles Sparling, who lived on Victoria Street right opposite the Normal School, agreed to look after the telescopes since they could be easily carried across to the school grounds to be set up for viewing. These were but temporary measures and in the summer of 1893, the Society decided to rent rooms from the Young Women's Christian Guild, 19 McGill Street. For most purposes, these quarters were adequate but some meetings drew large crowds and had to be held elsewhere. For example, an open meeting on April 18, 1893, attracted 150 members and visitors to the Physical Room at University College where C.A. Chant, a young Lecturer in Physics, spoke on the polarization of light. Dr Chant would later become the Head of Astronomy at the university, and the motive force behind the Society for the next six decades.

After only two years on McGill Street, the Society was offered new premises closer to the Observatory and to the university, in the Technical School at College and McCaul Streets. The Board Room served as the usual meeting place, with the adjoining Electrical Room suitable for lectures involving demonstrations. Another room in the school was set aside for storing the library and apparatus, and observations could be made from the grounds which afforded a good view of the horizon in all directions. But alas, the school enrollment grew to the point where no space could be spared and the Society had to move again in 1898.

The Canadian Institute, with broad similarities to the Astronomical Society in its goals, membership, and even some of its councillors, became the Society's new landlord. The Institute had managed to scrape together enough resources to erect a solid three-storey building at 58 Richmond Street East in 1877, which in subsequent years sheltered several of the more specialized societies under its roof. Its Board Room easily accommodated fifty people and so was suitable for most meetings, and there was extra space for the library. All in all, the arrangements were favourable for the Society, and the Institute only charged it a nominal rent, sufficient to cover costs of heating, light, and maintenance. In 1905, when the Institute moved uptown to 198 College Street, right on the university's doorstep, the RASC moved with them, and remained their tenant for forty more years. The yearly rent at the new location began at $100 which paid for the small room used as a library

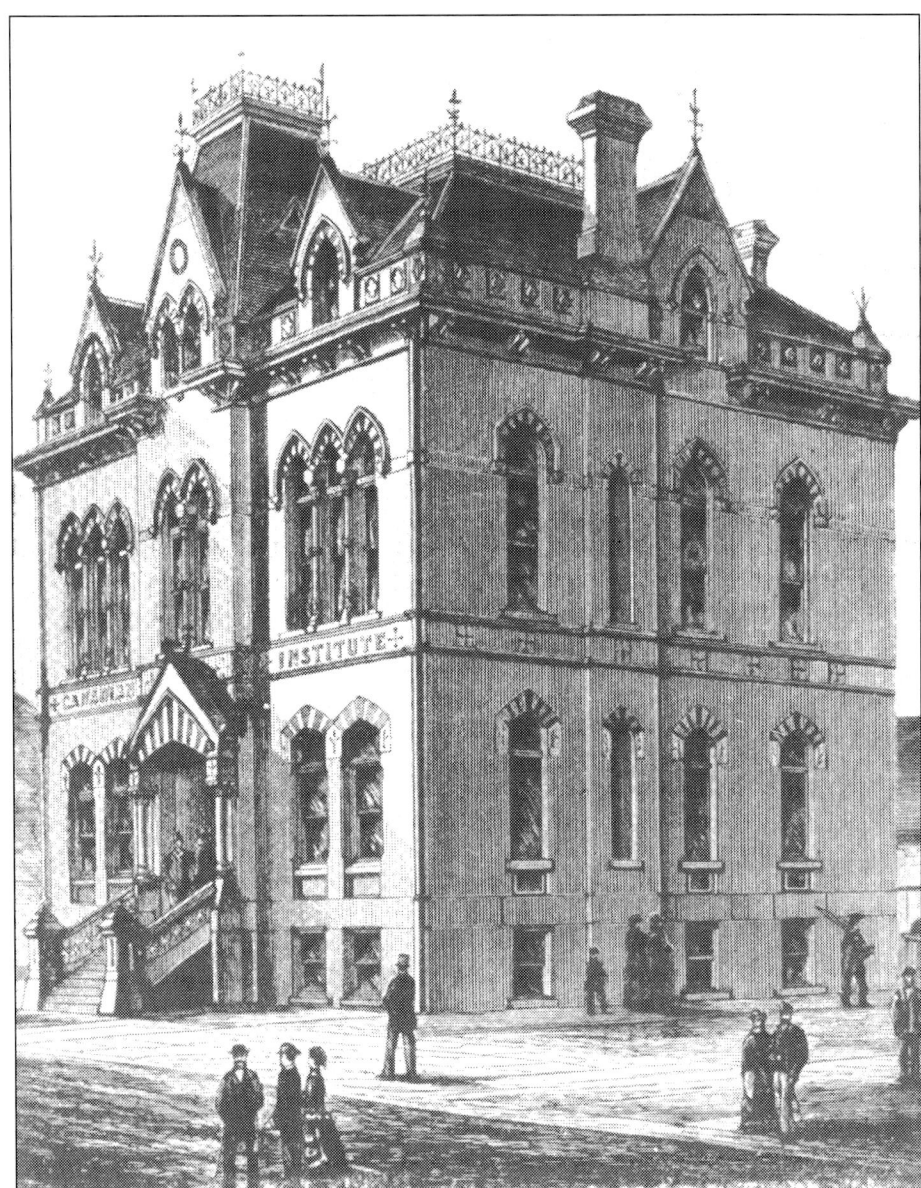

Canadian Institute buildings (above) at 58–60 Richmond Street East and (right) interior and exterior views of 198 College Street. These illustrations were previously published in the *Royal Canadian Institute Centennial Volume*, 1949.

and the fortnightly use of the lecture room.

The Society remained at 198 College Street long after it began to outgrow the space available. Even as early as 1908, "The Annual At-Home of the Society took place in its rooms at the Canadian Institute building, the accommodation being taxed to the limit." Thinking to the future, the Council set up a Building Fund in 1912. Occasional meetings were held at the University's Chemical Building (now demolished) and, on a regular basis from 1919 on, in what is now called the Sandford Fleming Building. The space at the Canadian Institute was used only for the library and office.

It was shortly after the move to College Street that the Society began to employ someone on a regular basis although there had been a period (1894–1900) during which Thomas Lindsay, as editor of the *Transactions*, was paid an honorarium of $50 per year. In 1906, the assistant librarian was paid $141 to be on duty for two hours each afternoon as well as on Tuesday evenings. Her hourly wage was thus about 25 cents. The following year an *Addressograph* machine was donated to aid in the work of mailing the new bimonthly *Journal* but by now the clerical duties as well as the routine library work required much more time. Consequently, starting in 1908, the assistant librarian's hours were increased to four per day and her annual salary to $250. A succession of women served the Society in this capacity for short spells until Miss Eva M. Budd began her long career with the RASC in 1912. Her title changed to assistant secretary-treasurer in the 1930s. Miss Budd often participated in meetings, reviewing articles or publications that had been received for the library, reading poetry, acting as scrutineer, and on at least one occasion, joining in an eclipse expedition. She retired in 1948 at age seventy-two, somewhat past her prime.

From time to time there was talk of using the Building Fund or trying to encourage some wealthy benefactor to provide a proper home for the RASC but these remained as mere dreams for some time yet. In all probability, the Society would have continued with the Institute indefinitely as a cost-saving measure, but in 1946 they were forced to make a change. The land at 198 College Street was in fact only leased from the university, and as the term of the lease was now up and there were plans for a new Chemistry Building (now known as the Wallberg Building) on the site, the RCI was required to vacate the building so that it could be demolished. The RCI purchased a building across the street, but with tenants already renting space in it, there was no room for the Society.

After two months of more or less frantic searching by the Society's building committee, satisfactory semipermanent quarters were found a few blocks north, at 3 Willcocks Street. The property was owned by the Christian Social Council of Canada, and the RASC rented offices on the second floor, with space for most of the library and a reading room. The premises were a great improvement, but the rent jumped threefold over what had been paid to the RCI. On top of this, there was a growing need for extra clerical assistance. There were complaints of unanswered correspondence, lack of attention to details, and a general inability to keep up with the mounting work load.

When Miss Budd retired, the Society felt able to pay more to her successor, Mrs E. Todd, probably with some increase in hours. Still the work continued to pile up and the Council realized they would soon have no choice but to make the secretarial position a full-time one. In the meanwhile, they had still another move to worry about.

Mr H.W. Barker, who had been treasurer and recorder for the Society, had also been on the Board of the Canadian Council of Churches. Because of this connection, he had been in a position to extend the Society's tenancy on Willcocks Street for two or three years beyond what would otherwise have been possible. Any reprieve was welcome as premises were exceedingly scarce in the early fifties but following Barker's death in 1951, there was no one to plead the Society's cause.

The task of finding new accommodation fell on the shoulders of the Society's incoming president, Professor J.F. Heard, who spent many hours searching the university neighbourhood for something suitable. Eventually, he discovered that premises at 15 Ross Street, a block south of College Street, were available on a three-year lease, and the move took place in the spring of 1953. Heard had recently been appointed the director of the David Dunlap Observatory, and whether in deference to his position, or simply from generous motives of good will, other members of his staff also helped greatly in the move. Each successive relocation faced by the Society was more difficult than the previous one. Besides the office files, furniture and library, there were hundreds of periodicals received in exchange for the Society's *Journal*, and back issues of the Society's own publications which kept piling up, and at each move these had to be organized, boxed and transported. William Hossack, the Society's librarian, had just completed his PhD work (the first in Astronomy at U. of T.), and he with the assistance of Gerald Longworth, the mechanical superintendent at the Observatory, carried out the tiresome task of sorting it all out. Then, when the Society's office secretary, Mrs Phelps, quit in May, the Observatory secretary, Miss Edna Fuller, filled in to see the Society through the summer, until Mrs Dorothy Williamson, wife of another U. of T. astronomer, assumed the position in the fall. No doubt Dr Chant, now well-on in his eighties, was proud to see the troops rallying round in the time of crisis.

Everyone realized that the Ross Street premises were really only a stopgap measure, and as unpalatable as another move was, they knew that a permanent home for the Society had become an urgent necessity. With the building fund now at $8,000, ownership seemed as though it might be within reach. Consequently the Council, in June 1955, instructed Heard, now past-president, and J.H. Horning, the treasurer, to investigate suitable permanent quarters for the Society's office and library.

On January 10, 1956, such a property was found, in a convenient location and within the Society's means. A large house at 252 College Street had come on the market as an estate sale, and while the asking price was $50,000, there was some hope that a much lower offer would be considered. Prompt action being necessary, Heard arranged that a group of the Executive Committee and some others inspect the property that evening. They decided that the property was admirably suited to the Society's needs, and an agreement was signed to purchase it for $32,000, with a down payment of $12,500, and a closing date in March. Fortunately, the Annual Meeting was only a week away, at which time the members ratified the action taken by the Executive Committee and authorized the use of all the special funds of the Society to whatever extent might be needed to meet the down payment. They also approved a financial campaign to seek donations both from members and from government and charitable foundations to furnish the new quarters suitably and to reduce the mortgage as much as possible. In light of a deficit of nearly $1,000 from the previous year, it was a bold decision, but undoubtedly it was the most auspicious move the Society had ever made.

Besides the premises to be occupied by the Society on the main floor, there was an apartment for the building caretakers. Another apartment was rented by a long-standing RASC member, Carl Reinhardt, who was conveniently on hand to supervise the many improvements and renovations carried out prior to moving date in May. Two more apartments on the third floor were available for suitable tenants. By making a separate side entrance, the basement was adapted for the use of the amateur telescope-makers group of the Toronto Centre, at the Centre's expense.

Though the funds of the Society were almost completely wiped out by the down payment, the response to the financial campaign was very encouraging. By the time of the next annual meeting, in February, 1957, the mortgage had been reduced from $20,000 to $13,000. Toronto Centre contributed $500 towards the furnishing of the library and reading room (facilities they were to use following regular meetings), Victoria Centre raised almost $1,000 for the building fund, and Reinhardt himself donated $2,000. Once the initial hurdle was passed, the advantages of the ownership became apparent: the rental income exceeded the combined building expenses and mortgage interest by about $1,000 per year. This favourable balance, along with further donations, a life-membership drive, and a lump sum of $2,000 realized from the sale of periodicals, enabled the Society to discharge the mortgage in 1961.

There are always hidden costs of course. Attending to tenant needs, collecting their rent, paying the caretaker, taxes, utilities and maintenance bills all added to the normal administrative burden of correspondence, meetings, agendas, minutes, notices, publications and membership records. *Handbook* sales were beginning to soar, entailing a lot of correspondence, invoicing and shipping. In recognition of these extra duties, the office position was now made a full-time one with the title of Executive Secretary. Mrs Williamson moved to the US in 1957, and was replaced for a short time by Miss Mary McCarthy, who was succeeded by Mrs Marie Fidler in 1958. When Marie left in 1972 to marry Sam Litchinsky, a prominent Calgary Centre member, Miss Rosemary Freeman assumed the position which she has held ever since. The Society has been extremely fortunate to have such dependable, dedicated executive secretaries to run its affairs.

The old house at 252 College Street really began to show its age in the early 1970s. A new furnace, installed at a cost of $1,300 contributed to a large

JOHN F. HEARD (1907–76) was born in St Thomas, Ontario, where Ken Chilton's father was among his school chums. He attended the University of Western Ontario (H.R. Kingston was one of his professors) and McGill where he took a course in astrophysics from A.V. Douglas and obtained his PhD in physics. He then went to London, England, for a second PhD and there met H.H. Plaskett who encouraged him to apply for a position at the soon-to-be opened DDO. Heard's qualifications and Plaskett's letter of reference secured him a position and, following a year of post-doctoral work at Yerkes Observatory near Chicago, he joined the staff in Toronto in 1935. For the first four months he wasn't paid at all, but made himself useful aligning and testing the telescope and putting the spectrograph together. He then assumed the position of demonstrator at a salary of $1,620 per year. During World War II he headed several air navigation schools attaining the rank of Squadron Leader with the RCAF. Returning to the University, he rose through the academic ranks and became head of the Astronomy Department and director of the Observatory in 1952, positions which he held until his retirement in 1965 following his first heart attack.

He originally joined the Montreal Centre of the RASC in 1930, and rejoined the Toronto Centre in 1935. He was a popular speaker in many Centres and gave more than one course of several lectures to Toronto Centre. He was on the Centre Council for thirty years including a term as President 1947–48, Nationally, he served as Vice-President and President 1949–54, Treasurer for most of the '60s, and chaired the Property and Constitution Committees at very demanding times. The Service Award in 1965 was small reward for his dedication.

Many of his delightful reminiscences were collected and published in the *Journal* following his death.

MARIE FIDLER was Executive Secretary of the Society from 1958 to 1972 and has been Editorial Assistant for the *Journal* ever since. During her years at the national office, she cheerfully handled a growing avalanche of work as *Handbook* sales nearly tripled and membership increased by over 70 percent. She proofread the *Journal*, arranged for advertisements in the publications, looked after the needs of three tenants, supervised the daily operation of the library and always took time to answer enquiries from school-children and to pass the time of day with visitors. Her joie de vivre, so well shown in this photograph, has won her countless friends in all parts of the country.

Marie's first marriage to Frank Fidler was blessed with two boys (and now three grandchildren). Her second marriage to Sam Litchinsky was an RASC romance which blossomed at the 1971 General Assembly in Hamilton and led to their wedding in Amherstburg, Ontario, the following year. They only had a year together before Sam's death, but Marie stayed on in Calgary, taking an active part in the Centre as Secretary. Following her return to Toronto in 1978, she continued her service to the Society as national Treasurer.

As a high school student at Toronto's Northern Secondary School, Marie won a gold medal for Pitman shorthand. In 1967 she was awarded Canada's Centennial Medal for her extensive volunteer work with hospitals and her dedication to the RASC, and in 1978 she received the Society's Service Award.

The photograph, by Glen Reed, shows Marie Fidler, arms raised in glee, at the Calgary General Assembly, 1976.

deficit in 1971. Plumbing, wiring, roofing all needed to be replaced, and very visible was a part of the library ceiling which was propped up with a piece of wood. The spacious verandah on occasion provided shelter for vagrants, and the attic a nesting place for squirrels. John Zarins, who had served with his wife as caretaker since 1956, died in 1972, and new occupants of the little apartment were less than eager to take on the problems of maintaining the building. News that the neighbouring Clarke Institute of Psychiatry was contemplating expansion, was all that was needed to convince the Council that the time had come to sell. For a while it seemed as though a favourable settlement would be reached with the Clarke Institute, but cut backs in provincial spending meant the curtailment of their plans, and cancellation of any interest they had in acquiring 252 College Street. (Eventually the Clarke Institute did purchase the property, though not from the RASC.) With some misgivings in the light of soaring inflation, the Society's headquarters was put up for sale in 1975. The national secretary, D.J. FitzGerald, and a small committee was appointed to deal with the problem of where to relocate, but a limit of $100,000 on the purchase price was imposed by Council.

The committee found no suitable properties for sale. There was certainly no advantage in moving to another old building where extensive renovations were required, and there were apparently no newer buildings zoned for commercial use in a convenient location. Finally with the signing of an agreement in the spring of 1976 to sell 252 College for the sum of $185,000, the committee was forced to make a decision. With time running out, they leased space at $7,500 per year in a new office building at 124 Merton Street, further from downtown than the Society had ever been, but still in midtown Toronto. It was a really lovely office, appointed with new furniture but less convenient for office workers or university people to drop in at lunch time to borrow a book or browse in the library, as they had done occasionally at the old location.

For six years, members of the property committee kept their eyes open and their ears to the ground, but every property they considered was found to be unsatisfactory for one reason or another. They had, in fact, formally got Council's permission to stop looking in 1982, but the announcement of a very large increase in the rent at Merton Street soon galvanized them into action once more. With a real sense of urgency now and a feeling that the economy was recovering from its severe recession, the decision was made in the spring of 1983 to buy a property at 136 Dupont Street, just a few doors from an identical one considered earlier. It was one of a row of three storey buildings with the main floor and basement zoned for commercial use, and the upper two floors designed as a two-bedroom apartment. The cost, $165,000, was somewhat less than the amount that had accumulated in the building fund, which allowed for legal and moving expenses. The space occupied by the Society on the main floor and basement was nearly the same as what had been available at Merton Street and on the main floor of the College

The only two properties ever owned by the RASC. Above is 252 College Street, headquarters from 1956–76, and right, the present home of the Society, purchased in 1983, at 136 Dupont Street.

CYRIL G. CLARK (1912-) succeeded against all odds. In England just before the First World War there was no government social assistance. Imagine the plight of a single mother with twin boys. Her only hope for work was as live-in help, but regrettably no employer was willing to take on a woman with more than one child. So it happened that Cyril Clark was separated from his twin brother at the age of four months and sent to London to Doctor Barnardo's Home for orphans. At age ten, the children had to move on, and Cyril chose to come to Canada. His destination was to be near Barrie, Ontario, but, in Cyril's words, "The farmer who had requested a boy had retired, so when I got off the train at Craighurst no one was there to meet me. The conductor set my trunk on the platform and the train pulled out!"

Fortunately the story has a happy ending. Cyril was cared for by a fine Christian family, married a girl from a nearby farm in 1933 and "lived happily ever after". After the first few years of married life in northern Ontario, he and Mary moved to Toronto in 1940 where he established an appliance business which was his livelihood until he sold it in the late 1970s. Presently, the Clarks live in Manilla, a village about an hour's drive from Toronto, and enjoy the winter months in Florida.

Cyril Clark joined the Toronto Centre in 1966 where he found the opportunity to share his life-long interest in astronomy with others at public starnights in city parks and on Saturday nights with visitors at the David Dunlap Observatory. His practical skills, good sense, and business experience were all appreciated as he served the Toronto Centre as Councillor and Second Vice-President and at the national level as Treasurer 1970-76 and Chairman of the Property Committee 1972-77. This was a particularly difficult period as the old building at 252 College Street required a lot of attention. Eventually he had a major role in the negotiations for its sale and the subsequent move to Merton Street. Even after 1977, he remained on the Property Committee and the Clarks many a time drove down from Manilla to help out with the move itself, with repairs and installations. Many of the homey touches around the present office are the result of their efforts. In appreciation for his commitment to the Society, Cyril Clark was awarded the Service Medal in 1987.

Cyril Clark on his back porch in Manilla, looking at a portrait of A.T. DeLury, a former RASC President and Manilla resident.

Street building. As a whole, it was much smaller than 252 College Street and was worth about half of the old property. Undoubtedly the value of the Society's assets shrank during the years of paying rent.

As a result of a very generous anonymous donation to the Society in 1981, it became possible to pay travel expenses for Centre representatives to come to the fall or winter Council meetings. Consequently, in the last few years, the attendance has grown beyond what can be comfortably accommodated at Dupont Street, and these two meetings are now held elsewhere, usually in the luxurious sixty-second floor boardroom of the firm of Smith, Lyons, where RASC member Michael Watson, currently the Society's second vice-president, practises law.

Many improvements have made the office procedures more efficient in recent years. The University of Toronto Press, who print the Society's publications, now mail out nearly all the bimonthly *Journals* and *Bulletins*, and ship most of the standing orders for the *Observer's Handbook*. The membership records are all computerized, as are the invoicing and accounting procedures. An electronic network links many of the Centres and some of the officers with headquarters and each other. Without these measures, additional help would surely have been needed, since the number of members served now is 50 percent greater than it was when the full-time executive-secretary's position was created.

As for the building at 136 Dupont Street, it continues for now to serve as office, library and a place for some committee meetings. But if the library is to grow, and if Council meetings are ever to return to headquarters, the Society will someday be on the move once again.

FRANKLIN C.J.T. LOEHDE (1936-) joined the Edmonton Centre while in junior high school and was soon attracted to all aspects of observing - meteors, auroras, lunar features and occultations, sunspots, and variable stars. He studied science at the University of Alberta, including Astronomy, and worked for two summers (1959-60) at the DAO. He was President of the Centre in 1961, during which time he helped to establish the Queen Elizabeth Planetarium and organized the first western GA. He was again Centre President in 1971-72.

A high-school teacher and administrator by profession, Loehde was the Centre's Director of Educational Activities for several years. A chapter on Astronomy in *Alberta - A Natural History* was his special Centennial project in 1967. He spoke a number of times on such topics as spectroscopy, on astronomy in Europe and Mexico and on the Edmonton Space Sciences Centre which he promoted enthusiastically as a member of the original Board of Directors and President of the Space Sciences Foundation. He arranged the "Donate-A-Star" program to raise funds for ESSC and organized the delivery of 16 000 promotional kits to classrooms across the province. Franklin's wife, Audrey, was a great help, serving as Social Convenor for many years. Her untimely death in 1989 was deeply felt by many in the Society.

At the national level, Loehde's first responsibility was as co-ordinator for lunar occultations on the Committee for Observational Activities in 1965. Since 1973, he has served continuously on the National Council, first as the representative of Edmonton Centre, then in the Presidential Offices and lately as Centre Rep once again. On receiving the Service Award in 1976, Franklin Loehde was commended for stimulating public appreciation of science especially among young people.

President Franklin Loehde with Dr Helen Hogg at the official opening of RASC Headquarters, 136 Dupont Street, in 1983.

M31 in Andromeda, a spiral galaxy somewhat like our own Milky Way, comprising hundreds of billions of stars.

CHAPTER 3

Giants and Dwarfs

This chapter is about the Society's populations – professional and amateur astronomers, women and men, youths and seniors, the dedicated few and the silent majority.

The title is not intended to be a putdown of any group, for astronomers know that even though giant stars shine brighter than dwarfs, it's the more plentiful main-sequence stars or dwarfs which determine a cluster's dynamics. Besides, giants and dwarfs are just at different stages in their evolution. Some professional astronomers, Bert Petrie, Doug Welch, Doug Gies and Pierre Bastien, for example, were once student members of the RASC. So were Richard Berry and Alan Dyer, later well-known Astronomy editors in the US, and Terence Dickinson, whose excellent books on the night sky are best sellers.

MEMBERSHIP

The tradition of the Society is that anyone with an interest in astronomy is eligible for membership. Leaflets explaining the Society's aims and activities were prepared as early as 1924 and these were sent out in reply to enquiries about membership and were distributed at public meetings and star nights. A French language version was printed subsequently.

Technically, all members must be elected by special resolution of the Council (i.e. 2/3 in favour) either at the national level for honorary or unattached members or by the Centre to which the applicant wishes to be attached. This is merely a formality, and a person is for all practical purposes considered a member once the fees are paid. There is, and always has been, provision for expulsion of a member whose continuance threatens the best interests of the Society, but this has very rarely been used. The Winnipeg Centre had to invoke this clause on two occasions to get rid of members who wanted to use the Society to expound the flat-earth "theory". (Only on the Prairies!)

Nearly every year from 1890 to 1905 complete membership lists with addresses were published in the Society's *Transactions*. In 1908 this information was included in the *Handbook*. Again in 1909, 1914, 1931 and 1936 lists were published in the *Journal*. But after that, the growing numbers made a separate printed roll the only practical way to provide the information. Such separate rolls were made available in 1958 and 1968. In recent years there have been calls for membership lists to be published again, but these suggestions have been consistently defeated by concerns that commercial interests might use copies for mailing out advertisements. Nonetheless, some Centres publish directories of their own members. The make-up of the membership in recent years can be gauged by two surveys. One study of 300 Ontario members was carried out by Dale Armstrong in 1984. He found that the median age was about thirty-five. Three-quarters had at least some college or university education, and so most members had incomes correspondingly higher than the general population. The people surveyed had been interested in astronomy for an average of eighteen years but the mean length of membership in the RASC was only 6.5 years. The other survey was completed just

Dr Andrew McKellar at the RASC meeting in Montreal, April, 1960.

ANDREW McKELLAR (1910-60) graduated from the University of British Columbia at age 20 and three years later received his PhD from the University of California. After post-doctoral work there and at the Massachusetts Institute of Technology, he returned to BC and accepted a position at the DAO following Plaskett's retirement in 1935. Like his colleague, R M Petrie who joined the DAO permanent staff at the same time, McKellar did research for the Canadian Navy during World War II, attaining the rank of Lieutenant Commander. Both men were elected Fellows of the RSC when they were only in their thirties. They shared in the design of the DAO's 112 cm telescope (later named for McKellar) and both were keen golfers. McKellar's main research interest was in the study of molecules in comets, stars and in interstellar space. He was probably the first astronomer anywhere to estimate the temperature of the 3° cosmic background radiation, long before its significance was appreciated and 25 years before it was measured at radio wavelengths by Penzias and Wilson. His work on the ratio of carbon isotopes in cool stars had important implications for stellar evolution.

McKellar joined the RASC in 1935 and at once took an active part in the Society, speaking at meetings in Victoria and Vancouver and to clubs and service organizations. He served on the Victoria Centre Council almost continuously from 1936-52 including two years as President and two as Honorary President. He served the national Society as 1st Vice-President and President. En route to special meetings of the RASC in Montreal in 1960, he spoke to a number of Centres and delivered his retiring Presidential Address on April 8, 1960. Less than a month later he died in Victoria as a result of an illness he had courageously lived with for 15 years. His son, ARW McKellar, is now an astronomer at NRC's Herzberg Institute of Astrophysics.

as this book was being written and comprises responses from 15 percent of the membership representing all segments of the Society. Some of the personal data was similar to the earlier results but now the median age was forty-four and the proportion with at least some college or university education had climbed to 87 percent.

CLASSES

By far the largest group within the organization has always been made up of the regular adult members who pay their fees annually. At various times they have been known as Active Members, Associates and Ordinary Members, the latter term still being operative.

Life Members form an increasingly important class. Anyone is welcome to join this group by paying the appropriate fee and life members now receive a large certificate suitable for framing. Since life members are normally on the roll until they die, while ordinary members typically belong for only a few years, the proportion of life members has steadily grown from about 2 percent in the 1920s through 5 percent in 1958 and 1968 to about 9 percent at present. There has traditionally been a provision whereby the Council can confer life membership on someone in recognition of specified meritorious service, but this has only rarely been done. Morris Altman, Donna Haley and Marie Fidler have been among the recipients for their years of service as auditor, solicitor and executive secretary respectively. Ian Shelton was recognized in this way following his discovery of Supernova 1987A.

The only other class recognized continuously since 1890 is Honorary Membership. This title is conferred by the Council "in recognition of noteworthy contributions to astronomy" though the wording of 1908 "scientific attainments of the highest order," more often guided the selection. At present this class comprises no more than fifteen

HONORARY MEMBERS OF THE RASC (in order of date of election)

- D. Kirkwood (1814–95)
- S. Fleming (1827–1915)
- S. Newcomb (1835–1909)
- E. Holden (1846–1914)
- W. Huggins (1824–1910)
- W. Christie (1845–1922)
- J. Morrison (1848–?)
- R. Ball (1840–1913)
- G. Darwin (1845–1912)
- S. Langley (1834–1906)
- E. Pickering (1846–1919)
- O. Struve (1819–1905)
- H. Vogel (1842–1907)
- J. Keeler (1857–1900)
- A. Downing (1850–1917)
- G. Hale (1868–1938)
- M. Loewy (1833–1907)
- C. Flammarion (1842–1925)
- J. Brashear (1840–1920)
- E. Barnard (1857–1923)
- S. Burnham (1838–1921)
- W. Campbell (1862–1938)
- W. Denning (1848–1931)
- T. Espin (1858–1934)
- E. Frost (1866–1935)
- K. Hirayama (1874–1943)
- W. Maunder (1851–1928)
- W. Monck (1839–1915)
- W. Pickering (1858–1938)
- F. Terby (1846–1911)
- S. Williams (1861–1938)
- A. Wolfer (1854–1931)
- O. Backlund (1846–1916)
- B. Baillaud (1848–1934)
- F. Dyson (1868–1939)
- P. Lowell (1855–1916)
- F. Schlesinger (1871–1943)
- H. Turner (1861–1930)
- W. Adams (1876–1956)
- H. Russell (1877–1957)
- E. Hertzsprung (1873–1967)
- H. Spencer Jones (1890–1960)
- H. Shapley (1885–1972)
- E. Hubble (1889–1953)
- H. Plaskett (1893–1980)
- W. Baade (1893–1960)
- J. Oort (1900–92)
- F. Stratton (1881–1960)
- O. Struve (1897–1963)
- B. Lindblad (1895–1965)
- I. Bowen (1898–1973)
- A. Unsold (1905–)
- S. Mitchell (1874–1960)
- J. Stebbins (1878–1966)
- G. Van Biesbroeck (1880–1974)
- B. Bok (1906–83)
- V. Ambartsumian (1908–)
- A. Joy (1882–1973)
- R. Woolley (1906–86)
- M. Schwarzschild (1912–)
- G. Clemence (1908–74)
- J. Shklovsky (1916–85)
- H.C. van de Hulst (1918–)
- B. Stromgren (1908–87)
- M. Ryle (1918–84)
- W. Iwanowska (1905-)
- L. Goldberg (1913–87)
- A. Sandage (1926–)
- A. Dollfus (1924–)
- R. Baldwin (1912-)
- F. Bateson (1909–)
- S. Hawking (1942–)
- R. Hanbury Brown (1916–)
- H. Hogg (1905–93)
- G. Reber (1911–)
- O. Gingerich (1930-)
- P. Moore
- R. Evans

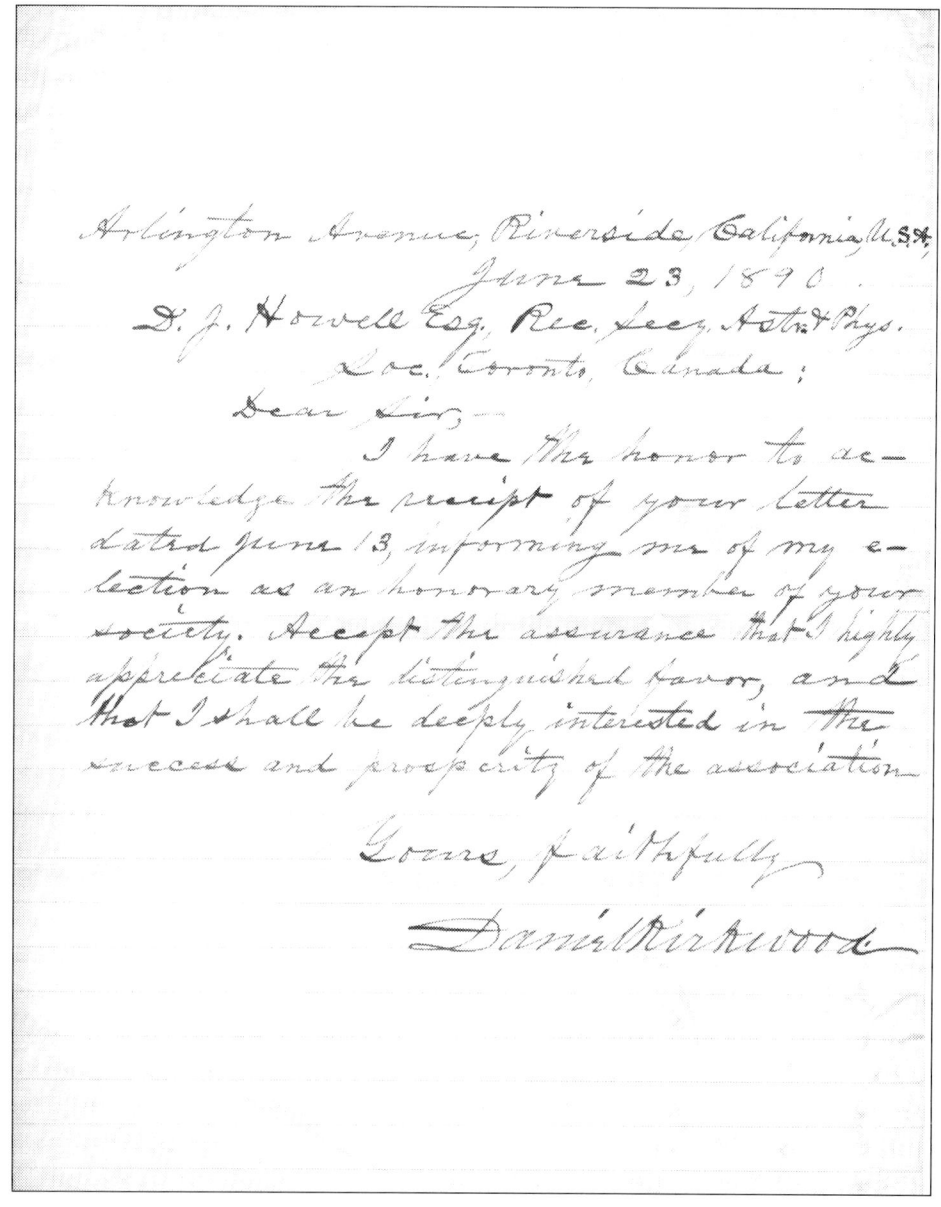

The first of many gracious letters of acceptance in the Society's archives.

ROBERT M. PETRIE (1906-66) came from Scotland to Canada with his parents when he was five years old. "Bert" joined the Victoria Centre at age 18 and within two years became Recorder and wrote his first paper "Variable Star Observing for Amateurs" for the Journal. Each summer during his undergraduate years he worked at the DAO observing and measuring spectra and determining orbits of spectroscopic binaries. On graduation from UBC in 1928, he secured a scholarship and went to the University of Michigan for his AM and PhD degrees. He worked for three years at the Detroit Observatory and then returned to the DAO in 1935. One of his long-term projects was the spectroscopic study of B-stars which led to a better understanding of the structure of the Galaxy. During the War he worked in civil defense and operations research and became Lieutenant-Commander in the Navy. An excellent administrator, he was appointed Dominion Astronomer in 1964 with responsibility for the proposed Queen Elizabeth II observatory project. At home and abroad, Petrie received many honours indicative of his distinguished career in Astronomy.

The RASC benefitted greatly from his talents. On many occasions he spoke to the Victoria Centre and he addressed several other Centres too. He published dozens of papers in the Journal and, as Associate Editor, instituted an informative and popular series called "Canadian Scientists Report". From 1936-41 he held office in the local Centre, including a term as President and was national Vice-President and President 1953-57. In spite of all the heavy demands on his time, Petrie was an avid sports fan and a keen golfer.

His second wife, the former Jean McDonald, was also an astronomer and his younger brother William, a physicist specializing in auroral research, was an active member of the RASC, serving at various times on the Councils of Victoria, Vancouver and Saskatoon Centres.

eminent international astronomers, though the limit and the title itself have varied over the years. In some cases these renowned figures have had little direct connection with the Society, but their willingness to be associated in this way has brought recognition and prestige to the Society. There was a period (from 1916 to 1944) when the Society elected no new honorary members and in fact all those who had been elected were dead by 1944. This lapse was probably due to the fact that the nomination of honorary members was not really anyone's responsibility. The situation was rectified in 1944 when the Constitution was revised and a standing committee comprising the president and two past-presidents was established for the purpose.

Of course many well-known foreign astronomers have chosen to join the Society as ordinary members and their support has also been important in publicizing the Society outside the country.

Corresponding Members and Fellows were recognized classes within the Society for many years but they ceased to exist by 1944. There has also been a desire from time to time to recognize those members who had made generous donations to the Society, and consequently there were a very few named Patrons in the early 1900s. Also, during a fund-raising campaign in 1954, two classes of donors were recognized as Sustaining Members and Patrons, but these descriptions were not formalized in the by-laws.

A Youth Membership category nowadays seems obvious but it was not always so. That is not to say that the Society used to ignore young people. There was, in fact, a sort of "Women, children and out-of-towners" class in effect from 1900 to 1908 by virtue of the fact that only men living in Toronto were required to pay the full fee, others paying half. However, the first attempts to form a separate Youth Membership class were not taken until 1931 when Victoria Centre in a generous act of leadership, made the following arrangements with the National Council:

> Until the age of 21, membership dues at this Centre are one-half of the regular dues. These Junior Members have all the privileges of the Society except the right to vote. The Centre receives no grant from Headquarters on account of the dues paid by these Junior Members. The General Society consequently profits equally from them as from the full members. If the Society is to flourish, there must be a continuing influx of youth and [its] enthusiasm.

Nonetheless, a lower-fee class of membership specifically for young people was not instituted nationally until the constitutional revision of 1944 formally made provision for full-time students at any educational institution, or other persons under twenty-one years of age, to become Junior Members of a Centre. The 1957 by-laws extended the class to unattached members but at the same time stipulated that only those enrolled full-time at an educational institution were eligible for Student Membership as it then became known.

Statistics on Student Membership are hard to find since the published membership lists do not indicate to which class the members belonged. However, in 1959 it was reported that 6.5 percent of the members were students and in 1964, as part of a study on the effect various proposed fee increases might have, it was indicated that there were 447 Student Members, slightly over 20 percent of the total. (Not surprisingly, amongst unattached members the figure was only about 8 percent, but in Kingston, twelve of its sixteen members were students, in Edmonton the proportion was nearly 40 percent and in Toronto 32 percent.) There seems to have been no policy change within the Society itself to account for this tremendous increase in student interest over a five-year interval; presumably the space age and the general expansion of universities were important factors.

Another important element was recalled by Dan Brunton, one of a keen group of students in the Ottawa Centre during the 1960s:

> Meteor observing was a passion for us at the time. It had all the right elements for a bunch of energetic, inquisitive teenagers; it was competitive, inexpensive, slightly off-beat ... and a wonderful excuse for staying up all night.

The Student Membership roll climbed even higher for a few years. The best evidence comes from lists of members elected at meetings of National Council. Of the new unattached members joining between 1966 and 1973, fully 40 percent were students. If the figures cited earlier are any indication, there would have been an even higher proportion of students among new Centre members. The boom came to a rapid end in 1973.

Councillors became aware that the Student Members got the same services as Ordinary Members but at a cost which the Society could not bear on a large scale. So in 1973 the by-laws were amended to do away with the Student Membership class and provide for Youth Members under eighteen instead. This, of course, had the immediate effect of requiring full fees from university students. The extent to which they continued to belong is unfortunately unknown.

The maximum age for Youth Members was restored to twenty-one in 1989. So far this policy has not brought a return of large numbers of Youth Members who presently comprise less than 5 percent of the total.

Senior Members got a separate class and a financial break for the first time in the 1989 by-laws when those at least sixty-five years of age could apply for membership at the same fee as Youth Members. This innovation was really just an obvious

New members are welcomed to Toronto Centre by Mary Anne Harrington (3rd from right).

Young members of Toronto Centre putting together an issue of 'Scope in the national office at 252 College Street.

Raymond Auclair seated at the head table at the 1989 General Assembly Banquet, with his wife Hélène.

Photo by Henry Lee

RAYMOND AUCLAIR (1952-), because of his career with the Canadian Coast Guard, has found it difficult to become attached to any Centre of the RASC. But even as an Unattached Member he has contributed substantially to the Society and as a result received the Service Award in 1989.

Captain Auclair joined the RASC in 1969 and became a Life Member in 1973. At least since 1981, he has attended most National Council meetings, formally as an observer, but in reality as a strong advocate of the interests of Unattached Members. He conducted a survey in 1981 to find out the prevailing attitudes and ideas of his Canadian counterparts and has since actively served on a number of national RASC Committees. Fluently bilingual, he has reviewed books for the *Journal*, contributed articles in English to the *Bulletin* and in French to *Québec Astronomique*, and spoken at a number of General Assemblies. Navigation is naturally one of his major interests, but whatever topic he turns to, his presentations always sparkle with his lively sense of humour. Even his unfortunate appendectomy during a trip to Soviet observatories has become a legendary story among those who heard his account.

In 1989 he broke new ground by organizing the first General Assembly to be hosted by Unattached Members. Volunteers came from the Cape Breton Astronomical Society and from the Coast Guard College where he was Chief of Training. His enthusiasm, energy and organizational skill made the event highly successful.

extension of the sort of privileges accorded by banks, theatres and many other organizations in the 1980s, but it was terminated in 1993 as the number of senior members began to grow rapidly.

Finally, there is a class called Associate Membership. Back in the 1890s, these were people who paid only half the normal fee because they lived outside Toronto and could not enjoy many of the benefits of membership. Then from 1900 to 1908, the designation applied to what is now called Ordinary Membership. For a long time, the term disappeared from the Society's vocabulary, but reappeared in 1969 with a different meaning which has been in use ever since. The purpose of Associate Membership now is to allow one or more family members of an ordinary, youth, or life member attached to a Centre to belong to that Centre without being a member of the Society. As recent survey results indicate, about 70 percent of members are married and about half of all members share their interest in astronomy with at least one other member of their families. Such people, for a reduced fee, get certain Centre privileges, but have no National rights. Associate Membership can also apply to members from one Centre who wish to have ties to another Centre but obviously do not want to have two or more full memberships in the Society.

UNATTACHED MEMBERS

There have always been members who lived too far from Toronto or any other organized Centre to attend meetings regularly. As just explained, these people were called Associates in the earliest years. After 1908, they were given no special status or reduced rates and were included in the list for Toronto, which was regarded as the headquarters of the Society and not really a Centre in the same sense as Ottawa, Hamilton or any of the others. Then in 1930 it was decided that each member would henceforth either be attached to a particular Centre of the member's choice, or else to a section known as Members-at-

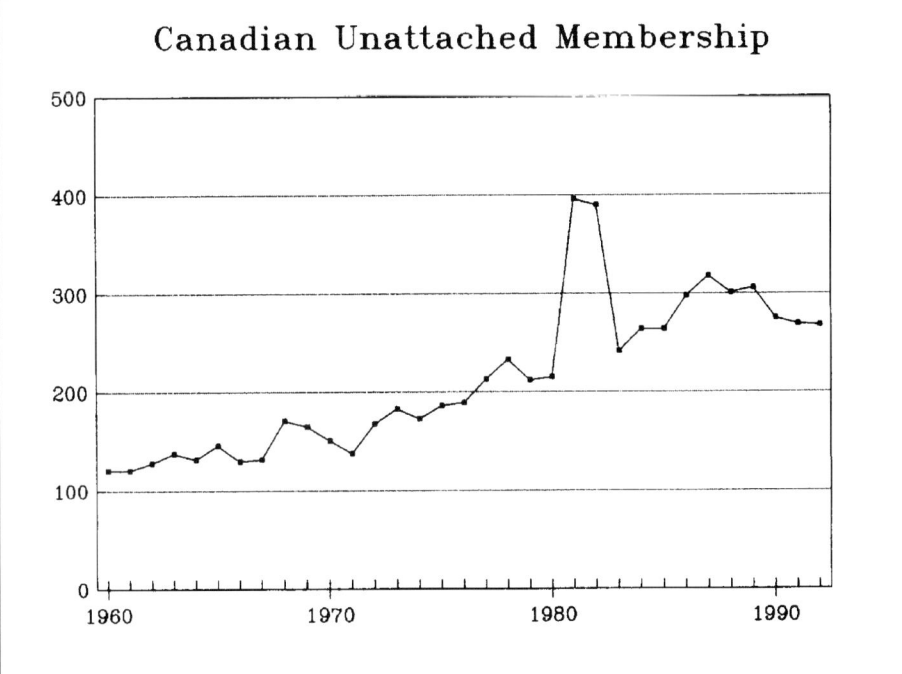

Large. The term Unattached Members came into use with the by-laws of 1944.

Although unattached members, by definition, do not attend Centre functions, some have taken an active and important role in the Society by contributing papers to the *Journal* or at General Assemblies. In fact, unattached members in Cape Breton Island, under the leadership of Captain Raymond Auclair, organized the 1989 General Assembly.

Committees have looked for ways to help unattached members feel more a part of the Society. During her presidency, Helen Hogg began efforts to send each unattached member annually some astronomical material such as star charts, posters and booklets. Centres were encouraged to hold meetings occasionally on weekends especially so that out-of-town members might attend. A survey of those unaffiliated with any Centre in 1979 led to the suggestion that membership directories and newsletters might be helpful but in the end the *National Newsletter* was considered the best means of communication.

The Unattached Members numbered about one hundred or 14 percent of the total when they were first counted in 1931, and the annual report of 1936 noted that they came from countries as widely scattered as Java, Palestine, the Argentine and Egypt as well as the principal European nations. There are now about 600 Unattached Members or 20 percent of the total and they represent even greater geographic diversity.

The accompanying graphs showing numbers of Unattached Members begin in 1960, the first year in which Life Members were included in the count. The figures in the Canadian graph for 1981 and 1982 (and therefore the totals for those years) are hard to believe. There seems to be no way to explain the dramatic jump from 1980 and the sudden drop in 1983. But the relatively strong increase in the

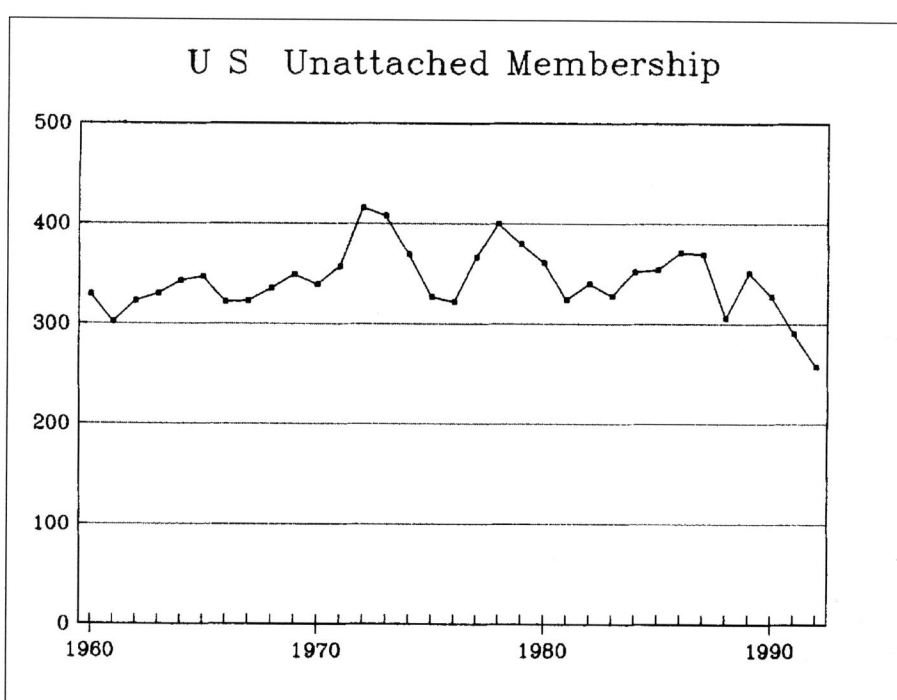
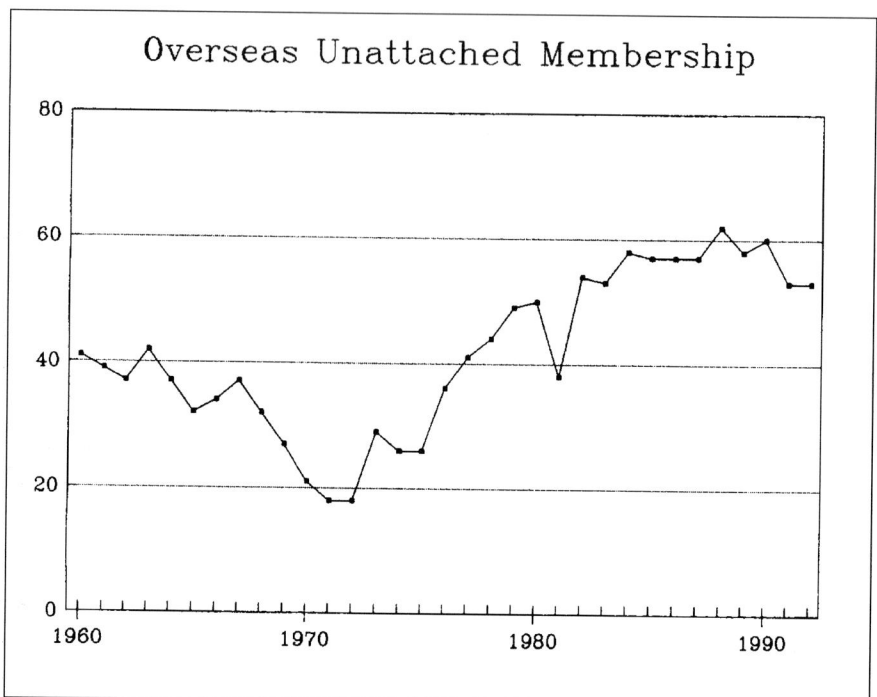

numbers of Canadian Unattached Members over the last thirty years is clear enough. With more and more new Centres starting up with the potential of absorbing unattached members, this is a surprising trend. American membership seems to be in recent decline though offset somewhat by an increase in the number of overseas members.

NUMBERS

The figure opposite shows the number of members in the Society as a whole. Prior to 1950 the data can only give an upper limit to the number of paid-up members at any particular time. Nowadays, the matter is cut and dried: the membership year ends on September 30 and members have until December 31 to pay their fees. But in the early years of the Society, there was apparently a great reluctance to reduce the roll by removing those who were in arrears. Certainly there were special circumstances.

During World War I, for instance, members serving overseas were excused from payment, and during the Depression, collection of dues was sometimes allowed to slip. With members in arrears for various lengths of time and some paying for more than one year at a time, it should be clear that exact membership statistics cannot be given. It was only after World War II, in 1948 to be exact, when the Society was in very bad financial straits, that a new policy of purging the lists of delinquent members was instituted.

The general trend, however, is clear enough – the Society has shown healthy growth throughout its history. There have been three periods when the rate of increase markedly outpaced the growth rate in the general population of Canada. During the first two decades, 1890–1910, when the Society was mainly a local club, personal contact was very easy and youthful enthusiasm saw almost limitless opportunities for expansion. The second surge occurred in the 1940s. World War II brought in a lot of new members from the Air Force who had used *The Observer's Handbook* in their navigation training. Undoubtedly Peter Millman and Jack Heard, both instructors in navigation, had a lot to do with encouraging their men to become members. Great public interest in Halley's Comet led to a rise in the '80s, with a remarkable 20 percent increase in numbers in 1985 alone. But unlike the earlier periods of rapid expansion, this gain in membership was followed by a decline to levels only moderately higher than those of the early 1980s.

Just as important and less well-known are data relating to the retention of Ordinary Members. On average, these people stay with the Society for four years with approximately an exponential drop-off in numbers with time. In other words, the half-life of an Ordinary Membership is about four years. Samples taken of various periods over the last eighty years show this statistic has been remarkably constant, with more variation from Centre to Centre than from time to time. Whether this is a good or bad retention rate is a matter for debate, but it does mean that the Society has had to be very active in recruiting new members since approximately 16 percent must be replaced each year just to maintain the status quo. The optimistic way of looking at this is to recognize that the large number of ex-members bears out the great success the Society has had in attracting the public to take an interest in astronomy. The fact that many do not stay for long is perhaps only to be expected. After all, personal growth always requires commitment and time, precious commodities for most people.

There have been instances where Centres lost the initiative needed to keep programs going. London and Windsor in the 1960s and Montreal in the 1970s saw just how quickly numbers can drop when little is done to retain old members or attract new ones.

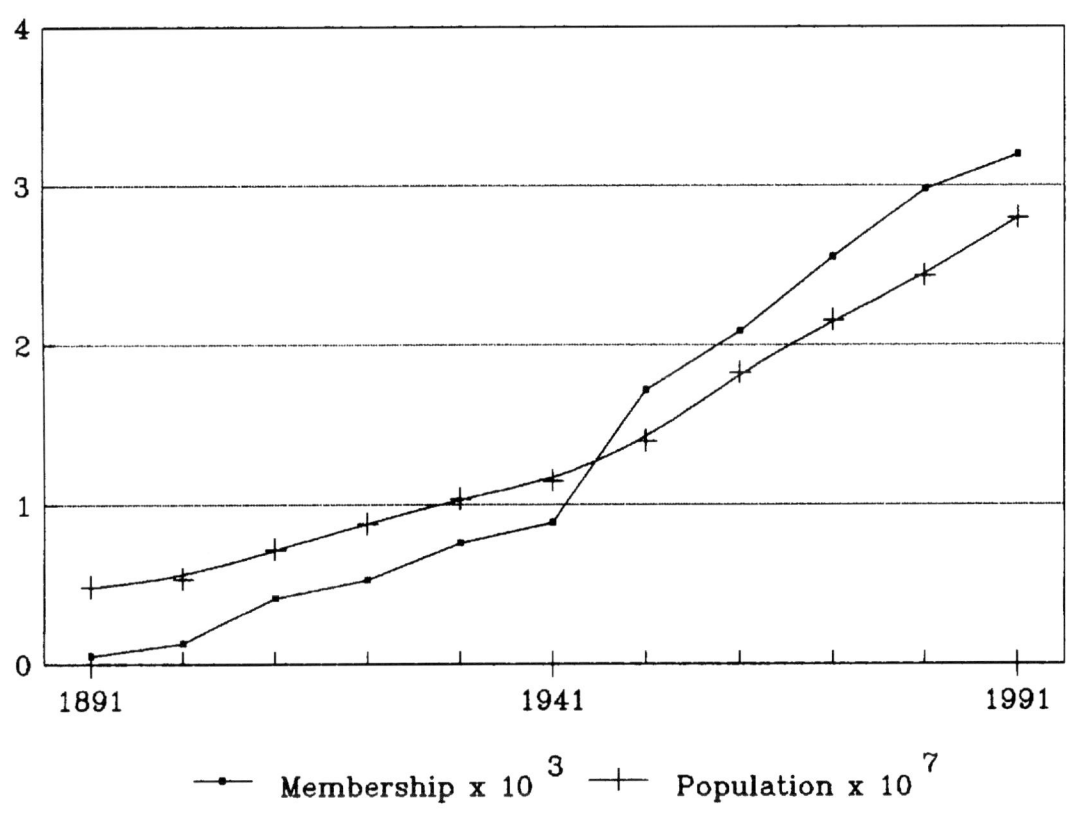

Much can be done to make new members feel they belong. Don Hladiuk and William Krosney of the Calgary Centre, as a result of a long-term planning meeting in 1984, recommended that a Centre membership committee establish a welcome session for all new members to be held at least once a year and that the committee should also produce a comprehensive up-to-date welcome package for all new members. As for retaining members, they recognized that education, goals and awards, socializing and fun were all important ingredients and

PETER M. MILLMAN (1906-90) was born in Toronto but spent most of his youth in Japan where his father was a missionary. Even as a boy, Peter had a keen interest in astronomy, especially Mars. He recalled getting his father to help him carry his telescope to the top of a Japanese mountain and also being impressed by a large rock, supposed to be a meteorite, in his town of Toyohashi.

Soon after beginning his studies at U of T in 1925, he joined the RASC and published his first paper in the *Journal* the following year. He earned his BA and the RASC Gold Medal in 1929 and then left for Harvard where he got his AM and PhD degrees. There he was greatly influenced by Dr Willard Fisher whom he credited with inspiring his lasting interest in meteors. During his years in the States, Peter married and he and his wife Peggy began their lifelong friendship with the Hoggs who were also Harvard students at the same time.

In 1933 Dr Millman returned to U of T where he lectured in Astronomy until 1941. He then enlisted in the RCAF, rising to the rank of Squadron Leader. (This later had astronomical spin-offs as he was able to organize airborne eclipse and meteor shower expeditions.) Following the War, Millman went to the DO in Ottawa, eventually becoming the Chief of the Stellar Physics Division. In 1955, he moved to NRC to head Upper Atmosphere Research until his retirement in 1971. In reality he never slowed down and continued his illustrious career, receiving many honours as an internationally recognized authority on meteors and meteorites.

In 1933, he began recruiting RASC members and other amateurs to observe meteors and started "Meteor News", a regular feature of the *Journal* for 25 years. He also contributed a section on meteors to the *Observer's Handbook* for 49 years. Peter Millman served for many years on the Councils of Toronto and Ottawa Centres including a term as President of the Ottawa Centre, and in various national offices including Librarian, President, and Honorary President. He was a popular speaker at Centre meetings across the country and was well-known even outside the Society for his radio series *This Universe of Space,* later published in book form in four countries, and for his weekly astronomy columns in *The Toronto Star*. In some small measure the Society tried to recognize his immense contribution by presenting him with the Service Award in 1967 and by posthumously naming its Endowment Fund for him.

recommended specific, well-defined responsibilities for each of the Centre's committees in these areas.

But when all is said and done, it's the personal touch that so often makes a difference. Here is what a new member in 1980, Becca Stone, had to say about her experience:

> On March 15, eleven of us gathered at IRO (Ottawa Centre's Observatory) for star-gazing. I am a new member and so was looking forward to seeing the telescope, site and the stars, but also to meet other members. Rob Dick kindly gave me a lift there and by the time we arrived, I was well briefed on the IRO's history. Doug George showed me the Orion Nebula — my first sight through a telescope of that size. Frank Roy was making hot chocolate for everyone and generous Frank ended up with none. ...
>
> I was greatly impressed with the Indian River Observatory. In particular it was the helpfulness and the warm welcome I received that impressed me the most.

We turn now to other groups of members, not designated by special membership classification, but who deserve special mention for their contributions to the Society — women, professional astronomers and award winners.

WOMEN

Since women have always been active in the Society in every way, a separate section outlining their contributions may appear redundant. The fact is, however, that women presently make up only about 8 percent of the total membership. Little has been done to understand the reasons for this small proportion, and no recent programs have been set up specifically to attract women to join the RASC. This is somewhat anomalous in the light of the Society's commitment to public education and the efforts of educators generally to encourage female students to share in the enjoyment of science.

Maybe when the next history of the Society is written, the interest and commitment of women will be so strong that current concern with such issues will seem quaint.

From the earliest years, the Society took pride in the fact that women were full participants in all its activities. In 1897, the year when Women's Institutes originated in Canada, President Paterson made this interesting observation about the Royal Astronomical Society (of London):

> Ladies are only admitted to the ordinary evening meetings [of the RAS] by special invitation of the President, sanctioned by the Council. ... I am happy to say that we, in this Toronto Society, have no such haughty rule of exclusiveness.

To be fair, the RAS had been established in 1820 when the practice of excluding women was generally accepted. Honorary Membership had been extended to some (Caroline Herschel and Mary Somerville being two notable examples), and by 1915 the old rule was abolished and women were admitted to regular membership.

In the RASC, women have always been regular members. But is their numerical strength growing or shrinking? At least five from a total of fifty-two signed the original roll in 1890, a proportion not much different from the present. By examining membership lists published in different years, it is possible to get an idea of how many women actually belonged at various times. In the early years, numbers were small and subject to variation between 9 percent and 19 percent with a mean value of about 15 percent. In more recent times, membership lists are available only for a few particular years, but with larger overall numbers, there should be less scatter from year to year. The trend seems disappointingly clear:

Year	1893–1914	1931	1936	1958	1968	1990
Women/Total	9% to 19%	16%	16%	10%	9%	8%

Since the sex of a member is not labelled on the membership records, some guesses had to be made.

What features of the Society did women find more attractive in years gone by, and what has happened in more recent times to deter them? The most obvious factor which affected women within the Society in the early years was the membership fee. From 1893 until 1908, women paid only half the regular fee, one dollar per year instead of two. Nothing in the minutes provides any rationalization for this, but it was very likely a pragmatic recognition that few women worked and those who did were paid less than men. It is doubtful that there was any conscious motivation specifically to encourage ladies to join, but the statistics could be interpreted to show that the lower fees had that effect. The proportion of women did drop in 1909 and 1914 to 12 percent and 9 percent respectively.

On the other hand, there were societal factors which may have been even more significant. There was a genuine interest in nature in the Victorian era, and this was considered quite an appropriate activity for women to engage in. A number of ladies showed great skill in making sketches at the telescope of planets and the lunar surface. A group in Simcoe, Ontario, using Dr J.J. Wadsworth's 32-cm telescope won considerable admiration for their work. They included Miss Stennett, Miss Beemer, and Miss Eva Brook, whose drawing of Saturn was used as the frontispiece to Volume 6 of the Society's *Transactions*. Mr Wallace of Orillia set up his telescope every clear night in the summer of 1897 and he noted that of 1,100 visitors, "fully 50 percent were ladies." In Toronto, Annie Gray observed occultations during a lunar eclipse in 1895, and several ladies observed the Leonid meteors in 1897. A special observing meeting for women was reported in 1898.

Mrs Annie Savigny was a Toronto member who took a special interest in dismantling the barricades

Louis Duchow and A.V. Douglas at the 1963 solar eclipse in Plessisville, Quebec.

ALLIE VIBERT DOUGLAS (1894-1988) was the first woman to be President of the RASC (1943-44), but this was only one of many distinctions she held. She was made a Member of the British Empire in 1918 for her work with the War Office in London, England, received honorary degrees from Queen's University in Kingston, Ontario, and from the University of Queensland, Australia, was President of the International Federation of University Women, and was selected in 1967 by the National Council of Jewish Women as the "Woman of the Century" for Ontario.

Allie Douglas (as she was known in her youth and to her closest friends) began her illustrious career in Montreal. She received her BA and MSc degrees from McGill and in 1921 went to Cambridge, where she studied Physics under Sir Arthur Eddington. (Many years later she wrote his biography.) She then returned to McGill where she lectured in Physics and Astrophysics and earned her PhD in 1925. In 1939 she was appointed Dean of Women at Queen's University, a position she held for twenty years. For much of this time, and until 1963, Dr Douglas was also Professor of Astronomy at Queen's and published many papers in international scientific journals and general periodicals.

Within the RASC, first in Montreal over a period of fifteen years, she was at various times Recorder, Secretary, or Treasurer, and she addressed the Centre nearly every year as well as speaking a number of times to the Ottawa Centre. After moving to Kingston and holding the Presidential national offices, Dr Douglas became the motive force in the formation of the Kingston Centre and was their Honorary President for many years. She contributed a long list of interesting papers to the *Journal*, many relating to the history of astronomy. She received the RASC Service Award in 1963.

HELEN HOGG (1905-93) was born Helen Sawyer in Lowell, Massachusetts. After earning her Bachelor's degree in 1926, she proceeded towards her doctorate in Astronomy under Harlow Shapley at Harvard in 1931. There she met a young Canadian, Frank Hogg, who was also doing post-graduate studies in Astronomy and they were married in 1930. A year later they set out in their Model A Ford for Victoria where a position awaited Frank at the Dominion Astrophysical Observatory.

Unfortunately, because of the depression, there was no opening at the DAO for Helen, but the Director, J S Plaskett, gave her the use of the 72-inch telescope to further the research she had begun at Harvard on variable stars in globular clusters. On some nights baby Sally, the first of her three children, slept in the dome while her mother took photographic plates.

In 1935, the Hoggs came to the new David Dunlap Observatory. They rose steadily through the academic ranks, with Frank becoming the Director from 1946 until his untimely death in 1951, and Helen becoming a full Professor in 1957. Dr Helen Hogg's international stature in her field grew as a result of her identification of hundreds of new variables and her many professional papers, notably the bibliographies of globular clusters.

A complete list of the honours and awards she received would fill this page. Many of them reflect her generous efforts to ensure that *The Stars Belong to Everyone* (to use the title of her popular book). Every week from 1951 to 1981 she prepared an astronomy column for *The Toronto Star* and characteristically handled the steady flow of correspondence from the public with kindness and care.

Within the Society itself, Dr Hogg spoke to numerous meetings from Victoria in 1931 to Halifax in 1980 and her many historical contributions to the *Journal* were always lively and fascinating. In the 1940s and 50s she served on the Toronto Centre Council prior to becoming the national President in 1957. Her myriad contributions were recognized in the presentation to her of the Service Award in 1967 and her election to a term as Honorary President from 1977-81. The Society's deep respect for her scientific work was marked in 1987 when she became the only Canadian among 15 eminent Honorary Members.

On becoming a Companion of the Order of Canada in 1976, Helen Sawyer Hogg receives congratulations from Governor General Jules Leger. Daughter Sally is on the left.

of tradition. In the course of a talk she gave in 1895 on the pleasures of the telescopic observation of nebulae, she had this to say:

> We men and women of the Earth bury our minds in things of matter, learn too little of the stars, gaze but seldom into the sublime stellar depths. Men fetter their eagle spirits, which fain would soar, by grovelling for some petty place or aim. We women weave our mental shrouds at pink luncheons, green teas or blue dinners. Would that we might rather ... search out new truths and endeavour to expand the field of knowledge.

Mrs Savigny contributed other papers to the Society, but perhaps most interesting was a report that she had communicated with the Council of Women in connection with the World's Fair in Paris. With this report she had forwarded a copy of the Society's *Transactions* as evidencing the part taken by women in Canada in scientific work.

Some of the ladies assumed administrative responsibilities within the Society. In 1892, Misses Gray, Pursey and Taylor were appointed assistants to the recording secretary, librarian, and treasurer, respectively. Sarah Taylor was in fact proposed to succeed Mr Lindsay as recording secretary, but she declined. The first female officer of the Society was Miss Elsie A. Dent. She had been nominated to serve on the Council for 1902, but withdrew her name. However, in 1904 she did accept the position of recording secretary, and continued in that office for over five years. Besides her duties of recording minutes and preparing bimonthly summaries of the meetings for publication in the *Journal*, Elsie Dent gave "Easy Star Lessons" at each regular meeting during 1905–06, and contributed a number of papers as well. Like Mrs Savigny she took a keen interest in women's work in astronomy and prepared a paper on this theme in 1901. Miss Dent spoke of the lack of scientific preparation, the systemic prejudice against "learned ladies," and "that instinctive defiance against intrusions into the happiness and comfort of family life." She acknowledged the encouragement male colleagues provided to those few women who had made a mark and gave several examples of highly regarded women who had made significant contributions to Astronomy in the past. Perhaps the most revealing part of the article was the information Miss Dent provided on contemporary scientific women – Lady Huggins and Mrs Maunder who collaborated with their husbands, Agnes Clerke and Dorothea Klumpke, famed for their own independent careers and Annie Sheepshanks and Catherine Bruce who, amongst others, gave generous financial support to Astronomy. Elsie Dent had written a lasting, informative, and forthright paper. She had done a considerable amount of thorough research and had evidently written to several astronomers for their own opinions. It was said that, "Miss Dent's paper fully established the right of our lady members to stand on an equal footing with the gentlemen in all the privileges and duties of membership." When Elsie Dent moved to Ottawa in February, 1910, members presented her with a gold bracelet set with amethysts and pearls as an indication of their esteem. Unfortunately, she and her sister Lillian, a kindergarten teacher in Toronto, disappeared from the RASC records after that date.

Whether social acceptability or lower fees was the more important factor in the enhanced level of women's membership during the Society's early years is anybody's guess. Perhaps other organizations without the fee bias will be able to settle the question. What does seem certain is that many of the Victorian gentlemen of the Society not only accepted the ladies as equals but went out of their way to encourage them.

Turning now to the 1930s, there were certainly no economic benefits in those very difficult economic times. Nonetheless for some reason the

Isabel K. Williamson (1907-) received the Chant Medal in 1948 and the Service Award in 1981, the only person to be doubly honoured in this way.

An RASC notice posted in the Sun Life building in Montreal, where Miss Williamson worked, originally attracted her to a series of introductory lectures in January, 1942. She immediately began to take an active part, and over the next 25 years made observations of auroras and meteors, organized others to do the same, and reported the results to appropriate scientific bodies. Miss Williamson built her own 15 cm reflector in 1947, but used the Centre's fine old refractor for lunar and planetary observation, rarely missing the weekly meetings. She participated in the AAVSO program of Nova searches from 1958 to '67 and was on their Council from '62 to '65. Dr Gartlein of Cornell University stated that her aurora reports were unequalled by any group in a large city and Dr Peter Millman paid tribute to the work carried out in 1946 under her direction: "I consider that the Giacobinid observations of the Montreal Centre are the outstanding example of meteor observations planned and carried out by amateurs for the purpose of producing scientific results of value."

Miss Williamson's enthusiasm led her to start a Messier Club in the '40s, the first of its kind in North America. Its purpose, to foster good observing skills through searching for, drawing or photographing all the Messier objects, lives on in the Society's Messier Certificate, awarded to members who complete the program.

From 1948 to 1971, Miss Williamson was the first Editor of the Centre's newsletter, *Skyward*, noted for its high quality, punctual appearance and lively style. For junior members she instituted "The Asteroid Club" which flourished from 1946-51. From 1942-50, Miss Williamson was the Centre's Recording Secretary, taking as part of her duties the publicizing and reporting of Centre activities in the newspapers, and from 1950-52 the Centre President. In recognition of her inspiring leadership, the Montreal Centre named their observatory for her in 1987. But perhaps the finest tribute that can be paid to her was written by David Levy in the dedication of his book *The Joy of Gazing* - "To Isabel K. Willamson, whose wisdom and teaching profoundly influenced a generation of amateur astronomers in Montreal."

Isabel Williamson presenting a paper in the University of Toronto's Hart House at the 1959 Annual Meeting. Seated beside her is the Chairman of the meeting, Peter Millman.

Photo by Geoffrey Gaherty, Jr.

number of members did not decline and the proportion of women was as high or higher than it had ever been. Perhaps people had more time for leisure activities, and were glad to have a free evening out watching the stars or attending a meeting. By now there were several Centres in the Society, and their varying degrees of success in attracting women may shed some light on what characteristics appealed to the ladies. In the 1930s, both London (with 37–38 percent women) and Winnipeg (with 28–27 percent) maintained high proportions of female members. Dr H.R. Kingston deserves a lot of the credit for many novel features which he introduced into the London Centre meetings. Contests on the recognition of constellations, book reviews, Handbook talks and occasional debates were arranged. Visiting lecturers were entertained at informal dinners at Wong's Cafe, social meetings with music and games were sometimes held at members' homes, and annual picnics at a member's farm out of town concluded with campfire talks. Whenever the weather allowed, observations were a part of every meeting. It all sounded like a lot of fun, and with a total membership of about fifty, these types of events were easier to organize than would have been the case in larger or smaller Centres. In Winnipeg, a Centre of comparable size, female members played a very active role. In 1927, Mrs E.L. Taylor became the first woman president of an RASC Centre, and she and other women took their turns along with the men in giving talks. When she visited Toronto that same year, Mrs Taylor spoke of the enthusiasm and good attendance at Winnipeg meetings. There were a number of visits to Winnipeg by distinguished astronomers in the 1930s, and these were always the occasion for a reception or luncheon. As in London, observing sessions were a regular feature of meetings whenever possible.

At the present time, Centres still have social events, and regular observing sessions, and women are more willing than ever before to accept responsibility on Centre Councils. In fact women are slightly more represented among Centre officers than among the general membership. So why are proportionately fewer women joining the Society now than in the past? As Mary Lou Whitehorne, the 1993 recipient of the Chant medal, recently observed:

> There are probably plenty of women who would like to be involved in the RASC but they must work for a living. They also bear most of the burden of raising children and homemaking in addition to bringing home the bacon. Is it any wonder that there is no time or energy left for anything else? Add to that the subtle (and not-so-subtle) sexism that infiltrates society in general and we have a very nice men's club indeed.

Women's accomplishments have received little specific attention within the RASC. The only Journal paper on "Women in Astronomy" of recent date was written by John Percy in 1981 and that was concerned with the decline in the professional community of female astronomers. RASC women have sometimes shown interest in this issue. During World War II, the Hamilton and Victoria Centres both held meetings entirely arranged by females. Madame Eugène Pettelier spoke to the Centre français in 1954 on "La femme canadienne-française et l'astronomie" and in recent years Sue Knight Sorensen spoke to the Kingston Centre and Myra Rutkowski to the Winnipeg and Calgary Centres on Women Astronomers and Women in Space.

PROFESSIONAL/AMATEUR MIX

The RASC is rare among astronomical societies in the importance of both amateurs and professionals to its history and continued well-being – perhaps the ASP comes closest in this respect. In general, RASC members of all varieties enjoy the friendliest of relationships but occasional rumblings are heard.

JACK L. LOCKE (1921-) received his undergraduate and graduate education at the University of Toronto, earning his PhD in molecular spectroscopy under H.L. Welsh and M.F. Crawford in 1949. He immediately joined the Stellar Physics Division at the Dominion Observatory, becoming head in 1955. Four years later he was put in charge of the construction and early stages of operation of the Dominion Radio Astrophysical Observatory in Penticton. In 1966 he resigned from the DO to pursue research in radio astronomy at NRC and the following year was one of the team who successfully linked Canada's two radio telescopes. Using long-base line interferometry, they determined an upper limit to the size of quasar 3C 273B as 0.02 arc-seconds. This pioneering achievement was recognized by the American Academy of Arts and Sciences in the presentation of their highest honour, the Rumford Premium. Locke's administrative talents were clearly evident during his term as the first Director of NRC's Herzberg Institute of Astrophysics, 1975-85, a post which put him in charge of all astronomical and astrophysical work carried out at government institutions, including NRC's participation in the Canada-France-Hawaii Telescope Corporation.

Jack Locke addressed the Ottawa Centre on a number of occasions; he served as their President 1954-55 and Honorary President 1979-89 and generously arranged suitable facilities for Centre meetings and library. At the national level, he participated in many General Asssemblies, was elected the Society's 2nd Vice-President in 1968 and was President 1972-74. He was presented with the Service Award in 1988 and was commended for the exemplary manner in which he encouraged the cooperation between amateur and professional astronomers.

Dr J.L. Locke in 1959 in front of the 26m radio telescope under construction at DRAO.

Professionals may wonder if their public lectures are merely a form of free entertainment, or if visitors' night at the observatory is anything more than an intrusion on valuable observing time. They may question why anyone would want to hunt down nebulae without using co-ordinates and setting circles, or what pleasure there could possibly be in looking at Saturn for the hundredth time. In the worst case, there may be some crank with a new theory of creation who could become an intolerable pest. For their part, some amateur members may think that the public lectures and papers in the *Journal* are there only to enhance the prestige of the expert. They may wonder how anyone calling himself an astronomer could fail to find the Dumbell Nebula with a pair of binoculars. Worst of all, some dedicated observer may wonder, as he mails off his thousandth variable star observation, if he is simply being "used" by the professional community as an unrecognized and unrewarded detector. Such attitudes as these are very sad, and fortunately seldom heard within the ranks of the Society. Truly, the way to get along is to recognize that every individual sees the universe through unique eyes. As Roy Bishop said in his presidential address, "What we bring, we find." No one way is right. Helen Hogg expressed the ideal philosophy in the title of her popular book, *The Stars Belong to Everyone*.

There are countless examples, some found in the pages of this book, of the deep interest of professional astronomers in the welfare of amateurs. By encouraging observations, answering questions, providing expertise and facilities, by willingly speaking at meetings, writing articles at a nontechnical level and by playing an active role in the Society, Canadian astronomers have been very generous supporters of the RASC and its activities. In fact, about a third of the RASC Centres owe their establishment to the direct action of professional scientists. Their impact has been out of all proportion to their numbers as they have never made up more than a few percent of the total membership. (Presently about three percent of RASC members also belong to CASCA.)

The RASC, per se, has been of limited direct benefit to professional astronomers, but in fairness, about all that any society can hope to do for professional scientists is to provide a forum for sharing information and perhaps to act in an advisory capacity to governments and funding agencies. The RASC *Journal* does publish professional research papers though that has never been its primary role. The Society has offered financial support to meetings of the AAAS, AAS and IAU and has contributed to scholarship funds in memory of several professional astronomers. The RASC Council has nominated astronomers to serve on advisory boards. At least up to the 1930s, delegates were sent annually to RSC meetings. Also five out of twenty-two members of the National Committee for Canada of the IAU were formerly appointed by the RASC Council. Now CASCA represents professional interests, and the RASC has no formal liaison with professional groups.

Nonetheless, amateur members do a lot which should be considered a benefit to their learned colleagues. They definitely ease the pressure on professional astronomers and their limited facilities through a tremendous range of public education activities. A few make genuine contributions to science by their discoveries and dedicated observations. Some find their niche in historical research and have significantly enhanced an appreciation of the development of astronomy in Canada. Many more help merely by sustaining the Society which over its life has brought Canadians and the Universe closer. The advantages to the profession of an educated public should be obvious. People who have been kept in the dark have no stake in the future. Informed people, on the other hand, feel like participants in the quest for discovery, and are far more likely to give moral and financial support to something they understand. Most professionals acknowledge their ultimate accountability to the taxpaying public and many realize that the RASC is an excellent bridge.

But can professional astronomers look to the RASC for any lasting legacies? Certainly there are the publications which have been a strong factor in bringing the work of Canadian astronomers to the attention of the world. Those who would see Canada's image enhanced abroad would do well to consider how they can strengthen the RASC *Journal*. The Gold Medal and more recently the Plaskett Medal are tangible evidence of the Society's desire to recognize the achievements of those about to embark on a career. Dr Heard recalled some other concrete contributions to the profession in 1976:

> Directly traceable to the Society have been the magnificent gifts of the Dunlap family, Walter Helm and Carl Reinhardt. The list continues to grow: Within the past few months a handsome bequest to the Frank Hogg Fellowship Fund has been received from the estate of the late Robert S. Evans, one time president of the Victoria Centre, whose will benefitted about equally the R.M. Petrie Fellowship and the Hogg Fund.

Mention could also be made of donations of equipment. In 1963, Evans presented the DAO with his 41-cm Cassegrain telescope with optics designed by Horace Dall of Dall Kirkham fame. DAO astronomers used this instrument for photometric work for several years and then loaned it to the University of Manitoba where it is still doing good work. The existence of the DAO itself may owe something to the Society which urged the government to establish the Observatory. The University of Alberta's first substantial telescope was donated by RASC member Cyril Wates. The Toronto Centre telescope makers donated to the University of Waterloo a coelostat which they had made for the 1963 eclipse.

Many Canadian planetariums have strong ties with the Society. From his first look at a star theatre

FREDERICK R. WILLIAMS joined the Vancouver Centre in 1929 and was their 2nd Vice-President in 1933-34. He moved to Victoria in 1949 where he served as Treasurer in such an exemplary manner over a period of fifteen years that he was presented with the Service Award in 1974. Unfortunately, not much is known about Williams, but George Ball recalled that he lived a quiet life as a bachelor in a residential hotel in Victoria.

This photo was taken on the occasion of the presentation by Robert Evans of his 41 cm Cassegrain telescope to the Dominion Astrophysical Observatory in 1963. Evans is on the extreme left of this picture, with D.G. Stoddart and J.C. Parsons on his left, facing the camera. Fred Williams, wearing a white hat, is immediately to the right of the telescope.

KENNETH O. WRIGHT (1911-) was born in Fort George, BC, but attended school and university in Toronto. He received the RASC Gold Medal on graduation in 1933 and went on to earn his MA the following year, during which time he helped the Toronto Centre by giving constellation talks and operating the slide projector at meetings and by organizing meteor observations. He moved to Ann Arbor to begin his PhD studies at the University of Michigan and joined the staff of the DAO in 1936, completing his Doctorate in 1940. He served the Victoria Centre as Councillor, Recorder, Vice-President, President (1945-46) and Honorary President and was the speaker at many meetings of the Centre and at their "Summer Evenings with the Stars". In 1961 he was elected national Vice-President and was President in 1964-66. While in office, he visited nearly all the Centres of the Society and spoke at General Assemblies. He tried to encourage more scientific content in the *Journal* and contributed dozens of papers himself.

Wright's primary research interest was in super-giant systems like Zeta Aurigae and in the relative abundances of chemical elements in solar-type stars. During his career he worked with several universities as lecturer, visiting professor, research associate or as a member of their governing boards. The Queen Elizabeth II observatory project took a great deal of his time and effort in the 1960s, and though it did not succeed, the 381 cm mirror eventually formed the basis of the Canada-France-Hawaii Telescope. On the occasion of Wright's retirement as Director of the DAO in 1976, a special symposium in his honour was attended by colleagues from Canada and abroad.

K.O. Wright, seen on the right, shows the DAO spectrograph to the Honorable Jean-Luc Pepin, Minister of Mines and Technical Surveys.

in Stuttgart in 1928, Chant was enthusiastic, and as other members of the Society began in the 1930s to visit the Adler Planetarium in Chicago and the Hayden Planetarium in New York, a movement for a Canadian facility began to grow. Speakers extolled the marvels of seeing the sky from inside the dome, articles were written and visitors like Roy K. Marshall of the Fels Planetarium in Philadelphia were invited to Canada to give public lectures in the hopes of arousing public interest. But the results were all too familiar to those who had witnessed the long struggle for an observatory in Toronto.

There were a few early examples of small projection instruments. One was used at Royal Roads College, near Victoria, and another at Central School, Forest Hill Village, Toronto, to train officers in navigation during World War II. A Montreal high school got one in May, 1945. Hamilton Centre raised over $1,000 to purchase a Spitz projector which they donated to McMaster University in 1949. The Nova Scotia Museum of Science began shows in Halifax in 1956. But the Queen Elizabeth II Planetarium in Edmonton was the first Canadian planetarium in a building of its own when it opened in 1960. Six months after it began operations, its director, Ian McLennan, was able to report:

> Over 10,000 people have attended shows presented with the assistance of members of the Edmonton Centre. The interest which has been aroused has resulted in a flood of requests for information regarding membership in the RASC. Naturally, this can only help the Society, and will, in time, be equally beneficial to the planetarium, as more people will be called upon to present star shows, handle classes, work on exhibits, etc.

Within the next few years, five large planetariums opened in major cities across the country and the Canadian Planetarium Association was formed. In fact the Association held its organizing meeting in the RASC Library at 252 College Street, Toronto, on May 20, 1965.

The indirect results of the Society's Planetarium Committee have already been described in chapter two. The Calgary Centennial Planetarium was the direct result of initiatives taken by the RASC in that city and will be discussed in chapter twelve.

By 1975, it was estimated that 750,000 people saw shows in Canadian planetariums each year. Huge successes like this by no means diminished the importance of the Society's public education programs. In fact all these facilities relied heavily on RASC members for their staffing needs, and all continue to enjoy the co-operation of the Society in hosting special events, such as seminars, courses, and Astronomy Days.

Throughout the Society's history, amateur members have enthusiastically endorsed and encouraged the work of the professionals. If anything, they may sometimes have been too anxious to prod the government to finance observatories or programs without understanding the full consequences. In 1964, for instance, members present at the Annual Meeting passed a motion giving full support for the construction of a large Canadian research telescope. Plans progressed and development of the 3.8-m Queen Elizabeth II Telescope began, but when the project was cancelled in 1968, few members realized that many in the professional community had felt there were better alternatives. In the end, the Society passed another motion deploring "the cutback in expenditure on astronomical research in Canada as evidenced by the cancellation of the QE II project." Again, when the closure of the Algonquin Radio Observatory was announced in 1987, and when Canada's involvement in Project Gemini was being discussed in 1991, members were only too eager to jump in. But professional members of the Society were able to explain the concerns for the fate of other government-funded programs and the need to be cautious. While the professional community should be encouraged to see this evidence of support from the amateur members, it should also heed the need for their continued presence in the RASC. Joint ventures between the RASC and CASCA, such as the Helen Sawyer Hogg Public Lecture and the Plaskett Medal are hopeful precedents for future cooperation.

AWARD WINNERS

During its first half-century, the Society offered only one award – the Gold Medal. Now it presents Membership and Messier Certificates, the Simon Newcomb and the Service Awards, the Chilton Prize and Chant Medal, and cosponsors the Plaskett Medal with CASCA. In addition, honours were given out at the Canada-Wide Science Fairs from 1962 to 1989, prizes are awarded in a variety of categories for displays at General Assemblies and many Centres have their own awards.

It was in 1905, the same year that Dr Chant inaugurated an undergraduate course in astronomy and physics at the University of Toronto, that the Society decided to institute a gold medal and to offer it annually to the university for presentation to the top student graduating with first class honours from that course. Ellis and Company, the same firm that produced the original RASC seal, handled the design and preparation. Even though the medals were only gold-plated, the cost to the Society represented a considerable outlay – $61.50 for the original die and $25 for each of the first several castings. As the very brief summary on the next page indicates, the great majority of the thirty-seven recipients over the next eighty-two years went on to make names for themselves in astronomy and allied fields.

By the 1960s, the awarding of the Gold Medal only to Toronto students was regarded as something of an anachronistic anomaly, at least in some quarters. Specialist courses in astronomy were now being offered at other universities and the list was growing as new universities sprouted all over the country. But

The following is a list of winners of the RASC Gold Medal. Seven of these people later became RASC Presidents and details of their lives can be found elsewhere in the book. Unless otherwise stated, the others in the list also made their careers in Astronomy. Wherever possible, their most recent career position is given.

1906 - W.E. Harper (see p. 77)
1907 - R.M. Motherwell (DO; died 1940)
1909 - R.K. Young (see p. 94)
1910 - R.J. McDiarmid (DO; deceased)
1911 - R.S. Sheppard (Edmonton School Superintendent; died 1967)
1913 - E.A. Hodgson (Head of Seismology, DO; died 1975)
1914 - G.S. Campbell (Teacher in sothwestern Ontario died in car accident before 1945)
1926 - F.S. Hogg (see p. 94)
1929 - P.M. Millman (see p. 44)
1931 - W.S. Armstrong (Principal of Campbellford High School, Ontario; died September, 1958)
1933 - K.O. Wright (see p. 54)
1935 - F.S. Patterson (After marriage, Dr F.S. Jones made her career in museums in US)
1937 - D.A. MacRae (retired Director of DDO)
1940 - W.F.M. Buscombe (Northwestern University, Evanston, Illinois)
1948 - R.W. Tanner (retired from NRC)
1949 - I. Halliday (see p. 145)
1956 - D.C. Morton (Director of Herzberg Institute of Astrophysics, Ottawa.)
1960 - C.R. Purton (DRAO, Penticton)
1961 - R.C. Henry (Johns Hopkins University)
1962 - J.R. Percy (see p. 9)
1963 - P.H. Reynolds (Geophysicist, Dalhousie University)
1965 - A.F.J. Moffat (Université de Montréal)
1966 - W.A. Sherwood (Max Planck Institut fur Radioastronomie, Bonn, Germany)
1967 - R.H. Chambers (Computer Services, U of Toronto)
1968 - P.G. Martin (Canadian Institute for Theoretical Astrophysics = CITA)
1970 - J. Kormendy (DAO)
1971 - B.F. Kinahan (Councillor for Metropolitan Toronto)
1972 - M.J. McCutcheon (Scientific Engineer with the High-Speed Networking Group, University of BC)
1976 - C. Rogers (DRAO, Penticton)
1977 - M. DeRobertis (York University)
1978 - D.R. Gies (Georgia State University)
1979 - N. Duric (University of New Mexico)
1981 - T.C. Box
1982 - M J Gaspar (Family physician in Barrie, ON)
1984 - G.D. Starkman (CITA)
1985 - G.A. Drukier (Institute of Astronomy, Cambridge University, England)
1987 - Man Hoi Lee (CITA)

Peter Millman (r), Gold Medallist of 1929, presents the Gold Medal to Richard Henry, 1961, in the RASC Library, 252 College Street.

even though formal suggestions for change were made and committees were formed to look into the matter, nothing happened until 1985. That year the Society asked Lloyd Higgs to chair a new committee with the mandate of examining possible ways to acknowledge the long-standing connections between the RASC and the University of Toronto and yet provide a national award for promising young Canadian astronomers. As Higgs explained, "The committee quickly focussed on the possibility that the Toronto Centre of the RASC could continue to offer the Gold Medal in ... modified form [to U. of T. students completing their final undergraduate year] while the Society could offer a new national award at the graduate level." Since the judging of such a graduate award would have to be the responsibility of professional astronomers, the Canadian Astronomical Society was at once contacted for their views. Before long an agreement was reached to institute a joint RASC/CASCA award with RASC paying for the medal and CASCA providing the judging expertise. This led to the presentation on May 30, 1988, of the first Plaskett Medal, named in recognition of J.S. Plaskett who played a pivotal role in the establishment of astrophysical research in Canada. The award is normally made annually to the graduate from a Canadian university who is judged to have submitted the most outstanding doctoral thesis in astronomy and astrophysics in the preceding two calendar years.

The recipients have been: Richard Gray, 1988; Peter Leonard, 1989; Paul Charbonneau, 1991; Eric Poisson, 1992; and Pierre Brassard, 1993. The Plaskett Medal is the only award now offered by the Society exclusively for professional or budding professional astronomers.

Membership Certificates and Service Awards are open to all members, amateur or professional. The Membership Certificate is a minor award intended to provide a polite, public "thank you" for service rendered to the Society or Centre; hundreds of members have been recognized in this way since the first certificates were presented to Ruth Northcott and Miriam Burland in 1961. The Service Award, in the form of a bronze medal, is presented on the recommendation of the Awards Committee for outstanding service to a Centre or to the national Society. Initially, the award was proposed as a means of honouring E.J.A. Kennedy who retired as national secretary after twenty-three years in that position but after due consideration, the decision was taken to make the award generally available. Centres were originally asked to pay for the cost of awards for their members and to restrict their nominations to one every three years on average, but by 1967 the national Society bore the expense. There have been over eighty recipients since the first presentation in 1959 (see list on next page), and each winner during the Society's first century has been accorded a special place somewhere in this book. The accomplishments of winners since 1990 are probably well known, but in any case properly belong to Volume II of the Society's history.

All the other awards are intended specifically to stimulate and reward amateur achievement in the Society. The Messier Certificate was started in 1980 on the suggestion of the Edmonton Centre, though Montreal's Isabel Williamson had organized what is generally acknowledged to be the first Messier Club way back in the 1940s. The basic objective, is to locate and identify all 110 objects in Messier's eighteenth century catalogue of star clusters and nebulae. David Levy, one of the outstanding "graduates" of the Montreal club, found the exercise a wonderful way to become familiar with the constellations, to learn about the variety of deep-sky objects, and to pick up good observing habits like keeping a log. About five certificates are approved each year by the National Council on recommendation of two Centre members. Two particularly remarkable feats should be mentioned – Kingston member Mark Sorensen's observation of all 110 objects with 11x80 binoculars, and David Levy's

WINNERS OF RASC SERVICE AWARDS

1959 - E.J.A. Kennedy and J.H. Horning (Officers)
1960 - C.M. Good (Montreal), F. Laforest (Montreal), F.L. Troyer (Toronto)
1961 - R. Peters (Victoria)
1962 - J. Asselin (Montréal), R.J. Clark (Vancouver), J. Ketchum (Toronto), J.A. Pearce (Victoria)
1963 - E.E. Bridgen (Montreal), M. Burland (Ottawa), A.V. Douglas (Kingston)
1964 - C.S. Beals (Ottawa), M.W. Burke-Gaffney (Halifax), W.T. Goddard (Hamilton)
1965 - D.C. Bawtenheimer (Windsor), J.F. Heard (Toronto), A.M. Crooker (Vancouver), E.S. Keeping (Edmonton), W.S. Mallory (Hamilton)
1966 - P. Lemieux (Montréal), S. Litchinsky (Calgary), W.J. McCallion (Hamilton), W.A. Warren (Montreal)
1967 - H.S. Hogg and R.J. Northcott (Toronto), P.M. Millman (Ottawa), H. Simard (Montréal), J.F. Wright (Vancouver)
1968 - D.R.P. Coats (Calgary/Winnipeg), G. Ball (Victoria), H. Fox (Hamilton)
1970 - N. Green (Hamilton), J.E. Kennedy (Saskatoon)
1971 - K.B. Meikeljohn (Calgary), J.N.R. Scatliff (Winnipeg)
1972 - V. Ramsay (Toronto), B.F. Shinn (Winnipeg), F. Schneider (Hamilton)
1973 - F.P. Lossing (Ottawa)
1974 - I. Halliday (Ottawa), F. Williams (Victoria)
1975 - D.J. FitzGerald (Toronto)
1976 - F. Loehde (Edmonton), K. Chilton (Hamilton), R. and P. Belfield (Winnipeg)
1977 - P. Marmet (Québec), J.R. Percy (Toronto), D. Russell (St Johns)
1978 - M. Fidler Litchinsky (Calgary/Toronto)
1979 - R. Noel de Tilly (Montréal), A.W. Scott (Toronto)
1980 - S.A. Mott (Ottawa)
1981 - L.E. Coallier (Montréal), A. Covington (Ottawa), H. Creighton (Toronto), I.K. Williamson (Montreal)
1982 - D.P. Hube (Edmonton), G.N. Patterson (Saskatoon)
1983 - C. Aikman (Victoria), L.A. Higgs (Ottawa)
1984 - L.V. Powis and J.A. Winger (Hamilton), P. Jedicke (London), H.N.A. Maclean (Niagara), C. Hallam and H. Lee (Windsor)
1985 - J.C. Fahrner (Calgary)
1986 - R. Brooks (Halifax), L. Enright (Kingston)
1987 - R.P. Broughton and C. Clark (Toronto)
1988 - A.H. Batten (Victoria), R.L. Bishop (Halifax), J.L. Locke (Ottawa), E. Orr (Hamilton)
1989 - R. Auclair (Unattached)
1990 - M. Grey (Ottawa)
1991 - D. Hladiuk (Calgary), I. McGregor (Toronto)
1992 - J.M. Frechette (Québec), M.S.F. Watson (Unattached)
1993 - D.A. Tindall (Halifax), E.J. Clinton (London)

single-handed record of 109 in one night – March 15–16, 1983.

For those who may not be inclined to observation, an award of a very different character is available. In 1978 the Halifax Centre proposed the creation of the Simon Newcomb Award to recognize creative writing in astronomy by nonprofessional members. The National Council approved the idea and three entries were submitted in that first year, 1979, and all three appeared in the *Journal*. The initial enthusiasm has unfortunately not been maintained. In recent years there have sometimes been no entries at all. The winners have been:

William J. Calnen (Halifax), 1979, "Astronomy at King's College, Windsor, Nova Scotia."
Christopher Rutkowski (Winnipeg), 1981, "What is Happening on the Moon? Lunar Transient Phenomena."
Philip Mozel (Toronto), 1982, "The Woodstock College Observatory."
Donald F. Trombino (USA), 1985, "Dr John William Draper – First to Photograph the Moon."
David Chapman (Halifax), 1986, "Recurrent Phenomena of Venus and the Venus/Earth Orbital Resonance."
Peter Jedicke (London), 1987, "Neutrinos, The Sun and Canada."

Some people defy categorizing – Phil Mozel was not only a Newcomb Prize winner but a recipient of a Messier Certificate too, and Peter Jedicke also received the Service Award.

The Chant Medal and Ken Chilton Prize are both awarded to amateur astronomers in recognition of their astronomical work. The Chilton Prize, established the year after Ken's untimely death in 1976, is awarded for a specific piece of work carried out or published during the year, while the Chant Medal is awarded for more extensive investigations of lasting value in astronomy or closely allied fields.

FRANK SISMAN (- 1960) joined the Hamilton Centre in 1935 and soon earned a reputation as an optical craftsman. By the time he received the Chant Medal for 1949, he had made five telescopes, many eyepieces, optical flats, a machine for grinding and polishing mirrors and a testing bench. His finished work was consistently of the very highest standard. Frank Sisman was always a very willing adviser to other amateur telescope-makers and did much to advance public interest in astronomy.

Unfortunately he suffered a stroke in the summer of 1949 and was still unable to attend the annual meeting the following year to receive the medal in person. He was apparently in ill health for the rest of his life.

(Clipping from Hamilton *Spectator*, January 14, 1950)

FRANK SISMAN

◀ David Chapman, a winner of the Simon Newcomb Award, stands beside the monument at the former astronomer's birthplace at Wallace Bridge, NS.

This 1948 photograph shows Keith Dalton with some of his shells and minerals – just two of his many carefully organized collections.

FREDRICK KEITH DALTON (1890-1975) joined the Toronto Centre in the early 1930s, participating in eclipse expeditions and writing reports for the *Journal*. After an absence of a few years, he rejoined the Society in 1944. Dalton spoke to the Centre a couple of times on *Astronomy as a Hobby*, and served on their Council for two years including a term as Recorder. He did a lot to bring astronomy to the public through talks and telescope demonstrations with his 10 cm Alvan Clark refractor, and prepared a manual for Boy Scouts and Girl Guides. Some of his papers in the *Journal* describe attachments he devised for his telescope to aid in finding Venus, Mercury and Polaris during daylight. Others, on tides and power generation, may seem to be a natural outgrowth of his professional occupation as an engineer with Ontario Hydro but in fact were just a facet of his deep and persistent curiosity to explore every aspect of nature. His experiments with radio communication before the advent of vacuum tubes and his investigations of the characteristics of different types of wood under fluorescence are two examples of his eclectic talents.

It was his three-part paper on *Microhardness Testing of Iron Meteorites* which proved especially significant and led to his receiving the Chant Medal for 1950. As a result of a long-standing interest in minerals, Dalton obtained a piece of the Canyon Diablo meteorite and tried to cut through it, breaking several hack-saw blades in the process. He thus became interested in investigating the hardness of meteorites with the use of a device called the Knoop indenter, well-known for testing metals. The advantage of this diamond-tipped instrument was its extremely small size which enabled measurements to be made on very small inclusions in meteorites.

The recipients of the Ken Chilton Prize have been:

Jack Newton, 1978 (see p. 61)
Warren Morrison, 1979 (see p. 149)
Craig McCaw, 1981, for outstanding astrophotography and development of cold camera techniques.
Chris Spratt, 1983 (see p. 97)
Philip Teece, 1988, for literary contributions to the promotion of amateur astronomy.
Lucien Kemble, 1989, for thousands of drawings of celestial objects including "Kemble's Cascade."
Doug George, 1990, for exceptional observing skill, and his codiscovery of Comet Skorichenko-George.

Newton, Morrison and Spratt continued with their work and became recipients of the Society's prestigious Chant Medal. This silver medal was first presented in 1940, the fiftieth anniversary of the founding of the Society, and was named in appreciation of C.A. Chant's great contributions to Canadian astronomy. The original hope had been to have Chant's profile engraved on the medal, but in the end, the cost of additional artwork was saved by having the traditional seal of the Society instead. The list of winners is as follows:

Bertram J. Topham, 1940	Frank J. DeKinder, 1955
H. Boyd Brydon, 1941	Maurice Drolet, 1956
W.G. Colgrove, 1942	Earl Milton, 1959
Cyril G. Wates, 1943	Raymond R. Thompson, 1967
Paul H. Nadeau, 1945	Rolf G. Meier, 1979
Isabel K. Williamson, 1948	David Levy, 1980
Frank Sisman, 1949	Warren Morrison, 1986
F. Keith Dalton, 1950	Damien Lemay, 1987
DeLisle Garneau, 1951	Chris Spratt, 1988
Jean Naubert, 1953	Jack B. Newton, 1989

Write-ups and photographs of all these winners are to be found elsewhere in the book. Readers will have to turn to the *Journal* to read about the accomplishments of Mary Lou Whitehorne, Chant medallist for 1993.

There was great interest in the Chant Medal at first and in fact by the time nominations closed in December, 1940, nine candidates had been proposed. While Topham got the award that first year, three of the others (Brydon, Colgrove and Sisman) eventually were recognized. It was not until 1944 that any procedure was developed for notifying the winner or for the actual presentation arrangements. Cyril Wates, for instance, first heard of his award when he got a call from a reporter asking about the medal he had won. A couple more years passed before the Council realized there was a need to remind Centres annually to submit nominations. Experience is a wonderful teacher, of course, and these and many other administrative details are now usually handled in a regular and efficient manner.

Four Chant medallists in 1993 (l to r) Jack Newton, Mary Lou Whitehorne, David Levy, Damien Lemay.

JACK NEWTON (1942-   ) joined the Winnipeg Centre in 1958 and became involved in the Moonwatch program. In 1969 he built a 32-cm reflector with the help of Frank Shinn and installed it in a three-metre domed observatory behind his home at the south end of Winnipeg. Newton started an astrophotography section of the Winnipeg Centre, served on the Centre Council and was President in 1970-72. But as a manager for Marks and Spencer stores he was transferred to suburban Toronto in 1973.

Once on Ontario, Newton built a new observatory and began experimenting with cooled emulsions and hypersensitized, gas-soaked films. His magnificent astrophotographs appeared regularly in the *National Newsletter* and in his book, *DeepSky Objects – A Photographic Guide for the Amateur*, published in 1977. The same year he received the Queen's Silver Jubilee Medal and in 1979 the RASC's Chilton Prize. During his six years as a member of Toronto Centre, Newton served as President in 1975-76.

Another business move took Newton to Victoria where he once again became active in Centre affairs, serving twice as Centre President, 1980-81 and 1990-91. With increasingly large telescopes and more sophisticated detectors, Newton kept pushing back the frontiers of what was observable. He is now able to reach magnitude 21.5 with a charge-coupled device (CCD) attached to his 63-cm telescope.

Besides being recognized as an outstanding astrophotographer, Jack Newton became well-known as a speaker and writer. A number of his articles appeared in the *Journal* and *Astronomy* magazine and two books, *The Cambridge Deep Sky Album*, and a *Guide to Amateur Astronomy* were co-authored with Philip Teece. Jack Newton received the Amateur Achievement Award of the ASP in 1988 and the Chant Medal of the RASC in 1989.

Jack Newton and his 40-cm telescope, Victoria, B.C.

The Pleiades (M45) is instantly recognizable because of a few bright stars yet most of the cluster members are fainter main sequence stars.

CHAPTER 4

Bright Lights

The officers of the Society are shown in the chart on pages 65 and 66. Except for the honorary president, their terms of office were first set down in 1944 when the Constitution stipulated that the president and vice-presidents were eligible for election to the same office at no more than three consecutive Annual Meetings. The two-year nonrenewable term for each of these offices, though it had been the practice since at least 1930, was not enacted until 1969. At the same time a three year term, renewable once, was prescribed for the other offices. Honorary presidents were appointed for an indefinite period until 1981 when a four- year term was instituted.

HONORARY PRESIDENTS

The honorary presidents of the Society can be conveniently considered in two groups. Ross, Harcourt, Grant, Ferguson, Simpson, McArthur, Drew, Porter and Dunlop were all appointed by virtue of their position as Ontario Ministers of Education. This tradition was no doubt started as a means of bringing the educational role of the Society to the attention of the provincial government but it probably didn't hurt in gaining its financial support as well. Perhaps the councillors of the Society had similar motives when they appointed Marc Boyer and W.E. van Steenburgh as honorary presidents in the 1960s. Both these men were Deputy Ministers of the federal Department of Mines and Technical Surveys which, through the Dominion Observatory, provided very generous annual grants to the Society. Whether or not there was actually any connection between the appointment of honorary presidents and their ability to provide grants, a conflict of interest would certainly be perceived to exist in today's political climate. It was just such a perception which led the Hon. Dr Pyne, Ontario's Minister of Education, to decline the honorary presidency in 1906 and which led to a nonpolitical candidate to be appointed by the Society.

The other honorary presidents have all been distinguished Canadian scientists with some previous ties to the Society. Several in fact have been RASC presidents. Their names are obvious from the chart of officers and they are written up elsewhere in the book. Arthur Covington, while never president, did receive the RASC Service Award for his significant contributions to the Society (see next page). The remaining honorary presidents for whom no biographical information is given elsewhere in this book were King, Klotz, Choquette and Herzberg. William F. King (1854–1916) was the Chief Astronomer for the Dominion Government when he was appointed honorary president in 1906. Later that year he became the first president of the Ottawa Centre. Otto Klotz (1852–1923) succeeded King as director of the Dominion Observatory in 1917, the position which he held when he became the Society's honorary president in 1919. He also succeeded King as president of the Ottawa Centre in 1909. Though Klotz confided to his diary that he disliked the Society's popular character, he did contribute over fifty papers to the *Journal*.

ARTHUR E. COVINGTON (1913-) joined the Vancouver Centre as a high-school student. He built and operated amateur radio station VE55 from 1930-32 and made his own 13 cm reflecting telescope, two hobbies which would later merge to make him Canada's pioneer radio astronomer.

After earning his BA and MA degrees from the University of British Columbia, he went in 1940 to the University of California in Berkeley. The War effort, however, brought him back to Canada and he accepted a position at NRC working on the development of radar. As he himself wrote, "When the opportunity to do other than radar investigations occurred at the cessation of hostilities, I presented a short account of the possibilities of research in cosmic noise to my immediate superior." His proposal was accepted and by 1946 Covington was doing his first solar microwave astronomy.

As a member of the Ottawa Centre, he published his first RASC paper in the *Journal* in 1951 and gave his first RASC speech to the Montreal Centre on the Reception of Radio Noise from the Sun, Moon and Galaxy. The same year, his wife was mentioned as a member of a string quartet playing for the Ottawa Centre Annual meeting. In 1954 he became the 1st Vice-President of the Ottawa Centre and the first Chairman of the Observers' Group. He was Centre President in 1956-57.

Over the years, dozens of papers by Arthur Covington appeared in the *Journal*. Most were naturally connected with his research at NRC, but many were of a historical nature. His concern for our astronomical heritage was a great asset to the RASC Historical Committee of which he was a member and Chairman for a number of years. He received the Service Award in 1981 and he was an active and interested Honorary President 1986-89.

Arthur Covington at the Goth Hill Radio Observatory near Ottawa, spring, 1971.

Mgr Charles-Philippe Choquette (1856–1947) taught physics at the Seminary of St-Hyacinthe, Quebec, where he later became the Superior. He organized and directed the Provincial Laboratory for Chemical Analysis and played an important role in the development of hydro-electric power in Quebec. He took a popular interest in astronomy and was the Montreal Centre's first president. Gerhard Herzberg (1904–) was awarded the Nobel Prize for Chemistry in 1971 for his contributions to the knowledge of molecular structure, particularly of free radicals. His studies have reached beyond the laboratory to include investigations of molecules in interstellar space, in planetary atmospheres and in comets. The Herzberg Institute of Astrophysics (HIA), a division of NRC established in 1975, is now the centre of federal government research in astronomy.

PRESIDENTS AND VICE-PRESIDENTS

It is obviously the prescribed duty of presidents to preside over meetings, preserving order and regulating proceedings. Until 1927, when a separate body began to take responsibility for the local meetings in Toronto, there were at least twenty biweekly regular meetings, an annual meeting and an At-Home, and some Council meetings each year. Merely attending all those meetings would have been a major commitment, but when coupled with the understanding that the president should exercise a general supervision over the interests of the Society including membership in all committees, the responsibilities could have made for a full-time job.

The ability of the early presidents to fulfill these duties varied greatly. Carpmael, first through work and later because of illness, rarely attended meetings, and Smith, by the time he became president, was too old to venture out at night. Plaskett and R.M. Stewart, coming from out of town, could only attend occasionally. So, by circumstance, some of the presidents were literally figureheads, with most of the work devolving upon the vice-presidents and other

	LIBRARIAN	RECORDER	SECRETARY	TREASURER	2ND V-PRES	1ST V-PRES	PRESIDENT	HON PRES
1890	A. MILLER G.G. PURSEY	D.J. HOWELL T. LINDSAY C. SPARLING	G. LUMSDEN	G.G. PURSEY D.J. HOWELL J. TODHUNTER	J. PATERSON	A. ELVINS L. SMITH	C. CARPMAEL	G. ROSS
1895	W.B. MUSSON	T. LINDSAY		C. SPARLING	F. STUPART C.A. CHANT	E. MEREDITH A. HARVEY E. MEREDITH F. STUPART	L. SMITH J. PATERSON A. HARVEY	
1900	Z COLLINS R. ATKINSON A. McFARLANE	J.E. MAYBEE J.E. WEBBER E. DENT	W.B. MUSSON J. COLLINS	J.E. MAYBEE	W.B. MUSSON	C.A. CHANT A.T. DeLURY	G. LUMSDEN F. STUPART C.A. CHANT	R. HARCOURT
1905	J. ELLIS K.M. CLIPSHAM A. SINCLAIR			G. RIDOUT C. SPARLING	L.B. STEWART		W.B. MUSSON	W.F. KING
1910	W.M. WUNDER	L. GILCHRIST			J. PLASKETT A.D. WATSON A.F. MILLER	L.B. STEWART J. PLASKETT A.D. WATSON	A.T. DeLURY L.B. STEWART J. PLASKETT	
1915	C.A. CHANT	W.E.W. JACKSON A.F. HUNTER	W.E.W. JACKSON	A.H. CREASE H.W. BARKER	W. BRUCE O. KLOTZ W. BRUCE	A.F. MILLER J.R. COLLINS	A.D. WATSON A.F. MILLER	J. PLASKETT O. KLOTZ
1920		J.A. PEARCE J.H. HORNING H.F. BALMER	A.F. HUNTER F.T. STANFORD		J. SATTERLY R.M. STEWART A.F. HUNTER	W.E.W. JACKSON W. BRUCE	J.R. COLLINS W.E.W. JACKSON R.M. STEWART	R.H. GRANT G.H. FERGUSON
1925		G.M. BRYCE E.J.A. KENNEDY	L. GILCHRIST		R.K. YOUNG C.P. CHOQUETTE	H.R. KINGSTON R.K. YOUNG	A.F. HUNTER W.F. HARPER	J. PLASKETT
1930	R.A. GRAY		R.A. GRAY	J.H. HORNING	R.E. DeLURY J.A. PEARCE	L. GILCHRIST R.E. DeLURY	H.R. KINGSTON R.K. YOUNG L. GILCHRIST	R.M. STEWART C.P. CHOQUETTE
1935	P.M. MILLMAN	R.H. COMBS	E.J.A. KENNEDY		F.S. HOGG	W. FINDLAY J.A. PEARCE	R.E. DeLURY W. FINDLAY	L.J. SIMPSON

	LIBRARIAN	RECORDER	SECRETARY	TREASURER	2ND V-PRES	1ST V-PRES	PRESIDENT	HON PRES
1940		H.W. BARKER			A.V. DOUGLAS D.S. AINSLIE A.E. JOHNS H.B. BRYDON	F.S. HOGG A.V. DOUGLAS A.E. JOHNS	J.A. PEARCE F.S. HOGG A.V. DOUGLAS	D. McARTHUR G.A. DREW
1945	D.W. BEST				J.W. CAMPBELL C.S. BEALS J.F. HEARD	H.B. BRYDON A. THOMSON C.S. BEALS	A.E. JOHNS J.W. CAMPBELL A. THOMSON	D. PORTER
1950	W.R. HOSSACK	F.K. DALTON F. TROYER			D. GARNEAU H.S. HOGG	J.F. HEARD R.M. PETRIE	C.S. BEALS J.F. HEARD	W.J. DUNLOP
1955	J.B. OKE W.T. TUTTE		J.E. KENNEDY		G.R. MAGEE P.M. MILLMAN R.J. NORTHCOTT	H.S. HOGG A. McKELLAR P.M. MILLMAN	R.M. PETRIE H.S. HOGG A. McKELLAR	
1960	L.A. CHESTER	L.V. POWIS	N. GREEN	W.R. HOSSACK J.F. HEARD	K.O. WRIGHT M.M. THOMSON J.E. KENNEDY	R.J. NORTHCOTT K.O. WRIGHT M.M. THOMSON	A. McK / PMM P.M. MILLMAN R.J. NORTHCOTT K.O. WRIGHT	M. BOYER V. STEENBURGH
1965	J.R. PERCY P. BROUGHTON			R. BROADFOOT J.F. HEARD	H. SIMARD J.L. LOCKE	J.E. KENNEDY H. SIMARD	M.M. THOMSON J.E. KENNEDY	C.S. BEALS
1970	H. CREIGHTON	P. ASHENHURST	C. HODGSON D. FITZGERALD	C. CLARK	H. KING R.J. LOCKHART J.R. PERCY	J.L. LOCKE J.D. FERNIE A.H. BATTEN	H. SIMARD J.L. LOCKE J.D. FERNIE	J.F. HEARD A.V. DOUGLAS G. HERZBERG
1975	F. TROYER	H. CREIGHTON	N. GREEN	P. BROUGHTON M. FIDLER	I. HALLIDAY F. LOEHDE	J.R. PERCY I. HALLIDAY	A.H. BATTEN J.R. PERCY	H.S. HOGG
1980	P. MOZEL	L. ENRIGHT	P. BROUGHTON		R.L. BISHOP M. GREY L.A. HIGGS	F. LOEHDE R.L. BISHOP M.GREY	I. HALLIDAY F. LOEHDE R.L. BISHOP	P.M. MILLMAN
1985	B. BEATTIE	H. LEE	D.A. TINDALL	B.R. CHOU K. MILLER	D. LEMAY P. BROUGHTON	L.A. HIGGS D. LEMAY	M. GREY L.A. HIGGS	A. COVINGTON I. HALLIDAY
1990					D.P. HUBE	P. BROUGHTON	D. LEMAY	

R Meldrum Stewart (1878-1954) was born into a missionary family in Gladstone, Manitoba. He probably appears as a child among the hundreds of his father's photographs in the Public Archives of Canada. Until he entered high school in Ontario at the age of 11, the only teacher "Bert" ever had was his father who instilled in him a love of classics, literature and astronomy. He did brilliantly at school and won many scholarships including the Gold Medal in Mathematics and Physics when he graduated from the University of Toronto in 1902.

Immediately on graduation, he was appointed by the federal government to the staff of the Chief Astronomer's Office, Department of the Interior. When the Dominion Observatory opened in 1905, Stewart was put in charge of the meridian-circle work and the time service. He became Assistant Director in 1918 and Director in 1924, a position which he held until his retirement in 1946.

Stewart served the Ottawa Centre from 1907 to 1914 successively as Treasurer, Vice-Chairman and Chairman and was their Honorary President from 1949-54. At the national level he was Vice-President and President from 1923-25. Throughout his career he willingly spoke at meetings of Toronto, Ottawa, Montreal and Winnipeg Centres. One of his early lectures to the Ottawa Centre in 1911 was on the Principle of Relativity. His papers in the *Journal* span the years 1907-49, with "The Early History of Astronomical Activity in the Canadian Public Service" appearing posthumously in 1971.

His kindness and modesty are evident in the remarks he made when he retired as RASC President in 1926. He expressed satisfaction that his successor, Mr Hunter, "who had done the work of the past two years would for the next year or two have the honour as well."

Lachlan Gilchrist (1875-1962) was the son of a pioneer family in Ontario's Bruce County. He attended Owen Sound Collegiate and the University of Toronto where he was a scholarship winner in mathematics and physics. Following graduation in 1904, he went to Chicago for his MA and PhD under Millikan and Michelson. Further graduate studies took him to Berlin, Cambridge and Oxford, and during the War he was in charge of X-ray work at the military hospital in Orpington, England. Gilchrist had been appointed Assistant Professor of Physics at the University of Toronto just before enlisting and returned to this position in 1919. He established geophysics at U of T and in the course of his career worked with the Geological Survey of Canada and was a consultant for British, American and Canadian governments, for Ontario Hydro and for Ontario and Quebec mining associations. He never married and lived in Victoria College's North House where he was a Don for many years. He was very popular because of his personality, his connections (W.L.M. King was one of his life-long friends), and his athletic prowess.

In spite of his active life and varied interests, Gilchrist found time for RASC affairs. Between 1909 and 1949, eight of his papers appeared in the *Journal*, indicative of a long-term attachment to the Society. He spoke to Toronto, Ottawa and London Centres and served as Vice-President and President of the Society between 1932 and 1935. Following this, he was on Toronto Centre Council for ten years.

Caught by the camera at an informal moment at the 1937 annual meeting of the RSC at Convocation Hall in Toronto, R.M. Stewart (left) borrows a light from Lachlan Gilchrist.

officers. Care had to be taken in fact, to ensure that vigorous vice-presidents were standing-by in Toronto when absentee presidents were in office. After 1927, national presidents no longer had direct responsibility for regular Toronto meetings and local affairs. Unlike the Centre presidents, the national presidents now had only the Annual Meeting and At-Home, and two or three Council meetings to attend each year. This had the effect of clearing the way for a remarkable succession of professional scientists from all parts of the country, a succession which lasted almost unbroken into the 1970s, when amateur members began once again to assume leadership roles and many professional astronomers diverted their attention to their own new Society, CASCA. But even during this middle period of the Society's life, it was difficult for presidents to travel from a distance to more than one or two meetings in a year. Certainly in the years before air travel, a president from Victoria, for example, could hardly be expected to travel across the continent by train to attend a one day meeting. Besides, the Society could not afford to pay travel expenses. Sometimes a president could fortuitously combine travel for some other reason with an RASC duty, but absentee presidents continued to be a fact of life off and on during most of this period.

Other than assuming the duties of president when circumstances required, the vice-presidents had no specific role until 1989 when the new by-laws required the first vice-president to be a member of the Constitution Committee and to chair the Publications Committee, and required that the second vice-president belong to the Finance Committee. These prescriptions so far seem to be working very well in giving prospective presidents a good understanding of how the RASC works.

As the chart of officers shows, the succession of two-year terms, second vice-president, first vice-president, president, has been the norm for most of the Society's history. However, a few vice-presidents never became president. With the exception of Ainslie whose services were required during World War II and Meredith and Lockhart who died in Office, all apparently asked to be relieved of further responsibilities.

The presidents and vice-presidents all have strengthened the Society in one way or another. Some who acted in name only brought prestige and international stature to the Society through association. Others became very involved in the routine work of the Society and contributed hundreds of hours of personal time towards its improvement.

The other officers, because of their prescribed duties, have always had their work cut out for them. Of course the one paid employee, now known as the executive secretary, has handled the main burden of the routine work for most of the Society's history, but the secretary, recorder, treasurer and the librarian could hardly have served the Society as figureheads. Because of the work involved at the RASC headquarters, these officers were consistently Torontonians until at least the 1970s when improved communications no longer made it essential for them to come to the office frequently. The librarian's position is really the only one remaining which must be handled by someone within driving distance of Toronto, though it continues to be advisable for at least one or two members of the executive to be nearby to meet unanticipated contingencies.

TREASURERS

After assorted attempts to define the treasurer's role in the early by-laws, the formal duties settled into a stable and constant statement from 1908 to 1968. The treasurer was to receive all monies due to the Society, to keep all books pertaining to his office, to make all payments, to prepare the cash and balance sheets and have them duly audited, and generally to act under the instructions of the Council. No payments were to be made unless first authorized by the Council, and all money coming into his possession

J. Henry Horning (1892-1978) had a 38-year career at Toronto's Oakwood Collegiate, starting as a teacher of Physics and Physical Education and retiring as Principal in 1958.

By coincidence, the school's first Principal was R.A. Gray who served the Society as Librarian and Secretary after his retirement in 1931, perhaps at Horning's suggestion. (Speaking of coincidences, the author also taught at Oakwood for several years, and among his colleagues were several RASC members; as well Sandra Garrie was the daughter of astronomer T.S. Jacobsen, and the principal at the time was Robert A Lawson, an Astronomy graduate from the University of Toronto.)

But to return to J.H. Horning. He joined the RASC in 1914, while attending the University of Toronto. A year later, newly graduated, he joined the army as a lieutenant, and served overseas in Europe and the Middle East. In fact some years later he gave a talk to the Toronto Centre, "Under the Northern Cross with Allenby in Palestine" which he illustrated with his own photographs. Some of his other talks included "Measuring Stellar Distances", "Measuring Planet and Star Diameters", and "The Meaning of Astronomical Terms". He served as the Society's Recorder in 1923, and was on the Toronto Centre Council in the 1930s, but his most remarkable contribution to the Society was his 30-year term as Treasurer - a record only surpassed by Chant as Editor.

Horning received the Service Award in 1960. As stated in his citation, "During lean years he urged caution but ... [in] the good years, he was the first to support bold and wise expenditures for the good of the Society." What better advice could guide any Treasurer?

J.H. Horning (l) presents the RASC Gold Medal to Donald C. Morton in 1956.

as treasurer was to be deposited in a chartered bank approved by the Council. Payments were to be made only by cheque signed by the treasurer and president. The only minor changes during this sixty year period provided that routine payments could be made without prior approval, and permitted some other designated officer to be the cosigner of cheques.

Undoubtedly, the executive secretary actually handled most of the book keeping, bill paying and deposits, even though the formal responsibility rested with the treasurer. Certainly it is a matter of record that in 1918 the assistant librarian (as the Society's employee was then known) assisted the treasurer with routine work, and in the 1930s her title became assistant secretary-treasurer to reflect those duties. The 1969 by-laws made this even clearer by stating that the deposit of monies, the safekeeping of securities and the disbursement of funds was *under the control* of the treasurer, subject to the direction of the Council, and in the 1989 version, the treasurer was required to supervise deposits, disbursements and safekeeping. He or she is now required to prepare and present statements for the Council three times per year, to chair the Finance Committee (charged with preparing an annual budget, administering the special Funds and investing the assets of the Society), and to belong to the Executive and Property Committees. The treasurer's position is perhaps the most sensitive on Council. Everyone rightly expects the treasurer to be completely familiar with all aspects of the RASC's finances, and if reports are not ready and explanations not convincing, councillors and members are inclined to be less understanding than they would be towards other officers with less precisely defined portfolios.

SECRETARIES

At various times this officer has been called by different names: corresponding secretary, general secretary, national secretary, and probably some unofficial and less complimentary ones as well. It has always been his duty (never hers) to conduct the official correspondence of the Society, and to report on such communications at the first opportunity.

In the early decades of the Society, correspondence played a far more significant role than it does now. Indeed, there was a whole class of "Corresponding members" comprising foreign astronomers whose opinions and contributions were solicited and out-of-town amateurs anxious to communicate their observations to the Toronto Society.

Since 1908, the secretary has also been charged with the responsibility of issuing notices calling meetings of the Society and Council, and since 1969, also sending copies of minutes to every councillor. In practice, routine matters are normally left to the executive secretary and the preparation of the minutes to the recorder, though it is still the secretary's responsibility to see that the mailings of notices, agendas, ballots and minutes are carried out as required in the by-laws. Also, in recent years, the secretary has been an important force in keeping track of various policies and procedures to ensure the smooth and consistent operation of the Society.

RECORDERS

The positions of recorder and secretary are somewhat synergistic. In fact until 1900 the recorder was called the recording secretary (in contrast to the corresponding secretary), and had the duties of keeping minutes and notifying members of meetings. This latter responsibility was assumed by the secretary in 1908, and ever since then the recorder's one important duty has been the preparation and submission of accurate minutes of Council and of the Society. Up to 1927, this duty included summarizing all the regular meetings of what would now be called the Toronto Centre, and preparing accounts of these meetings for publication in the *Journal*. It was a large

E.J. ARTHUR KENNEDY (1880-?) was introduced to astronomy by J R Collins and joined the Society in 1910. He spoke occasionally at meetings and served in various offices beginning in 1928, when he became Secretary of the Toronto Board (as it was called before becoming a Centre). Kennedy also served on the 1932 Eclipse Committee, was elected the Toronto Section's Vice-Chairman for 1935-36 and Chairman for 1937-38. When he retired as National Secretary in 1958, a position he had held continuously for 23 years, he was ready to step down. He had not been in the best of health and was unable to keep up with all the day-to-day work required of his office. E.J.A. Kennedy was regarded with affection by the Council, who named him Honorary Secretary and presented him with the first Service Award. A corsage was sent for Mrs Kennedy as they embarked on a European holiday in 1959, and a tea was arranged for his 80th birthday the following year.

Not a lot is known of his personal life. He first appeared in Toronto directories in 1898 as a clerk. By 1909 he was a book-keeper for a confectioner and eventually he had his own candy business. He was still living in suburban Toronto in 1966, but by 1968 his name no longer was included in the RASC membership roll.

JOHN ROBINETTE COLLINS (1865-1957) was a devotee of astronomy for most of his long life. He observed the transit of Venus in 1882, and recalled his experience 71 years later at a meeting of the Toronto Centre when plans were being laid for the observation of a transit of Mercury. He was educated in Toronto and at Chautauqua College in New York State, and remained an avid reader of science throughout his life. John Collins was a charter member of the A&P Society in 1890 and his brother, Zoro, joined in 1893. They both took a special interest in optical design and published a number of papers on this and on other topics in the *Transactions* and in the *Journal*. Every Saturday, from 1900 to 1948, J.R. Collins ran a column on Astronomy in the Toronto *Telegram*. In 1905, he went on the Canadian government eclipse expedition to Labrador and in 1932 was chairman of the RASC Eclipse Committee. He served as Secretary of the Society from 1902 to 1917, and then was elected Vice-President and President. Following these responsibilities he was very active in local Toronto affairs, being the first Chairman of the Board in charge of meetings. Collins was fond of travel and took at least three extended motor trips in the 1930s, visiting western Centres and several U S observatories. Besides his dedication to the RASC, he was a President of a Geological Society and a fellow of the AAAS and the American Geographical Society.

In writing his obituary, E.J.A. Kennedy recalled him as "a kindly gentleman, always carrying out his duties with great diligence and care. With his keen sense of humour and good memory, he was always ready with a story or an interesting and amusing sidelight on the lives of the great scientists of his day."

In this group photo, taken at Louiseville, Quebec, the site chosen for viewing the August 31, 1932 eclipse, Kennedy is at the far left of the back row. The complete key identifies those in the back row as Kennedy, Simpson, Beatty, Mrs Benson, Ireton, Benson, McKone, Clipsham, Barker, Paterson, Kingston Jr., Carrick, Maybe, Collins, ____, Thomson, and those in the front row as Hepburn, Connor, ____, Mrs Paterson, Miss Benson, Mrs Kingston, Mrs McKone, Miss Budd, Mrs Collins, Mrs Connor, ____.

In this recent photo, Rosemary Freeman, executive secretary, works at her desk in the national office.

responsibility, though it was also expected of each Centre's recorder. Now, however, the recorder has usually only four meetings per year to worry about, three of the National Council and one Annual Meeting. Still, with Council meetings lasting several hours, the task of preparing accurate minutes is important and time-consuming.

Though the membership traditionally elected Recorders, the by-laws of 1989 changed the process so that the Council now appoints a recorder for a specified term not exceeding three years, but renewable any number of times.

LIBRARIANS

The library will be discussed in chapter five where its importance to the Society will be evident. The librarian is and always was responsible for the safekeeping and proper condition of the books, maps, publications and archives belonging to the Society. Beginning in 1900 he was also required to take charge of the Exchange List and to oversee the distribution of the Society's publications to those on the list as well as to the members generally. This latter duty was surely a formality, as the actual mailing of the Journals was always done by the executive secretary (formerly assistant librarian). The 1969 by-laws no longer made the librarian responsible for the distribution of the Journal but did require him to maintain an up-to-date catalogue of all items comprising the library and to make new acquisitions subject to terms and provisions determined by the Library Committee of which he was chairman.

Like the recorder, the librarian is no longer elected by the membership but is appointed by the Council for a term of up to three years, renewable at will.

OTHER OFFICERS

A curator was also named as an elected officer in the 1900 and 1908 by-laws, though not in the 1944 or subsequent versions. D.J. Howell held the position for the first two years and Robert S. Duncan from 1902 to 1943. It was their duty to be in charge of all the Society's property except that entrusted to the librarian's care. Judging by occasional reports from the curator, he supervised the maintenance and lending of telescopes and slides. A celestial globe, meteorites and several telescopes which once belonged to the Society are now nowhere to be found. At what point these items disappeared is not known, but they may have gradually gone after the curator's position was abolished. Before 1900 and since 1944, this equipment and material was theoretically the librarian's responsibility, but without an up-to-date inventory it would have been difficult for a succession of librarians to keep track. As far as the telescopes were concerned it is not hard to imagine that repairs were needed which volunteers may have offered to carry out; perhaps through illness or neglect, the work was not done and the constantly changing Councils soon forgot that the equipment had ever belonged to the Society. Realistically, equipment cannot be expected to last forever and the telescopes did serve a useful purpose for many years. Still it is with regret and embarrassment that one realizes, for instance, that such heirlooms as the speculum which once belonged to Sir Adam Wilson have vanished.

The editor of the Journal has only been considered an officer of the Society since 1969, and the other editors only since 1989. They are named in the chapter on Publications.

The executive secretary was included in the list of officers from 1969–89, a rather anomalous situation since the by-laws stipulated that no officer was to receive any remuneration for services. Presumably the executive secretaries of the time were grateful that the by-laws were not always rigidly followed! Their names are found in chapter two.

CHAPTER 5

Keeping a Balance

Normal stars exist in a state of equilibrium with the tendency of hot internal gases to expand balanced by gravitational attraction. Change can occur gradually but if these opposing forces become markedly unequal, catastrophe will result. The Society too has evolved slowly, constrained by the need to balance its desire to spend and thus expand its sphere of influence with the cautious retention of its limited resources.

INCOME AND EXPENDITURE

In the diagram shown here, the widths of the various sections indicate, at five-year intervals, the proportion of the national Society's revenue and expenses attributed to various sources and activities. The first feature which should be noted is that the Society is a nonprofit organization. Revenue and expense are always within a few percent of each other, and while there is a profit in some years, the deficit in other years just about cancels it out. In other words, the RASC fully applies its revenues to fulfill its objectives.

On the expense side, the proportions devoted to various purposes have been rather constant. Within a few percentage points, publications have always accounted for 60 percent of the expenses, salaries have grown from 10 percent (when the help was only part-time) to a 20 percent share, and other office expenses have consistently been about 10 percent. Library expenses, which at one time claimed about 5 percent of the budget have become insignificant, and rent paid out, while never a large portion of expenses, is nonexistent now and in those years when the Society occupied its own building. Travel costs have begun to make inroads in the last few years. Before 1982, the only travel assistance the Society could afford was a grant of half the fare for a few delegates to attend the annual General Assembly. Then a donation of $30,000 changed the situation so that travel expenses of Officers could be paid and assistance provided for Centre representatives to attend Council meetings.

On the revenue side, there have been some remarkable trends over the years. Government grants, for a long time the main source of Society funding, have now become a small proportion of

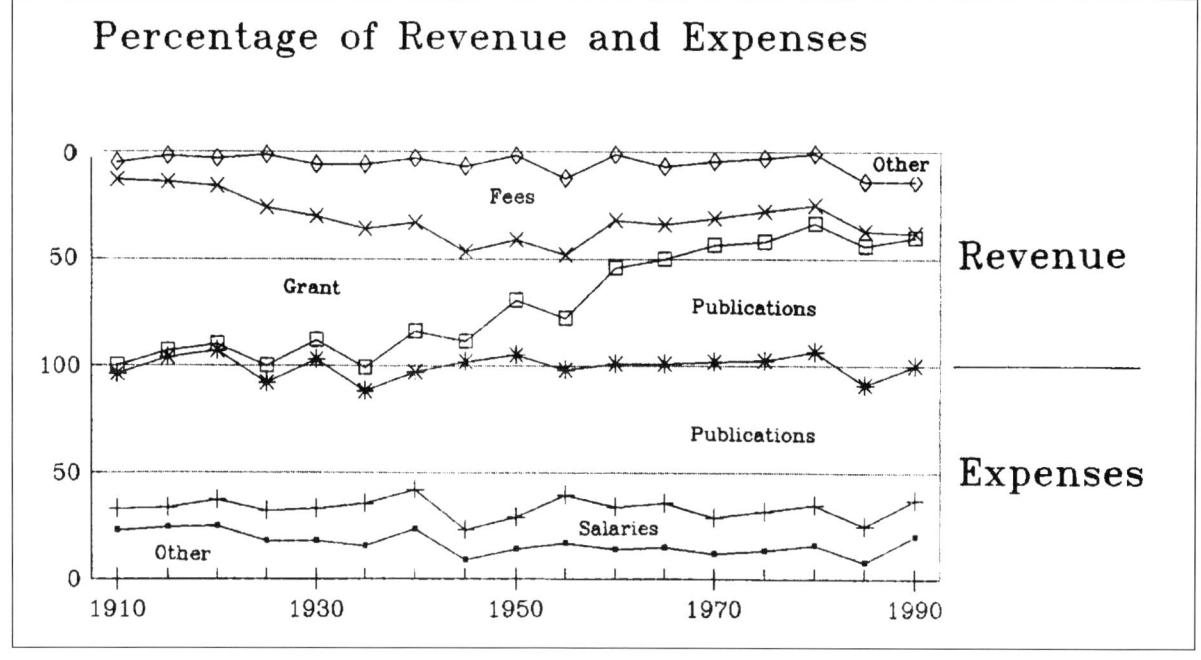

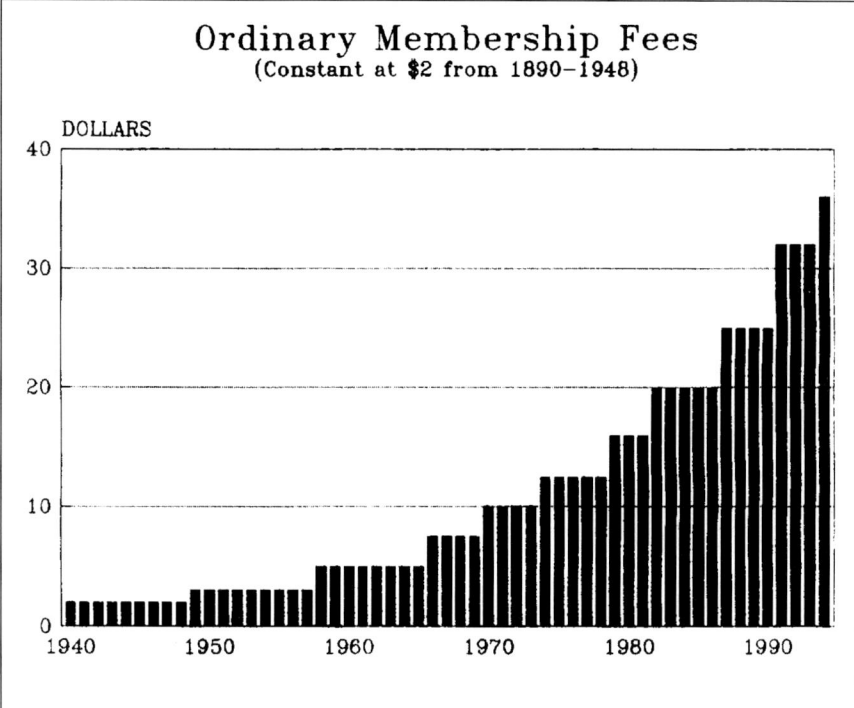

the operating budget. Fees played an increasingly important role until about 1950 when they accounted for about 40 percent of revenue. Since then they have become less important so that they now provide for only 25 percent of income. In this connection, it should be pointed out that during the period 1947–68, the national Society claimed 66 percent of attached members' fees. Since then it has received only 60 percent with the balance retained by the Centres. Revenue from publications has grown from insignificance in the first decades to the dominant position of over 50 percent in recent years. More will be said about publications in chapter seven.

Income on investments has been significant since 1982 when the interest earned by the Endowment Fund was first credited to the general operations of the Society. However, in future years, the hope is that most if not all of this interest will be retained within the Fund. Rental income from the apartment above the national office has also brought in a modest income, more than sufficient to pay the operating costs of the building. Still another small but important source of revenue in recent years has been the sale of promotional items. Originally, lapel pins and insignia, suitable for sewing onto a blazer, were made available in 1957. In the last couple of years, thanks to the work of Cathy Hall, other items have been added including stickers, key chains, golf shirts and tuques. Besides bringing in a few hundred dollars each year, these items provide good publicity for the RASC.

The fee structure for the past fifty years is shown in the figure. During the earlier period, 1890–1948, regular members paid $2. There were a few temporary measures, for instance when out-of-town members and women paid only $1, and the life membership fee increased from $20 to $25 in 1908, but on the whole there were no significant changes until 1949. While fee increases apparently out-paced the consumer price index from then until about 1980, they have lagged a bit recently. Salaries, of course, have increased faster than the CPI over the long term.

GRANTS

Without government grants, it must be obvious that the Society, had it survived at all, would have been a very different organization from what it is today. Members' fees could certainly have never supported the publications on a continuing basis, and the publications did not begin to generate significant revenue for decades. Indeed, it was not until about 1925 that the income from fees was sufficient even to pay the salary, rent, library and office expenses. There are interesting parallels with the financing of professional astronomical research in Canada, which, until the establishment of the DDO in 1935, was almost entirely provided directly by the government. This typically Canadian tradition of funding arts and science, unfettered by the needs of the military or other powerful interest groups, should be a matter of national pride. It is a tradition which has benefitted the RASC for nearly a century.

The A&P Society of Toronto, encouraged by the willingness of the Ontario Minister of Education, the Honorable George Ross, to become their honorary president, decided early in 1893 to apply for funding through his department. The fact that the Society's corresponding secretary, George Lumsden, was also the Provincial Secretary (akin to a deputy minister) might have also had something to do with their resolve. Within a few months, word was received that the provincial legislature had approved an annual grant of $200. This was raised to $300 a few years later, and then in 1905, there was a substantial increase. C.A. Chant, RASC president at the time, has told the charming story of how this came about:

> One morning I walked over to the Parliament Buildings to see the Provincial Treasurer. The Hon. Colonel Matheson was in and I was admitted to his office without delay. I told him who I was and said I had come to ask about money. He said our grant was in the Estimates and we should receive it. "But," said I, "we think we should have more." He walked into his inner office and brought out our new volume, a copy which had been sent to each member of the Legislature. "How much do you want?" he asked. I replied, "We think we should have $500." "Five hundred additional?" "Oh," said I, "we are modest, but we think we should have a grant of $500." He said he would consult his colleagues, but as I was leaving he remarked, "I think your publication very creditable." A few weeks later there came addressed to me a check for six hundred dollars!

Those were the days when six copies of ten-page applications were not required by the bureaucracy! Except for some years in the 1920s, the Provincial Department of Education continued the $600 grant until 1933. The Depression nearly brought cancellation, but lobbying of Premier Hepburn kept the grant coming though at a reduced rate of $500, a level which was maintained until 1956. Then, prompted by the financial needs relating to the newly purchased building, the Society sought and received an amount of $1,000 annually. This continued when the Ontario Department of University Affairs assumed responsibility in 1965 and through 1974, when the Ministry of Culture and Recreation provided the grant. The years 1975–83 marked the last phase of provincial support with grants rising from $1,000 to $1,600 by the final year. One bit of a break which the Society continues to enjoy as a nonprofit organization is exemption from sales tax on some stationery and printing supplies.

The City of Toronto also provided some financial aid in the form of a $100 annual grant from 1898 to 1933. According to the minutes of 1898, this resulted from an interview which Society President Arthur Harvey had with a committee of City Council. The grant was made "for the purpose of providing opportunities for the general public to view celestial objects with the telescope." In more recent years, the RASC has obtained indirect support from the City through tax relief. As a registered charitable organization and property owner, the Society pays no property tax on the portion of the building which it occupies.

The largest grants, however, came through the federal government and its agencies. No approach was made to Ottawa for funding until 1906, by which time a Centre was planned in the capital city and the Society's name reflected its national aspirations. As the minutes record, "A delegation from the Society ... waited on the Hon. Sir Wilfrid Laurier, Premier of Canada, and asked him to consider the request of the Society that pecuniary assistance be given to enable the Society to publish more elaborate and more frequent reports." A $1,000 annual grant was approved for 1906 and 1907, and the Society, with this adequate financial backing, at once launched publication of its *Handbook* and bimonthly *Journal*. The amount was doubled in 1908, and except for temporary decreases during World War I, this $2,000 annual grant was maintained until 1933. The Great Depression which saw the end of the municipal grant and the reduction in provincial support, also brought about a 20 percent cut in federal funding. By 1949, the situation had become rather desperate. RASC expenses were progressively exceeding revenue; Toronto Centre donated $100 to help out, and $500 had to be borrowed from the Building Fund. Fees were increased from $2 to $3 while rebates to Centres were held at $1. Delinquent members were culled from the rolls. The price of The Observer's Handbook was increased from 25 to 40 cents, numbers of the Journal were combined so that two months were issued as one and advertisements were solicited for the publications. Binding of periodicals in the Society's library was suspended. The treasurer, J.H. Horning, led lobbying efforts to get the federal grant restored to its pre-Depression level. Members across the country wrote letters and visited their Members of Parliament and in the end the $2,000 annual grant was resumed in 1949. The history of federal support since then is summarized in the table below. After a long trend of increasing grants, it will be seen that nothing was received in 1987. That year, the Council, on the editor's recommendation, made the decision not to ask for a grant since the Society had experienced substantial and growing surpluses in the previous three years.

GRANTS RECEIVED

Year	Source	Amount
1953–54	Minister of Finance	$3,000
1955	Minister of Mines & Technical Surveys through the Dominion Observatory	3,000
1956–60	" " " "	3,500
1961–66	" Dominion Astronomer	3,500
1967–69	Department of Energy, Mines & Resources through the Director of Observatories	5,000
1970–71	National Research Council Committee on Grants and Scholarships	5,000
1972–74	NRC C'ttee on Scientific Publications	5,000
1975	" " " " "	6,000
1976	" " " " "	6,500
1977	" " " " "	7,000
1978	" " " " "	7,150
1979	Natural Sciences and Engineering Research Council (NSERC)	8,000
1980	NSERC	$7,000
1981	"	8,000
1982	"	9,000
1983–85	"	10,000
1986	"	10,400
1987	"	0
1988–89	"	2,000
1990	"	2,400
1991	"	10,000
1992–93	"	5,000

As vital as these grants have been to the RASC, the Canadian public has certainly been repaid manyfold for this fiscal encouragement. The countless volunteer hours which have gone into the careful preparation of the publications have brought the work of Canadians to the attention of thousands of readers at home and abroad. As for the provincial and municipal support, meetings of all kinds have enriched the lives of campers and city-folk, young and old, from Victoria to Saint John's. Though these intangibles do not show on a balance sheet, they are the real life and spirit of the Society.

FUNDS AND DONATIONS

From time to time the Society provided for special needs by establishing funds segregated from the general operating account. The first to be established was a Building Fund in 1912, necessitated by the cramped quarters at 198 College Street. The Society had accumulated assets of $3,411 by the end of 1911, and transferred $1,372 of this to establish the Fund the following year, investing the proceeds in 4.5 percent bonds. By the time the Fund was used towards the purchase of 252 College Street in 1956, it amounted to $8,908.

In 1925, a life membership account was set up with $50 which had been received from two new life members that year. The general practice for the first thirty years seems to have been to credit the interest earned on these deposits to the general operating account. By the end of 1955, the life membership account amounted to $2,543, $543 of which went to pay for the building the following year. From 1957 on, the practice has been to maintain only a book-keeping account called "Unappropriated Life Membership Fees," of which a certain percentage, currently 5 percent, is paid out each year to the operating account. This has the effect of smoothing out revenue from life memberships so that the financial statement does not fluctuate through boom and bust according to the number of new life members in any particular year. The interest earned by these fees is still credited to the general account, but is not shown separately.

The third fund to be established had a creative origin thanks to the initiative of President W.E. Harper. Starting in the fall of 1929, certain members of the Society began to write a series of short fortnightly articles on popular astronomy for Southam newspapers across Canada. These were somewhat like syndicated columns, except that a variety of authors took part under Harper's editorship and the $25 per column fee went entirely to the Society. Some other newspapers purchased rights to the series, and so at the end of the project's two year span, $1,357 had been added to the coffers of the Society. Council decided in 1932 that these monies should be kept in a segregated fund to be known as the Southam Press Fund. Except for interest which accrued, no new amounts were added to the fund. Occasional withdrawals paid for striking the Chant Medal, the Gold Medal, a film on Solar Prominences and a $500 contribution to the F.S. Hogg Memorial Scholarship Fund. The Southam Fund was closed in 1956 when the entire amount of $1,951 was used towards the purchase of 252 College Street. Following the acquisition of this property, the only surviving fund was the Life Membership Reserve, but as explained already, that was no longer kept as a separate account. A Building Account was also set up in the books in which revenue from rent and

WILLIAM EDMUND HARPER (1878-1940) made outstanding contributions to the RASC. Other than C.A. Chant, no professional astronomer before him had been more devoted to the Society and to the popularization of his science. He gave over 100 radio talks, many of which were published in a number of newspapers across the country. Harper put all the royalties from these into the Southam Fund which eventually became an important source of capital for the purchase of the Society's own headquarters. He wrote over 80 papers for the *Journal* as well as a great many notes, reviews, seminar abstracts and texts of his radio talks. Harper served on the Councils of Ottawa and of Victoria Centres and spoke not only at meetings of these Centres but also in Vancouver, Edmonton and Hamilton. He was President of Victoria Centre in 1922 and national President in 1928-29.

He grew up near Owen Sound, Ontario, attended the University of Toronto and, on graduation in 1906, was the first winner of the RASC Gold Medal. He immediately joined the staff of the Dominion Observatory in Ottawa and in 1919 moved to the DAO in Victoria where he subsequently became Director. During his lifetime, he was credited with computing orbits for more spectroscopic binaries than any other astronomer in the world and was also noted for calibrating the absolute luminosities of thousands of stars of spectral type A. He was a councillor of the AAS and a Fellow of the RSC.

At the opening of the DDO in 1935, Harper was awarded an Honorary Doctorate by the University of Toronto. He is seen here (rear right) on that occasion with co-recipients Harlow Shapley of Harvard on his right, and former Astronomer Royal Sir Frank Dyson in the foreground. DDO Director C A Chant, Harper's teacher thirty years earlier, is at the right front.

expenses connected with the property were shown separately from the general account in the financial statements.

One of the tenants at 252 College Street was Carl Reinhardt, a retired mining geologist who took a great interest in the Society. He donated $2,000 to help reduce the mortgage on the property, and also gave $1,000 to help build up the library. Following his death in 1962, when he bequeathed a further $1,000 for the same purpose, a Library Fund of $2,000 was established. This was amalgamated with the general account in 1970.

Because the RASC is a registered nonprofit organization, generous members and friends have not only been able to support the Society through gifts and donations, but have had the added incentive of reducing their personal income taxes. Contributions, large and small make a significant difference to the ability of the Society to carry out its programs.

By far the largest bequest in the Society's history came from the estate of Walter J. Helm of Port Hope in 1960. Mr Helm had been vice-president and general manager of the Midland Loan and Savings Company and was a member of the RASC since 1925. According to the terms of his will, approximately $400,000 was designated, the interest from which was first to be paid out to certain of his descendants for their respective lifetimes, and following their deaths was to be transferred to the RASC in order to create a fund to be known as the Walter Helm Endowment Fund. However, the will directed that the clear annual income from this fund must be used by the Society for the general purposes of the David Dunlap Observatory; so the Society only benefits to the extent that it can be said to be supporting professional astronomy at the DDO. At the end of 1992, the Walter Helm Endowment Fund stood at $175,000 and the DDO received $17,500 during the year. (A separate provision of Helm's will in 1960 also established the $100,000 Walter Helm Scholarship Fund at the University of Toronto.)

An ongoing concern of the Society is the provision of some suitable way of honouring deceased members who have played a prominent role in its work. The passing of Professor Ruth Northcott in 1969 led to the establishment of a Fund in her memory. Contributions were sought from her many friends both within and from outside the Society, and an amount of $4,544 was raised by the end of 1970. This was augmented by proceeds from her estate in the next few years. With accumulating interest and a transfer of $10,000 from surplus revenue in 1988, the Northcott Fund increased to about $38,000 by the end of 1990. The success of this fund's operation, and its continuing existence, can be credited to the careful way it was set up and the specific guidelines which were established. Payments out of the fund have gone towards library acquisitions, to paying a share of the costs of the Ruth J. Northcott Lecture (held approximately every two years at the General Assembly), to helping Centres with special publications, such as Centre histories and a French language almanac and as a grant to the Society of $1,000 to assist with the publication of the seventy-fifth anniversary issue of the *Journal*.

The death in 1977 of another past-president, J.F. Heard, also led to a fund named in his honour. In this case, the contributions were earmarked to pay the publishing costs of several short "Reminiscences" which Heard had written over the years, and the fund was retired in 1979, once its purpose had been achieved.

Less successful was the way the Society dealt in 1975 with a bequest of $10,000 from the estate of E.R. Paterson who had been a member of the Montreal Centre since 1925. With the best of intentions, this very generous amount was put aside in the E.R. Paterson Fund. The original plan was that income from the fund would be used to pay for library acquisitions. However, the move of national office in 1976 raised the recurring issue of what

JOHN EDWARD KENNEDY (1916-) graduated from Queen's University in 1937 and earned his MSc at McGill in 1942. He became a Professor of Physics at the University of New Brunswick, where Lloyd Higgs and Mary Grey (later RASC Presidents) were among his students. Through friendship with Peter Millman, he worked at the DO in the summer of 1953 and was soon organizing observations of meteor showers at UNB and participating in eclipse expeditions. His important work in uncovering the history of astronomy in Canada also began in those years when he brought attention to the pioneering work of Brydone Jack who established an observatory at UNB in 1851.

In 1959, the Kennedys moved to Toronto and Ed (as his friends know him) began work at the Defence Research Medical Laboratories. Fortunately for the RASC, he was the right person at the right time and place to asssume the position of Secretary. His strong and efficient presence ensured that the Society operated in a business-like manner. As a member and Chairman of the Editing Committee, his insistence on high standards and his reassurance were a vital help to the Editor, Ruth Northcott.

Immediately following his six-year term as Secretary, Kennedy embarked on the ten-year round of Presidential Offices. During these years and subsequently, he addressed nearly every Centre of the Society; he spoke frequently at General Assemblies, and contributed many papers to the *Journal* and other international publications.

Professor Kennedy taught Physics and was Assistant Dean of Arts and Science at the University of Saskatchewan. A founding member of the Saskatoon Centre in 1969, he presently represents the Award on the National Council. He received the Service Centre in 1970 for his guidance and untiring efforts in the interests of the Society.

RUTH J. NORTHCOTT (1913-1969), as a student at the University of Toronto, produced such meticulous lab reports that Professor R.K. Young suggested she join the Astronomy department as a computer. This she did after earning her MA in 1935. In the early years she compiled meteor and variable star observations from amateurs, and, working with R.E. Williamson, produced what was probably the first radio map in galactic co-ordinates, based on Reber's observations. But throughout her academic career, her main research interest was radial velocities and spectroscopic binaries.

Professor Northcott taught an Extension Course for teachers for nearly 20 years and was completely responsible for the laboratory work for astronomy students at the University. She took a rare personal interest in her students, photographing her classes and keeping in touch with many graduates.

Miss Northcott joined the RASC in 1935 and held office in the Toronto Centre as Recorder, Vice-Chairman and ultimately Chairman in 1943. She frequently spoke at meetings on a wide variety of topics, served on the Public Star Night Committee, answered queries from the question box, reviewed current astronomical happenings at each meeting, and gave a series of short lectures intended to acquaint new members with the use of the *Handbook*.

At the National level, Professor Northcott carried out her work as Assistant Editor and later as Editor of both the *Journal* and the *Handbook* with the utmost care. She strove to see that everything reflected the highest possible standard and brought the circulation of the *Handbook* to unsurpassed levels. On top of these heavy responsibilities, she served during the years 1959-65 as Vice-President and President of the Society. The Society presented her with the Service Award in 1967.

J.E. Kennedy watches as RASC President Ruth Northcott presents the first cheque from the Walter Helm Endowment Fund to DDO Director J.F. Heard, July 11, 1962.

should be done with the library and in the absence of any clear direction, expenses were kept to a minimum and the Paterson Fund got very little use. In 1984, it was amalgamated with two other funds into an Endowment Fund which was designated to provide for long-term growth of the Society, a purpose with which Mr Paterson would surely have been in full agreement.

The proceeds from the sale of 252 College Street in 1976 amounted to $185,000, and only about $7,500 per year was needed initially to pay the rent at the new address on Merton Street. To some members, especially those from outside Toronto who were not used to such high property values, this seemed like a wonderful windfall. In fact, over the twenty year period the Society had been at 252 College Street, the value of this asset had increased at a compound rate of just over 9 percent per annum, an excellent return but hardly spectacular. Certainly, if ever the Society was again to enjoy the benefits of ownership, a new building would have to be found soon before rent and inflation eroded the value of this new-found capital. So the Society again set up a Building Fund from the net proceeds of the sale with $85,000 retained in an 11 percent mortgage, and the balance (about $90,000) mainly in investment certificates at 9 percent. As an appeasement to those who felt that Centres should have some claim on these assets, a committee was formed to examine the finances of the Society. One of their recommendations, adopted by the Council, was that $35,000 should be set aside as a Special Projects Fund to give Centres access to money for capital expenses such as observatories, telescopes and other equipment. From the point of view of the Property Committee, siphoning off this amount of money made the task of finding another building that much more difficult. However, they eventually succeeded in 1983 without exceeding the amount which had accumulated in the Building Fund, and the Special Projects Fund did pay out over $8,000 for some worthwhile undertakings in the years of its existence, 1978–1983. These included loans and grants ranging from $100 to the Niagara Centre for duplicating equipment to $1,000 which went to the Saskatoon Centre towards a 41-cm telescope. Council has always insisted that a Centre present financial statements and a clear commitment, monetary and otherwise, to a project before a Special Projects grant would be considered.

The RASC got a wonderful boost in 1981 when it received a gift of $30,000. This has been referred to previously as an anonymous donation which allowed for travel support enabling officers and Centre representatives to attend meetings on a more regular basis. It is now possible to give credit where credit is due. The donor whose generosity has made a significant difference in the effective operation of the Society is none other than pioneer radio astronomer, Grote Reber.

Finally, following the purchase of 136 Dupont Street, the small balance remaining in the Building Fund was amalgamated with the Paterson Fund, the Special Projects Fund and Reber's donation to form an Endowment Fund which amounted to $105,600 by the end of 1984. Since the interest was credited to the general operating account, the Endowment Fund grew only when principal was added to it. Ten thousand dollars was transferred from surplus revenue in 1988, and other donations brought the balance up to nearly $119,000 at the end of 1990. Following the death of Peter M. Millman that year, the fund was renamed in his honour. A separate Centennial Fund was established in 1988 with a start-up of $5,000, also paid out of the operating surplus that year, and subsequent donations and interest brought that Fund to a total of just over $10,000 by the end of 1990. The purpose of the Endowment Fund was to provide for the long-term growth of the Society, while the Centennial Fund was intended to finance special projects at both the National and Centre level. *The Beginner's Observing Guide* and this book are two major projects that have benefitted from this latter source.

While the Funds are not large enough yet to allow the Society to fulfill its objectives as it might wish, there is certainly a firm foundation on which to build. Combined with the long-standing tradition of extensive contribution of time and expertise by volunteers, the Society is well poised for a second century of even greater accomplishments.

Grote Reber addresses the 1977 General Assembly in the auditorium of the McLaughlin Planetarium, Toronto.

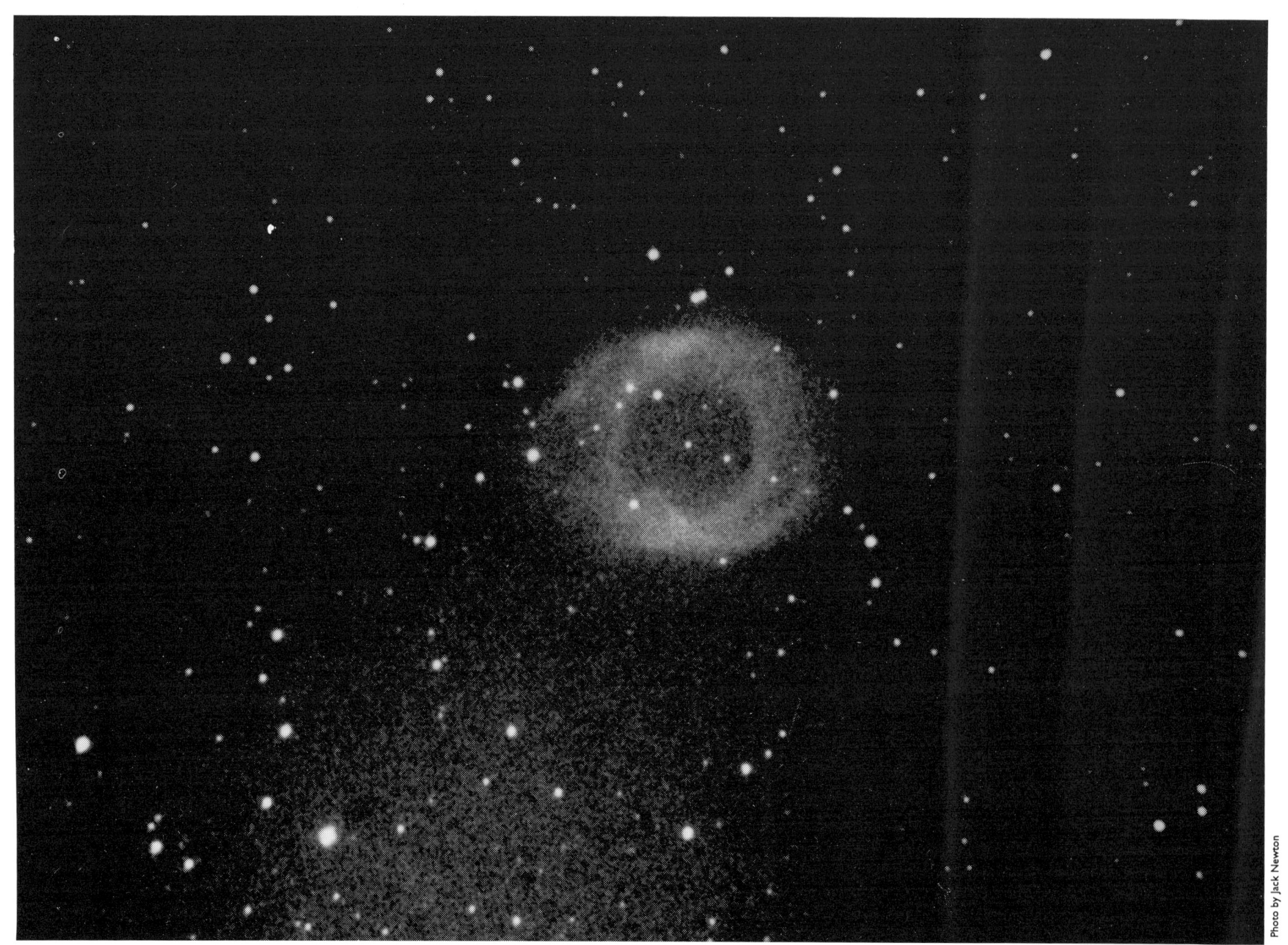

The "Helix Nebula" (NGC 7293) in Aquarius is an example of a shell of gas ejected from the hot central star.

CHAPTER 6

Of Volumes and Space

The story of the Society's library can be likened to the more dramatic phases of stellar evolution. Unlike the Society's finances which have kept in equilibrium within fairly close limits, the library has at times expanded to the point where major portions of it have been cast off, forcing the remaining core into a new phase of its life.

Starting from nothing in 1890, the Society's library grew rapidly. By 1901, the librarian was able to claim, probably with justification, that the collection was "by far the best of its kind in Canada." At that time it contained some 1,000 astronomical and physical works, comprising 165 books, 500 transactions and reports, and 363 maps and charts. Amazingly, no books or periodicals had been purchased until 1897, and even then only a small amount of money was spent. Practically all of the journals had been obtained in exchange for the Society's own *Transactions*. These exchanges included *Publications* of the ASP, *Monthly Notices* of the RAS, the *Journal* of the BAA, *Proceedings* of the Royal Society and of the RSC, a nearly complete set of the *Publications* of the United States Naval Observatory, volumes from the Cincinnati Observatory and the Lick Observatory, and periodicals such as *The Sidereal Messenger, Astronomy and Astrophysics,* and *Scientific American*. Prominent members, Lady Wilson and Larratt Smith donated many of the books and maps, and distinguished astronomers from England, France and the United States including Isaac Roberts, M. Loewy, E.C. Pickering, and G.W. Hill contributed photographs, annals and works.

The members of the A&P Society of Toronto were evidently very enthusiastic about their splendid collection. The librarian stated in 1892 that some of the works were in continual circulation and members frequently reported at meetings on what they had been reading. In 1896, though the borrowing rate only averaged about one item per member per year, the reading room was used frequently. Successive librarians put in many hours of work; rules were drawn up for borrowing, catalogues were prepared, published and updated. On top of this, the responsibility for distributing the Society's own publications became part of the librarian's duties starting in 1900. No wonder the first paid employee of the Society was the assistant librarian.

The young Society had set its sights very high. The generosity of individuals, observatories and sister societies was truly remarkable, but it was also the fine calibre of the *Transactions* of the A&P Society and the extensive and impressive list of those to whom it was sent that attracted the attention of others and prompted their willingness to help. By the early 1900s, the Society had a tiger by the tail, though no one seemed to realize it.

Books are far from being the most costly item in a library's operation. Books take up space which must be rented or otherwise provided. Shelves must be built, periodicals bound, new books catalogued, adequate insurance maintained, decisions made and borrowers hounded. For the sort of library the Society aspired to have, all this was too much for volunteer help and a very small budget. Had the

Society been really serious about maintaining a first-class library, it would have had to make a large financial commitment which its relatively small membership base could hardly have supported. Momentum kept the collection growing, however. In 1911, Librarian Angus Sinclair prepared a 31-page catalogue of the books and R.S. Duncan a six page list of lantern slides belonging to the Society. The catalogue was divided as follows:

- General treatises and text books, 43 authors
- Bibliography and history, 30 volumes
- Spherical Astronomy and geodesy, 7 volumes
- Theoretical Astronomy and celestial mechanics, 22 volumes
- Practical astronomy, instruments and methods, 30 volumes
- Observations, descriptive astronomy, astrophysics, 76 volumes
- Ancient astronomy, 5 volumes
- Works of physics, 63 volumes
- Works on mathematics, 34 volumes

For reasons which will be explained later, most of the above material is still in the Society's collection, but very little of the following material (from 1911) is:

- Bound periodicals, 21 titles
- Reports of learned societies and institutions, 36 titles
- Publications of Observatories, 60 observatories represented - Almanacs, tables, star charts and catalogues, over 100 volumes - Monographs (Bound) 213 items

The best of the lantern slides including photographs of solar, stellar and lunar subjects, planets, comets, meteorites, portraits, spectra, aurorae and miscellaneous views have been saved but unfortunately a set of the 1905 Labrador eclipse expedition and Shackleton's expedition to the South Pole have disappeared.

Having enough space to comfortably house the collection and some tables and chairs has always been a problem. In the very earliest days, when the meetings were held from house to house, the librarian kept the books at his own home. That must have been a real imposition on Mr Pursey and his family. When the Society began to rent its own accommodation, one room was generally for meetings and another for the library, but unless arrangements were made to have someone in attendance, members still could not easily get access to the books except at the time of meetings. Borrowing was easier from 1898 to 1905, when the Canadian Institute and the Society shared a reading room, the Society's collection being kept in separate bookcases from those of the Institute. Members were thus able to use the library whenever the Institute was open. After the move to 198 College Street, even though the RASC continued to rent space from the RCI, the two organizations apparently kept their libraries separate.

Relentlessly over the next three decades the periodicals from other societies and observatories kept piling up; every year 600 more were added requiring another three metres of shelf space. One third came from the British Empire, including Canada, one third from the United States, and one third from other countries. Some subscriptions to periodicals and some new books were purchased, but as there were only about a dozen appropriate astronomy books published each year, it was not too hard to keep that aspect of the library up-to-date. These purchases and some costs for binding put the direct annual library expense in the $200 range, but as always there were the hidden costs of rent, salary and so on. One year, 1920, saw an expenditure of $412 but that paid for a complete set of *Astronomisches Nachrichten* from 1823 to 1919 (209 volumes), some

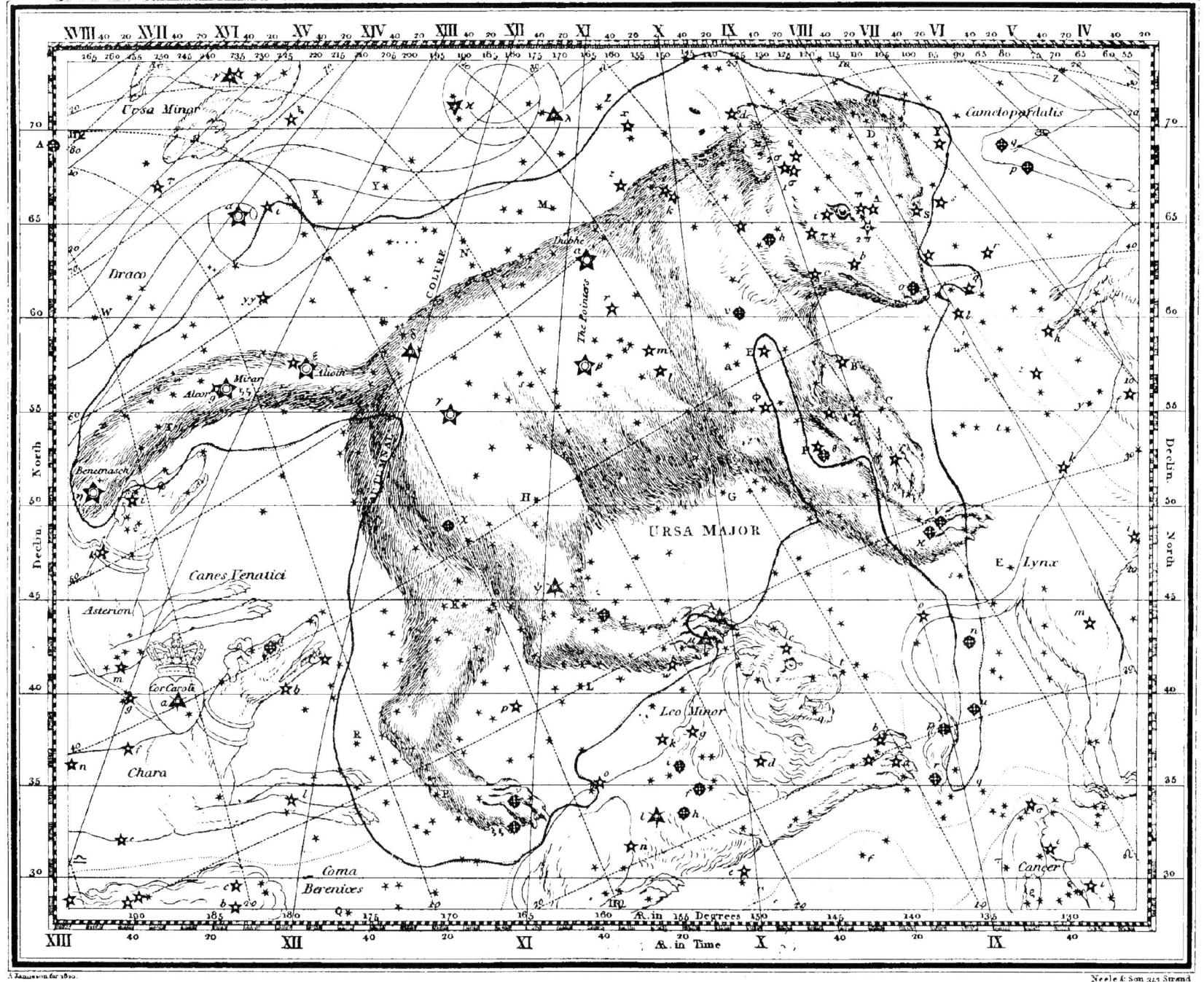

A plate from the RASC's copy of *Jamieson's Celestial Atlas*, 1822.

early volumes of the *Bulletin de la Société Astronomique de France*, and thirty-two books! By 1923 the collection was valued at $10,000.

The library continued to get moderate usage; on average about 100 items were borrowed each year, a figure pretty static from the 1890s to the 1930s even though membership in Toronto had doubled. Reading between the lines, it seems that the RASC Library at 198 College Street became a virtual storeroom, what with the growing numbers of the Society's own *Journals* as well as all those received on exchange. The library became increasingly difficult to use. Some books inevitably were "borrowed" without being signed out. The continual complaint of librarians throughout this period was that the space was woefully inadequate.

Librarian R.A. Gray, in his report for 1932, made a suggestion which signalled a willingness to change. He noted that "many valuable reports ... seldom or never consulted ... might be of great value to a class of reader of a more advanced type than the general calibre of our members; space taken up by these reports might be used for works of a more popular character." Perhaps he was already thinking of the needs of the newly announced David Dunlap Observatory. Perhaps Dr Chant had planted the germ of the idea in Gray's mind. In any case, what Gray suggested soon came to pass. Though one may lament the marvellous part of the collection which slipped from the Society's control, the technical material which went to the Observatory was undoubtedly better catalogued, cared for, and more widely read than it could have been in the RASC. Yet it was all still available for serious RASC members to use.

Initially, the books and periodicals were lent to the university's Department of Astronomy and were kept on the main campus, where they were still convenient for members to consult on occasion. As plans for the DDO progressed, however, it became clear that the real need for the library was at the Observatory, some twenty-five kilometres north in Richmond Hill. There had originally been plans for a special endowment to provide a research library there but those hopes were dashed. It seems that Colonel Reuben Wells Leonard of St. Catharines had agreed to contribute $25,000 for a library, payable when the plans for the observatory were fully matured. But unfortunately Leonard died about two weeks before a public announcement of Mrs Dunlap's magnificent gift could be made. There was no mention of a bequest for this purpose in Leonard's will and so, of course, those plans fell through. Consequently Chant proposed to the Society that the technical portion of its library be moved to the observatory where the collection would be catalogued, periodical sets completed where possible, and the whole made available to members of the Society. Though the Observatory was an inconvenient location for most members, the Council nonetheless unanimously approved the loan for a renewable term of three years and the Board of Governors of the university cordially accepted the proposal. The list of serials comprised over 100 titles, and in all, three and a half tons of bound volumes and pamphlets were transferred in 1935 to form the nucleus of the DDO library. As technical publications continued to arrive at the Society, the practice was to leave them out on the library tables for a month or so before shipping them off to the Observatory, a policy which is still adhered to, though on an infrequent basis and smaller scale. The loan was extended for many years until finally, in 1961, an agreement was reached whereby the University of Toronto purchased the books from the Society. Though the university librarian estimated their value at $8,700, the Society, mindful of the great contribution made by Chant and others at the DDO, accepted only $2,000.

The removal of the technical periodicals from the Society's rooms at 198 College Street made the rest of the collection more accessible to members. During 1937, the library was redecorated and refurbished, and the books were reclassified, thanks to the hard work of the librarian, Peter Millman, and the assistant secretary, Eva Budd. Among the more popular periodicals still retained by the Society were *Popular Astronomy, The Telescope, Publications* of the ASP, the *Journal* of the BAA, *L'Astronomie*, and *Southern Stars*. All of these continued to be received in exchange for the Society's own publications. In addition, each year a few books were added, either through purchase or donation. By 1944, the library collection was valued at $4,000 for insurance purposes. Comparison with the $10,000 figure used in 1921 gives an independent confirmation of the value of the portion that was transferred to the DDO.

During Millman's term as librarian (1936–46), several measures were taken to stimulate wider use. Hours at the national office were increased to include some afternoons, evenings, and Saturday mornings. A number of books were actually taken to Toronto Centre meetings so that members could borrow them without even walking over to 198 College Street. Some of the Centres were loaned groups of books for a couple of months at a time making access easier for their members and borrowing by mail was instituted with members paying postage in both directions. This latter provision was not really satisfactory, however, as no up-to-date catalogue of the available books was published between 1911 and 1970, and so members could hardly be expected to request books if they did not know what was on the shelves. Nonetheless, the measures did have moderate success. Circulation reached a high point in 1944 when 210 items were loaned, though on a per member basis, there had been a definite decline in library use over the years.

The premises on Willcocks Street to which the Society moved in 1946 provided good facilities and a spacious reading room for the library but the subsequent location on Ross Street which the RASC occupied from 1953 to 1956 was less adequate. Once again, in 1953, the Society was forced by constraints of space to dispose of some of its collection. This time, considerable meteorological material was given to the Meteorological Service on indefinite loan and many of the older periodicals having little to do with astronomy, were sold for $82. In addition, the exchange list was cut back. Seventy-eight organizations were removed from the list, though most of them became paying subscribers to the *Journal*.

From 1953 to 1968, the library saw its greatest use in its entire history with the annual borrowing rate reaching a high of 447 in 1960. The chief reason for this success was that the national office was open for a while after meetings of the Toronto Centre. Both the Ross Street location and the house at 252 College which the Society purchased in 1956 were just a couple of blocks away from the university buildings where the lectures were held. So it was very convenient for members who were interested to walk over to the national office, have a cup of tea or coffee, and select a book or two to read in the weeks between meetings.

Another boost for the library came in the form of a bequest of fifty books and $1,000 from the estate of the late Carl Reinhardt in 1963. Other donations of books and money were made over the years but this substantial gift was the beginning of a Library Fund. Though Council decided to eliminate the Fund as a separate entity in 1970, the understanding was that the money was still to be available for Library purposes.

The increased circulation which took place in the 1960s was not confined entirely to Toronto members. Work was progressing on a catalogue. Bev Oke made a listing of one hundred of the most recently published books available in 1958, and he also tried recommending a few specific titles in brief articles in the *Journal*. Borrowings by mail increased to forty-two in 1960 and seventy in 1963. When the complete Library Catalogue became available as a thirty-eight-page booklet in 1970, copies were sent to each Centre and to each Unattached member. Soon growing numbers of members from all parts of the country were using the mailing service. Unfortunately, customs difficulties precluded borrowing by members in the States and overseas, but requests from remote parts of Canada were especially gratifying. For these isolated members, the Society was providing a really unique service. A supplementary catalogue was printed in 1974, but a long postal strike the following year reduced circulation to 125, still a respectable figure.

Another aspect of the library's collection, slides and films, also attracted more attention in the 1960s. Of course, glass lantern slides had been an important part of the library since the 1890s but in the 1960s, excellent 35-mm slides in black and white and in colour became commercially available, illustrating all sorts of celestial objects. The Society purchased hundreds of these over the next few years, including the earliest space photographs of the Moon, taken by Ranger VII, and of Mars by Mariner IV. Other slides contributed by members included interesting series taken on solar eclipse expeditions. In 1970, about fifty slides per week were being borrowed on the average.

The RASC also purchased a few films over the years. One on Solar Prominences, bought in 1948, four others added in 1960, including *Universe*, and Ranger VII's flight in 1965, ending with its crash on the lunar surface were most popular. However, the high initial cost, and the amount of time and effort required to repair damaged films meant that relatively few were acquired.

There was a sharp decline in borrowing by Toronto members after 1968. That year the Centre began holding its regular meetings in the McLaughlin Planetarium and the increased distance to the national office meant the end of postmeeting visits to the library. A few members still dropped in at 252 College Street during the day, but the move to Merton Street in 1976 really spelled the end of usage by Toronto members. Librarians did try opening the office on a few Saturdays, both at Merton Street and later at Dupont Street, but the response hardly warranted their time.

From all quarters, interest in the library dried up in the 1970s; postal strikes and increased postal rates played a role, as did the closure of the library for some months during the move of the national office in 1976. Only twenty-four requests in total arrived that year. The move also saw the disposal of excess sets of the *Journal*. St John's was one Centre which took the opportunity to get a nearly complete set free.

A very important consideration in the decline of library use was the lack of new books. After 1975 almost no purchases were made. There was a discouraging feeling that the Society could not possibly keep up with the ever-increasing number of astronomy books. Many members found their local public libraries adequate, at least for the popular works, and several Centres were building their own collections of more specialized items. And of course it was always easy to rationalize the lack of purchases as a money-saving measure. A chicken-and-egg situation developed where members lost interest because the collection was not being renewed, and successive librarians were reluctant to spend much money when there appeared to be so little interest.

Something had to be done. There was no point in maintaining a room full of books which almost no one used. About 1984 the Historical Committee came up with a plan that may yet prove to be the salvation of the Society's library. The committee recommended, and the Council agreed, that the focus of the library would be the history of astronomy with special attention given to its development in Canada. The Library and Historical Committees met jointly and agreed on 253 books which were either duplicates or which were not compatible with the library's specialized purpose. These books were removed to make way for new acquisitions of more appropriate titles, and ninety-eight of those removed were sent to Centres which requested them. At present, some new books of a historical nature are purchased, but much more needs to be done.

There were several sound reasons behind the recommendation for the library's renaissance. In the first place, the Society had a truly fine collection of historical books which had been acquired during a century of existence, and the prospect of building on this firm foundation provided a definite goal which seemed manageable in terms of the Society's resources. Universities across the land were being hard hit by financial restraint, and few if any had resources to devote to library books on such "peripheral" topics as the history of astronomy. At the same time, there was increasing academic interest in the history of science, including astronomy, and there seemed to be a possibility that the RASC could build an important reputation as a centre for studies of this kind. Building the collection and the reputation would take years, and the library would probably never be used by large numbers of members, but it was felt that serving a small group well was to be preferred to the alternative of trying unsuccessfully to serve everyone. The success of this limited venture is by no means assured; it will depend greatly on the devotion of future librarians and the support of their committees and the Historical Committee.

PETER BROUGHTON (1940–), the author of this book, received the Service Award in 1987. He is seated in the Society's Library at 136 Dupont Street — an appropriate pose as his first RASC post was as librarian. It was partly as a result of his experience with this fine library that he developed a special fondness for the history of astronomy.

The first issue of each of the Society's long-standing publications. The *National Newsletter* which started in 1970 didn't have a title page.

CHAPTER 7

The Three-Body Problem

Over three hundred years ago, Isaac Newton proved that if there were only two bodies in the Universe, the Sun and the Earth for instance, and if their positions and velocities were known at some specific instant, then their future course could be determined exactly, and their orbits could be described as well-defined curves – ellipses in the case of the Earth and Sun. The problem of predicting the future course of three interacting bodies kept many of the best mathematical minds occupied in the Eighteenth Century, but in the end it turned out to be an impossible problem to solve exactly. All that can be done in general is to approximate a solution by proceeding in short steps, calculating how each body would move if the others were fixed, then taking the newly found positions and velocities as a start for the next step, and so on. Until the advent of electronic computers in the middle of the twentieth century, the task was excruciatingly time-consuming and susceptible to error.

For most of the Society's life, there were two publications, the Journal and The Observer's Handbook, and they came out predictably, year after year, with internal changes but no strong external forces pulling them from their accustomed course. The introduction of a third publication in 1970 coincided with profound changes in the dynamics of the Society which made the future far more difficult to predict.

THE JOURNAL, ITS PREDECESSORS AND ITS EDITORS

The Journal traces its ancestry back to the year of the Society's Provincial incorporation, but just as the name of the organization changed over the years, so did the title of its publications. The various names and editors are as follows:

Years	Title	Editor
1890–93	Transactions of the Astronomical and Physical Society of Toronto	(Lumsden)
1894–99	" " " "	Lindsay
1900	Transactions of the Toronto Astronomical Society	"
1901	" " " "	Harvey
1902–03	RASC Selected Papers and Proceedings	"
1904–05	RASC Transactions (Including Selected Papers & Proceedings)	Chant
1907–56	The Journal of the RASC	"
1957–69	" " " "	Northcott
1970–75	" " " "	Halliday
1976–80	" " " "	Higgs
1981–83	" " " "	Batten
1984–88	Journal of the RASC/Journal de la SRAC	"
1989–	" " " "	Tatum

Lumsden was never named as editor, but was expected to fulfill editorial duties as the Society's corresponding secretary. Harvey, who only edited a couple of volumes, was nearly seventy when he asked to step down due to poor health. Lindsay, Chant and Northcott all died in office. Fortunately the Society has not been so hard on its more recent

ALAN H. BATTEN (1933-) received his BSc degree from St Andrew's University in Scotland and his PhD from the University of Manchester. He came to Canada in 1959, originally on a Post-Doctoral Fellowship at the Dominion Astrophysical Observatory, and remained there for his entire professional career, retiring as a Senior Research Officer in 1991.

He soon settled into life in Victoria, marrying in 1960 and joining the RASC Centre in 1962. This was the start of a long and very active association with the Society, including a term as President of the Centre 1970-72 and of the national Society 1976-78, and as Editor of the *Journal* 1980-88. During his term as President, he visited all 18 Centres, speaking in French when appropriate and always delighting his audience with his talent as a raconteur. Alan Batten received the Service Award in 1988 for his extensive and energetic contribution to the well-being of the Society.

Professionally, he took a special interest in radial velocities and binary stars, publishing over 100 papers and a book entitled *Binary and Multiple Systems of Stars*. His very significant work was marked by a DSc from St Andrew's in 1974 and election to the Royal Society of Canada in 1977. He then began to delve into the history of the famous Struve family, and this ultimately led to another book, *Resolute and Undertaking Characters* in 1988. Several of his papers on famous astronomers have enriched the pages of the *Journal* for both amateur and professional readers.

Dr Batten's administrative abilities have also benefited ASP, CASCA, of which he was President 1972-74, and the IAU of which he was Vice-President 1985-91. He has organized a number of international symposia and headed the organizing committee for the IAU General Assembly in Montreal in 1979.

Alan Batten at the end of his term as RASC President in 1978. He is holding a mounted portion of the Bruderheim meteorite presented to him by the Edmonton Centre in appreciation of his service to the Society.

editors, and in fact a motion was passed in 1984 by which editors were to be appointed for five-year terms with renewals possible.

Though the main burden of duty has always been on the editor's shoulders, there were associate and/or assistant editors from 1907 to 1963 (generally the directors of the DO, DAO and Meteorological Service), and an Editing Committee, book review editor and assistant to the editor in the more recent years. With the exception of the present assistant to the editor, Marie Fidler, who has received a small honorarium since 1976, and Thomas Lindsay who was paid $50 per year back in the 1890s, these people have never received any remuneration for their work.

The *Transactions* came out annually until 1902. Combined issues of the publications were printed in 1902–03 and 1904–05. There was no issue in 1906, but the papers which accumulated during that year gave the new journal a good start when it began in 1907. The names *Canadian Astronomical Journal* and *Canadian Astronomical Record* were suggested, but in the end the Council decided on *The Journal of the Royal Astronomical Society of Canada*. The first issue of the *Handbook* also came out the same year, and both were Chant's responsibility for the rest of his life. In fact, the editorial work of the *Journal* and the *Handbook* was only separated in 1970, the same year in which the *National Newsletter* first appeared. It now seems almost miraculous that one volunteer could have served alone as editor where there are now three people, each devoting a tremendous amount of talent and energy to their respective publications – the *Journal*, *Observer's Handbook* and *Bulletin*. Yet Dr Chant for fifty-two years and Miss Northcott for the following fourteen years did carry out these editorial duties in an exemplary manner. It should also be remembered that from 1915 through 1947, the *Journal* was issued ten times per year, i.e. monthly except during the summer. Council "temporarily" reduced the *Journal* to a bimonthly starting in 1948 as a cost-saving from the United States or Britain, sometimes contributed papers which were read at meetings and then published. These tended to give the *Transactions* a place on the international stage.

The *Journal* began with a fresh approach and appeared on a bimonthly schedule, starting in 1907. In addition to the reports of meetings at Toronto and the other Centres, the *Journal* now began to include formal papers which were not read at any meetings. Such an expanded scale of publication was only made possible by the prospect of a good supply of papers from the astronomers at the newly established Dominion Observatory, the federal government grant, and the initiative and enthusiasm of Editor Chant. The by-line which appeared on the title page announced that the *Journal* was "Devoted to the Advancement of Astronomy and Allied Sciences." Though the emphasis was on original papers, Chant made clear in his 1907 editorial (the only one he ever wrote) that:

> There are few technical articles on astronomy, which, if clearly written, have not a real value to the amateur, while the work of the latter is always of interest to his professional brother. There will be room for both in the pages of the Journal. If all unite, the result will be highly creditable to Canadian science.

That statement has been the guiding principle of the *Journal* ever since.

Turning the professional aspect to good account, Chant argued for the continuance of the federal grant during World War I by pointing to the part the *Journal* played in bringing recognition "to our Dominion by the scientific world abroad." In fact for many years the *Journal* of the RASC was the only scientific periodical in Canada issued on a regular schedule. On the completion of the twenty-fifth volume, Chant again noted that its "widespread circulation ... has not only united those interested in astronomy in Canada but has also supplied a bond of sympathy with those of like mind all over the world."

The *Journal* developed an even more professional image under Ruth Northcott's direction. The cover changed from its old gray colour to a new, bright shade of green. Abstracts in both English and French began to appear with each paper. Standardized styles were adopted in figures, tables, references, preferred spellings, abbreviations and in the use of numerals. The referee process was begun in 1961. In her report as editor in 1965, Northcott stated that "approximately 10 percent of manuscripts are rejected, 25 percent are printed essentially as submitted and the remainder require a wide range in both time and effort before our standards for publication are met." (For whatever reasons, the rejection rate at present is considerably higher.)

Policy concerning the content of the *Journal* was discussed in the 1960s not only by the National Council of the RASC but also at meetings of the National Committee for Canada of the IAU. Following these discussions, a revised statement was published in 1966:

> The *Journal* aims to provide a means of publication for the professional astronomer and to keep the non-professional astronomer well-informed. Each manuscript is reviewed by one or more referees. Papers in the following categories will be considered:
>
> (a) Research papers on astronomy and related topics. These should emphasize methods and results and long tabulations with extensive calculations should be avoided.
> (b) Non-technical articles and reviews of progress in various branches of astronomy. These should be authoritative, of general appeal, and contain a minimum of highly specialized technical material.

To commemorate the 500th anniversary of Copernicus' birth, the Polish Societies undertook to finance a solar telescope at the Manitoba Museum's Planetarium in 1973. The 41 cm heliostat mirror, to which Frank Shinn is pointing in this photo, reflects the sun's image down a telescope (enclosed behind the memorial plaque) into a room in the subterranean planetarium.

B. Franklyn Shinn (1911-) had two careers. He was first of all a musician. Having graduated from the Royal Academy of Music and the London College of Music in England, he became a church organist and owner and director of the Shinn Conservatory of Music until 1968 when he began his second career as Assistant Director (and later Director) of the Planetarium at the Manitoba Museum of Man and Nature.

Frank Shinn joined the Winnipeg Centre at the invitation of Professor Robert Lockhart in 1954 and soon became a vital force in the Society. He originated the Centre's newsletter (later called *Winnicentrics*) and edited it for many years; he organized observing nights in Winnipeg parks, spoke at many meetings and was President 1963-66. Hosting two General Assemblies, in '66 and '74, kept Frank and his wife Florence very busy.

As an observer, he was involved in the Moonwatch project starting in the late '50s, contributed to the visual meteor program and to AAVSO in the '60s and later took part in many grazing occultation expeditions. He was an expert in the design and construction of telescopes and models to illustrate such things as comet orbits and binary stars. The heliostat at the Planetarium was largely his conception. His work on a 32 cm mirror for Jack Newton was documented in a film about the project. In 1972 Frank Shinn received the Service Award. In 1978, he moved to Vancouver Island and served on the Council of Victoria Centre. From 1978-80 he edited the *National Newsletter* and wrote many articles for it himself. In retirement he continued to help as Assistant Editor and as a member of the Editorial staff of the *Newsletter*.

(c) Brief reports of original observations and new scientific projects undertaken by members.

While many members said that most of the papers were "over their heads," many professionals found that the guidelines restricted their needs to be detailed and technical. Consequently, in 1969, the National Committee for Canada of the IAU presented a report on the feasibility of publishing a professional Canadian journal of astronomy. Opinion was divided on the merits of this, on the commercial prospects of a new publication, and on its effect on the *Journal* of the RASC, and in the end the report was tabled. There still is no Canadian journal especially for professional astronomy, and there is not likely to be one with costs soaring, many international places to publish research and talk of electronic journals being the way of the future. CASCA has, however, published a newsletter, *Cassiopeia*, since 1973.

Successive editors have struggled to bridge the gap between general readers and those with a scientific background. In 1978, Editor Lloyd Higgs indicated that, "In addition to original research papers, [the Journal] welcomes contributions of an historical, biographical or educational nature, of general interest to the astronomical community." Alan Batten saw the *Journal* as "a meeting place for amateurs and professionals of a kind that no other publication … can quite fulfil." Jeremy Tatum expressed the hope, in a recent editorial that amateurs will take

> great pride that they are able to support the publication of a nationally and internationally recognized Journal that gives the professionals an opportunity to publish their work in this country, and recognize that the great support that amateurs and professionals have given to each other through the Journal is in large measure responsible for the high esteem in

which the Society is rightly held. ... The professionals must also recognize and appreciate the support they are receiving from the Journal.

What do amateurs themselves have to say? Frank Shinn, a prominent Winnipeg Centre member wrote this in 1976:

> The first few copies [of the Journal] sailed completely over my head. After a few months – or was it years – I began to get some idea of what it was all about. Then, in fear and trembling, at National Assemblies I began to ask some of the experts in certain fields questions about some aspect of their specialty that was puzzling me. I found that I need have no fear. They were only too ready to help with answers, and I began to realize that ... they were willing to treat me as if somehow I had earned the right to be considered a colleague.

David Levy made the point, in 1989, that:

> the Journal is a door to the world of professional astronomy. We may not understand all the words and formulae, but its pages force us to examine the style and content of a professional contribution to science. We cannot get that out of the other publications.

Furthermore, it would be a mistake to think of amateur members solely as consumers of professional papers. The *Journal* and other Society publications provided many budding authors with a first opportunity to get into print. Some have built on this experience and developed wider reputations. The proportion of papers by amateurs peaked at about 30 percent in the 1940s and is now closer to 10 percent. On the other hand, the proportion devoted to original research has climbed from a low point of 25 percent in the 1930s to about 60 percent recently.

CHRISTOPHER E. SPRATT (1942-) joined the Niagara Falls Centre in 1961 where he began making variable star estimates for AAVSO with a 6 cm refractor. Following his graduation from the Niagara Parks School of Horticulture, his work took him to many parts of Canada. He was at various times an unattached member, and belonged to Centres in Ottawa, London and Victoria where he is currently employed as a gardener by the University.

Wherever he went, Spratt continued his variable star work. He was the top observer for AAVSO in 1976-77 and by 1982 he had reported over 30 000 observations to that organization. In recognition of this remarkable accomplishment and his highly-regarded computational work with comet ephemerides, the Society awarded Chris Spratt the Chilton Prize in 1983. The same year the International Halley Watch named him the Co-ordinator of Visual Observations for Northwestern North America. He continued to amass observations of variable stars but found comets and asteroids increasingly fascinating.

His observations of these solar system objects have appeared frequently in *IAU Circulars*, *The International Comet Quarterly* and *Sky and Telescope* and his numerous papers on asteroids, comets and meteorites have formed a major contribution to the *Journal*. He has written many popular articles for Centre and national newsletters, local newspapers and *Astronomy* magazine. He has a large personal meteorite collection. Chris Spratt was honoured with the Chant Medal in 1988. Furthermore, at his suggestion, a number of asteroids were named after prominent Canadian astronomers and the tribute was reciprocated when asteroid "Sprattia" was "Christened" in 1991. The photo (by Alice Newton) shows Chris Spratt (on the left) being congratulated by Dave Balam, who discovered the asteroid and suggested the name.

Education has at times been recognized as the common goal of authors and readers of the *Journal* either in the form of self-development, or in a more formal sense. As long ago as 1911, English astronomer Arthur Hinks wrote in his popular book *Astronomy* "To keep abreast of recent work, the student must refer to the periodicals. ... The *Journal* of the BAA, the *Publications* of the ASP and the *Journal* of the RASC are all full of interesting matter." Thinking more pragmatically of the needs of educators, John Percy began a regular feature called "Education Notes" in 1977, which over many issues has provided much useful information for teachers at all levels. The NRC Associate Committee on Astronomy in fact subsidized this feature at a rate of $500 per year. But education has been no more successful in bringing consistency to the *Journal* than any of its other themes.

Seeking common bonds between the diverse interests of its readers will no doubt be a continuing feature of the *Journal*. In the long run, the quest for the ideal mix itself may be the chemistry that keeps the Society vibrant.

THE SCOPE OF THE JOURNAL

Some attempt will now be made to summarize the substance of one hundred years and 40,000 pages of published material. Indices appear at the end of each year; a general index compiled by W.E. Harper was published covering the period 1890–1931 and another for the period 1932–1966 was edited by Ruth Northcott. In the latter case, 500 copies were printed at a cost of about $2,500 and sold to members at $5 each to members and at $8 to others. These indices have been very valuable in the present survey.

First of all there have always been features, or columns or departments which have run for several years, some still continuing. Editorials have been written sporadically, and "Education Notes," already mentioned, began in 1977 with John Percy's initiative and continued intermittently under the direction of Serge Demers, David DuPuy and Roy Bishop. Brief biographies of the authors of articles in each issue appear on a page headed "About Our Authors". This feature began in 1964 and was handled successively by Miriam Burland and Marie Fidler. Book reviews have been included in most numbers since 1915, and in recent years a Book Review Editor has been responsible for seeking out appropriate reviewers and co-ordinating their contributions. And ever since the *Journal* celebrated three quarters of a century, Alan Batten has made a judicious selection of old excerpts entitled "Seventy-Five Years Ago." Short contributions under the heading "Notes" have run since 1958, this section of the *Journal* growing out of an earlier column called "Notes and Queries" which was directed by Chant from 1908 until his death. In its early form, members, especially amateurs, were invited to submit questions and Chant agreed to try to find answers but the queries soon dried up and the editor was often left to prepare notes himself on topics he hoped would be of interest.

Several other departments have come and gone. Helen Hogg's popular selections under the heading "Out of Old Books" (1946–66) came about as a by-product of her very extensive bibliographic searches for references in the literature to globular clusters, variable stars and novae. A very valuable outcome of one of these columns was J.B. Tyrrell's presentation to the DDO of William Wales' original manuscript written at Fort Churchill in 1769. Another *Journal* feature of very long standing was Peter Millman's "Meteor News" which comprised 131 columns between 1934 and 1973, and his "IGC Visual Meteor Program Progress Reports" from 1961–67. These columns, and of course Millman's enthusiasm for the subject, undoubtedly had a strong influence on the reputation Canadians built in meteor research.

Some outside organizations have had regular reports in the pages of the *Journal*. Between 1907 and

1966, forty-two meetings of the American Astronomical Society were summarized by Canadians who attended. Meetings of the American Association of Variable Star Observers were officially reported in the *Journal* from 1937–46 and 1952–66, and a column called "Variable Star Notes" prepared by AAVSO secretaries Mayall and Mattei, ran from 1952 to 1981. These notes included items of current significance to observers such as unusual light curves and lists of observers and their estimates. Reports of AAVSO meetings and the lists had been published in *Popular Astronomy* until 1936; the AAVSO's own *Quarterly Reports* did not begin until 1947 and these were in the form of typed sheets. The AAVSO did pay $50 per year to the RASC towards publishing costs of "Variable Star Notes." Abstracts of papers presented by Canadian professional astronomers at CASCA meetings have been published since the inception of that organization in 1971 and their printing costs are now fully covered by the authors.

There were other columns which provided some insight into what Canadian astronomers were doing professionally. All the major observatories sent in Notes: The DO for the years 1907–20, 1933–4, and 1953–69; the DAO for 1919–20, 1928–40, 1953–76, and 1979–82; the DDO 1952–73. There were seismographic records published from 1907–20, Notes from the Meteorological Service from 1909–25, and magnetic observations spanning 1910–21. Other reports from government agencies and various university departments of astronomy appeared sporadically in the 1960s and 1970s.

Other features of long standing were designed to keep members informed of recent scientific developments. These appeared under a variety of titles. "Astronomy and Astrophysics Progress" began in 1893, while "Notes on Astrophysics" and "Astronomical Notes, News and Queries" began with the *Journal* in 1907. This became simply "Astronomical Notes" in 1921 and comprised

Thomson, on the left, presenting a certificate to John Patterson who had preceded him as Director of the Meteorological Service.

ANDREW THOMSON (1893-1974) was the last of a line of Directors of the Canadian Meteorological Service to play an important part in the RASC. Thomson grew up near Owen Sound, Ontario, and attended the University of Toronto, earning his MA in 1916. After a year at Harvard and a short time working as "mathematical aide" to Thomas Edison, he was employed as a physicist in the Department of Terrestrial Magnetism at the Carnegie Institute, 1918-22. During this time he went to Brazil for the 1919 solar eclipse and was at sea for two years investigating atmospheric electricity. His career then took him to Samoa where he directed a meteorological observatory. Work in New Zealand, Germany and Norway followed before his return to Canada in 1931. He was with the Meteorological Service of Canada from then until his retirement in 1959, the last 13 years as Director. He was a Fellow of the RSC, Vice-President of the American Meteorological Society and of the Royal Meteorological Society, received an honorary DSc from McGill, and was named to the Order of the British Empire.

Thomson seems to have become interested in the Society through the 1932 Solar eclipse. He was certainly part of the group at Louiseville, Quebec, and he joined the RASC later that year. He spoke to the Toronto Centre a few times in the 1930s, was elected to the General Council in 1936 and was national Vice-President and President 1947-50. While on a business trip to the Pacific Coast he visited the five western Centres from Winnipeg to Victoria. Several papers of his appeared in the *Journal*, including his Presidential Address, "The Unknown Country", a fascinating look at Canada's Arctic.

"brief reviews of interesting and important papers appearing in the different scientific periodicals." From 1928 to 1931, R.E. DeLury prepared a column called "News and Comments.". Then for a long time, from 1932 until 1956, current astronomical news items of this sort were absent from the Journal. Starting up again during Ruth Northcott's regime, a long series of papers ran under the heading "Canadian Scientists Report". In the first of these, R.M. Petrie wrote, "The articles have been planned to present authoritative and yet non-technical reviews and accounts of scientific work in which Canadians are taking part. ... It is the privilege and duty of our Society to help all members to an appreciation of [this] growth through the pages of the Journal." A new department called "Advances in Astronomy" was prepared by Ian Halliday from 1965 to 1972, and Ian McGregor tried "Canadian Astronomical News" a couple of times in the 1983 National Newsletter.

As for Society activities per se, minutes of Annual Meetings for 1890–1901 appeared in the Transactions and for 1912–1990 in the Journal. Annual Reports, lists of officers and reports from individual Centres were also included in the Journal until 1959. By this time the reports were virtually using up one of the six numbers each year, so it was decided that these reports would hereafter come out annually in a separate booklet known as the Supplement to the Journal. Reports of General Assemblies and RASC Papers, presented at the General Assemblies, continued in the Journal from 1959 to 1972. Citations of award winners and newly elected honorary members and obituaries of prominent members have also been traditionally in the Journal as have presidential addresses and speeches given by Hogg and Northcott lecturers.

All this notwithstanding, formal papers have always been the dominant feature of the Journal. Naturally, Canadian professional astronomers contributed strongly. In some of the early numbers all the formal papers were by astronomers at the DO. Not surprisingly, with so few professionals to draw from, there were times when the editor's well ran dry. In the early 1920s for instance, sometimes two numbers of the Journal were combined under one cover but Chant was usually able to fill the bucket somehow. Other publishers were willing to let him reprint occasional articles of theirs and he chose some outstanding ones. So it was that papers by such famous figures as Einstein, Eddington and Jeans, American astronomers Curtis, Shapley and Russell, physicists Millikan and E.O. Lawrence, and many others including P.M.S. Blackett, J.C. Smuts, C. Stormer and M.N. Saha are found in the pages of the Journal. (Recent policy has been a strict avoidance of any such reprints. For instance, Geoffrey Burbidge's Petrie Lecture, delivered to CASCA in 1979, came out in Nature soon thereafter and consequently never appeared in the Journal.)

Occasionally the flow went the other way. The editor of the Astrophysical Journal sought permission to reprint Chant's 1904 article on the reflecting power of glass and mirrors. J.S. Plaskett's paper, "Some Recent Interesting Developments in Astronomy" from 1911 and L.B. Stewart's "The Form and Constitution of the Earth" from 1913 were selected by the Smithsonian Institution for inclusion in their publications. A.D. Watson's 1916 paper on Horrox was reprinted in the English Mechanic, and Annie Cannon's paper of the same year describing the beginnings of stellar classification was used by Shapley in his Source Book in Astronomy. Recently, a number of international periodicals have reprinted papers from the Journal, a gratifying indication of the stature and value of the RASC's work.

Of course, not all papers by famous foreign astronomers were reprints. In some cases, lectures delivered in Canada by outstanding visitors such as Russell, Barnard, Olivier, Peebles and Christy, were published. But there were also original contributions from such well-known astronomers as Swings,

Pawsey, Kuiper, van de Kamp, and amateurs of the stature of Denning and Haas. These are but examples to illustrate the *Journal*'s international flavour.

Some other famous astronomers were drawn into contributing to the *Journal* by controversy. A.F. Miller made unfortunate allegations in 1925, based entirely on second-hand sources, that observations of Zeta Cancri by Flammarion were claimed by Otto Struve as his own. This brought swift and rather caustic replies from Otto's two grandsons, Georg and Otto, themselves world-famous astronomers. Though Chant published his regrets, Miller was unrepentant. Another example was Boyd Brydon's 1933 paper, "The Illusions of Monsieur Antoniadi" which in reality was a review of a chapter entitled "The Illusion of the Canals" in Antoniadi's book on Mars. Brydon was convinced that there were canals on Mars, and in unwarranted and inflammatory remarks dismissed Antoniadi as having "nothing to point to as proof of his ability as an observer except his failure to see things that others have seen." Chant surprisingly allowed this to be published, but did add an editorial note, "The pages of this Journal are open for discussion but not controversy. ... Space will be found for a suitable article supporting M. Antoniadi's views." Indeed, Antoniadi did reply himself, with an article "The Delusions of an Amateur," and W.H. Pickering also wrote because he felt he had been misquoted in the course of the debate. If there is anything good to be said about these episodes it is that the Struves and Antoniadi held no grudges. The younger Otto Struve later accepted Honorary Membership in the Society, and Antoniadi later published papers in the *Journal*.

Amateurs were not always the ones on the losing end of an argument. R.E. DeLury's paper, "Haze Effects on Measurements of Solar Rotation," published in 1916, led to rebuttals by W.S. Adams and H.H. Plaskett and counter-rebuttals by DeLury over the next three years. Professor J.W. Campbell, in his published presidential address "The Problems

ALBERT DURRANT WATSON (1859-1926) combined a very successful medical practice with many other interests. He was well-known as a poet; in fact he wrote a verse of O Canada which is still used in many churches. Spiritualism, occult phenomena and psychical research also attracted him.

Watson was born at Dixie, near Toronto, and studied at the Toronto Normal School and at Victoria University. In 1885, he married Sarah Clare, a daughter of the original Secretary-Treasurer of the Toronto Astronomical Club. Watson joined the Society in 1892 and became an active member as a speaker and writer. Several poems of his and a number of papers of a historical nature and on calendar reform appeared in the *Journal*. Watson was on the Council as early as 1902 and was elected Vice-President in 1912. He was RASC President 1916-17.

O Canada

Almighty Love, by Thy mysterious power,
    In wisdom guide, with faith and freedom dower,
Be ours a nation evermore
    That no oppression blights,
Where justice rules from shore to shore,
From lakes to northern lights.
May love alone for wrong atone;
    Lord of the lands, make Canada thine own!
Lord of the lands, make Canada thine own.

ALFRED TENNYSON DELURY (1864-1951) and his brother **RALPH EMERSON DELURY (1878-1956)** were both Presidents of the RASC. Another brother, Justin, became Provincial Geologist for Manitoba and spoke on more than one occasion to the Winnipeg Centre. Their father was a shoemaker by trade who emigrated from Ireland to settle in the village of Manilla, about 110 km northeast of Toronto. A fourth brother, Daniel, who became a legislator in Minnesota, recalled that their father was an excellent story teller and a poet by nature. No doubt this accounts for the literary names of the boys, and A.T.'s engrossing interest in Irish writers whose works he collected. Like many of his contemporaries, Alf (as his family called him) went into teaching following his high school years. He then attended the University of Toronto and was in the same class as Chant, both men graduating in 1890. After some short-term posts elsewhere, he returned to Toronto as a Lecturer in Mathematics at the University in the fall of 1892. He got his MA in 1902 and during his career he became Head of the Department of Mathematics in 1919 and Dean of Arts from 1922 until his retirement in 1934. His nephew, D.B. DeLury, was Head of Math at U of T from 1958-68.

In his early years of lecturing, A.T. DeLury taught some astronomy courses within the Mathematics Department. The papers he read at meetings of the Society and which were published in the *Journal* encompassed such topics as comet orbits, parallax, aberration and relativity, and biographical pieces on Poincaré and G.H. Darwin. He was a very popular speaker and hundreds attended his talks.

He and Ralph both retired to Manilla where a park and community centre now perpetuate the DeLury name. A.T. DeLury maintained his interest in the Society and even in retirement came to the city to give an occasional lecture to Toronto Centre.

Ralph DeLury, like his older brother, attended the University of Toronto where he was prominent in various organizations and athletics. An all-round student, he played football, basketball and hockey and was the cartoonist for the campus newspaper. He received his BA in 1903, MA in 1904, and his PhD in 1907, his thesis topic being "The Rate of Oxidation in Arsenious Acid." During this period, he wrote a 444-page book, *A General Method of Calculation in Kinetics*. After a year at Princeton on a research scholarship he joined the staff of the Dominion Observatory, Ottawa, where he spent his entire career on solar investigations. One might suppose from this illustrious start, that Dr DeLury was embarked on a brilliant career. In fact, much of his work was marked by controversy.

Ralph DeLury presented many papers to the Society, starting with two on convection as a factor in stellar variation in 1909 and continuing up to 1940; for six years, 1928-34, he wrote the "News and Comments" column in the *Journal*. In some of his papers he presented ideas on the association of sunspots with everything from agriculture to zoology. He was, however, not merely playing with statistics as many others before and since have tried. He believed that at sunspot maximum an increase in ultra-violet radiation from the faculae, or bright spots, would result in increased ionization of the Earth's atmosphere which would in turn lead to increased cloudiness and rainfall. These meteorological effects, he reasoned, would have an effect on crops and economics. The newspapers of course latched onto the sensational claims. But in spite of this, or perhaps because of it, the Society elected Ralph DeLury Second Vice-President in 1932, which led to his Presidency in 1936-37. He had also served the Ottawa Centre as President in 1921-22.

A.T. DeLury, nattily attired for a University of Toronto Hockey game. (U of T Archives). Photos of Ralph DeLury and of the family are reproduced through the courtesy of the Manilla Public Library.

of Space Travel" (1948) was less than enthusiastic about the prospects for space travel as outlined in Willy Ley's book *Rockets and Space Travel*. Though there was nothing personal in Campbell's remarks, Ley was not impressed and wrote a long letter to the editor clarifying and correcting some of Campbell's statements. On other occasions, astronomers simply let controversial papers pass without rebuttal. Such was the case in the 1950s when Millman and Beals refused to be drawn into a controversy with American meteoriticists Leonard and LaPaz following papers they had published in the *Journal*. A tactful way of handling such problems was exemplified by a pair of papers appearing in 1980. Jean René Roy's paper, "Comments on the Astronomical Alignments at Callanish, Lewis" was accepted for publication on the understanding that Gerald Hawkins would be given the opportunity to write a rebuttal. This he did and both papers appeared together in the *Journal*.

While both famous and factious contributions have flavoured the pages of the *Journal*, more staple fare has provided the bulk of its menu. Summarizing this may be about as futile as attempting to describe the lifetime diet of a centenarian. Jeremy Tatum found that even the contents of current *Journals* was difficult to classify. Asked the proportion of papers devoted to original research, he came up with 65 percent and for the proportion by Canadian authors, 74 percent. Easier to answer was the fraction by female authors (5 percent) and the fraction of papers in French (4 percent).

French first appeared in the *Journal* in 1941 with "Mars, La Planète Etrange" by Albéric Boivin. A few other contributions by Quebec members came out in the 1940s, but then there were almost no French papers until 1976 when Editor Lloyd Higgs began to actively encourage submissions in our country's "other" official language. Other small steps were taken to make the *Journal* more truly bilingual. Biographical notes of Francophone authors were published in both French and English, and Abstracts of papers appeared in both languages. The introduction of a bilingual cover in 1984 also provided the opportunity to spruce up the image of the *Journal* by including a photograph on the cover.

Some representative papers by Canadian scientists might include the following, with one arbitrarily selected from each decade.

- "Light Pressure" by G.F. Hull in 1901. (Hull was born and educated in Ontario and was a member of the A&P Society for a while, but his illustrious career was in the States.)
- "An Extraordinary Meteoric Display" by C.A. Chant in 1913. (This was the subject of further study by J.A. O'Keefe in 1959.)
- "Recent Investigations on the Auroral Spectrum" by H.J.C. Ireton in 1925. As Ian Halliday pointed out in a review of significant papers on Solar System Astronomy from the first twenty-five years of the *Journal*, Ireton summarized the work leading up to the historic identification at the University of Toronto of the auroral green line.
- "The Dimensions and Structure of the Galactic System" by J.S. Plaskett in 1936. The great importance of Plaskett's work in understanding the structure and rotation of the Milky Way was highlighted in the republishing of the diagram from this paper with the proceedings of the 1990 "Kingston" Conference on "The Age and Evolution of the Galactic Disk and Halo."
- "Wave Length Standards for Radial Velocity Determinations" by R.M. Petrie, published in several parts in 1946.
- "Some Characteristics of 10.7 cm Solar Noise" by A.E. Covington in 1951 was based on observations of the Sun dating back to 1947 – the first radio astronomy done in Canada.
- "The Galaxies of the Local Group," a beautifully illustrated 1968 paper in two parts by Sidney van den Bergh. It was reprinted as a seventy-four-page

CARLYLE S. BEALS (1899-1979) received his education in his native Nova Scotia and earned his post-graduate degrees at Toronto and in London, England. For one year he taught Physics at his alma mater, Acadia University, and then in 1927 joined the staff of the DAO. Using a microphotometer of his own design (a device which produces a graphical profile of spectral lines), he was able to examine the complex nature of interstellar lines and to study the motion of discrete clouds about the Galactic centre. He also was very interested in stars with emission lines in their spectra and was an authority on particular classes of these bodies known as Wolf-Rayet and P Cygni stars.

On arriving in Victoria, Beals at once joined the RASC and in subsequent years was Secretary-Treasurer, Vice-President and President of the Centre. During his years at the DAO, he spoke regularly to the Centre and sometimes to Vancouver Centre as well. His wife Miriam, an accomplished pianist, occasionally provided the entertainment at dinner meetings.

In 1946, Beals was appointed Dominion Astronomer and the family moved to Ottawa. In spite of the heavy demands of his new position, Beals found time to host open houses at the DO, to speak at meetings of the Society in Ottawa and other Centres and to serve as RASC Vice-President and President. As the DO now included geophysics as a major component of its operations, Beals appropriately became very interested in impact craters, both terrestrial and lunar. He received several honorary degrees, medals and other distinctions in recognition of his important work in astronomy, geophysics and administration. He is the only Canadian to have been President of the American Astronomical Society. The RASC presented him with the Service Award in 1964 and CASCA established an Award in his name in 1981.

Dr C.S. Beals with the sun-dial presented to him on 21 February, 1964 by the members of the National Committee for Canada of the IAU.

booklet and sold through the DDO at $2.40 per copy.
- "Current Trends in Astronomical Spectroscopy" by K.O. Wright was based on an invited address before an international conference in the Netherlands in 1975.
- "The Liquid-Mirror Telescope as a Viable Astronomical Tool" by E.F. Borra in 1982 was, as Alan Batten pointed out, the first modern paper on a concept now beginning to produce results.

Over the last few decades, the *Journal* has also served as the medium for proceedings of a number of astronomical conferences, including the (US) National Science Foundation Conference on Binary Stars (1957), A Symposium on Space jointly sponsored by the Canadian Association of Physicists and Section III of the RSC (1960), an IAU Colloquium on the Orbital and Physical Parameters of Double Stars (1973), a symposium in honour of K.O. Wright (1977), the "Kingston" conference held at the DAO in 1989 and sponsored by the Canadian Institute of Theoretical Astrophysics (CITA).

In spite of the oft-heard complaint of many amateur members that the *Journal* is full of abstruse analyses of astrophysical arcana, the *Journal* has always provided opportunities for astronomers, both professional and amateur, to publish nontechnical papers aimed at a general audience as well as papers of a more cultural aspect dealing with the history, philosophy, or sociology of their science. In a sense, these papers are nearly all written by amateurs since the authors usually have no formal educational qualifications in the peripheral disciplines involved. The *Journal* has also traditionally welcomed well-written papers on astronomical topics by amateur astronomers. A sampling of both could include professional members, like Alan Batten who has contributed many outstanding papers of an historical and philosophical nature and amateur astronomers like Mary Lou Whitehorne and Chris Spratt who have carried out research of professional standards. It is the eclectic nature of the *Journal* which makes it unique in an age of specialization. It is the sharing of space by amateurs and professionals which makes it exceptional. Is there any other publication in the world where a high school student would write the lead article and a Nobel Prize-winner would be the author of a paper a few pages later?

Who reads the *Journal*? A recent survey suggests that the vast majority of members read it at least occasionally. (6 percent say they never read it.) Moreover the *Journal* is of some importance on the international professional scene. In 1990, eighty-five organizations or institutions exchanged their publications for ours, and 302 were paid subscribers – 52 in Canada, 166 in the USA and 84 from other countries. International citation indices published over the last fifteen years show that *Journal* papers were cited an average of 112 times a year. While this puts the *Journal* well down in the ranking order, with only about 25 percent of cited publications doing poorer, it's not bad considering that many of the papers in the *Journal* are not intended to be used as a basis for further research.

PRINTING, FINANCING, AND ADVERTISING

The Society's printers were Brough and Caswell for 1890–1, who supplied 500 copies of the forty-page *Transactions* for $95. The following year, Rowsell and Hutchison were engaged on a lower tender of $76 for 300 much fatter copies thanks to the influence of Charles Sparling, the Society's recorder and later treasurer, and an employee of the firm. Carswell and Company looked after the Society's needs for a couple of years and then the firm of Z.M. Collins, a prominent member of the Society, did the printing from 1901 until he moved from Toronto in 1919. Ever since that date, the Society's publications have been handled by the University of Toronto Press.

The Press had originally been organized for the purpose of printing examination papers, stationery and other supplies, and was located in the basement of the university library, though by 1919 it had begun to print other specialized journals and some books. At first, the choice looked bad though postwar conditions may have been the culprit. The paper used was of very poor quality and the number of misprints was appalling. But within a year, things improved greatly, and the U. of T. Press has generally done a first-rate job and has been very accommodating towards special needs and flexible schedules. In 1973, the RASC reached an agreement with Xerox Corporation to have them microfilm past, present, and future *Journals*. A free microfilm is supplied to the Society along with a percentage of any sales revenue.

The price of individual copies of the *Journal* went from 25 cents to 50 cents in 1948, and has gone up over the years to the present price of $12 per copy. Annual subscriptions, for libraries and institutions, cost close to the price of regular membership fees until 1965 when the price of a subscription was set at $12, including *The Observer's Handbook*. The history of subscription prices has been upward ever since, the latest increase being from $60 to $72, approved in 1990.

There is another source of *Journal* revenue which may surprise those unfamiliar with periodicals struggling to survive on small circulation. It has become commonplace for academic publications to expect their contributors (or the contributor's employer) to pay a fee towards the costs of printing, editing and distribution. In the case of the *Journal*, these "page charges" were instituted in 1962 at a rate of $10 per page. Since then, authors have been levied ever increasing amounts, with the latest increase, from $50 to $60 per page, being approved in 1990. Recognizing, however, that these fees would deter many excellent contributions, the Council decided that members without institutional backing would be exempt from paying page charges.

Reprints of individual *Journal* papers, stapled in covers bearing the title and the author's name have always been available to authors who were willing to pay for them at cost. While these provide no net revenue to the RASC, the Society has, from time to time, reprinted papers or series of papers from the *Journal* in the hopes that they could be sold as separate booklets. The most ambitious project of this kind was the republishing in 1920 of Robert Grant Haliburton's "The Festival of the Dead" – a virtual treatise of 126 pages. The author was the son of the well-known Nova-Scotian judge and creator of "Sam Slick" and was himself a judge and author. The paper had originally been presented to and published by the Nova Scotian Institute of Natural Science in 1863 and contained much information on the connection of the calendar of primitive peoples with the Pleiades. Chant, and apparently the rest of the Council, were persuaded that it was a Canadian work of lasting importance that was so scarce as to need reprinting. So the Society undertook to do this, not only in instalments in the *Journal* but then as a separate book which was offered for sale at $1.00 postpaid. Judging from the large stock of these which had to be disposed of when the Society moved in 1976, the project was not particularly successful. Happier results can be reported from the reprinting of a series of articles by Dr B. Haurwitz on "The Physical State of the Upper Atmosphere" as a ninety-six-page booklet. This was so well-received that two lots of reprints were issued. The second edition was updated by the author and 1,500 copies were made available at 75 cents apiece. As it was wartime, and the booklet was one of the most important available anywhere on the subject, meteorologists and instructors in aeronautics found it particularly valuable.

Two other examples of useful reprints, these dating from 1937, were "Telescope Mountings for Amateur Builders" by H.B. Brydon, a forty-eight-page booklet which sold for 25 cents and P.M. Millman's "General Instructions for Meteor Observing," an eighteen-page brochure which was available for 10 cents. Still another important reprint was in fact the RASC's celebration of Canada's Centennial in 1967. Thanks to John Percy's suggestion and Ruth Northcott's persistence and determination, several authors co-operated in writing a series of review papers entitled, *Astronomy in Canada, Yesterday, Today and Tomorrow*. These appeared in an enlarged edition of 132 pages in the October *Journal* that year, and were simultaneously printed as a paperback with an attractive cover and sold to many libraries and individuals across the country for $2.00, the normal price at the time for a single copy of the *Journal*. Within a couple of months 1,350 copies were sold.

From time to time, between 1920 and 1960, ads appeared either on the inside or outside cover of the *Journal* for such things as telescopes, telescope-making supplies and publications.

OBSERVER'S HANDBOOK

It was June, 1990, and Roy Bishop was, as usual, devoting most of the summer to the preparation of the *Observer's Handbook* for the following year. Co-ordinating 236 pages of information and data from over thirty contributors is enough to test the organizational skills, perseverance, tact and care of any editor, but to the world outside he gave not the slightest sign of frustration or stress. The phone rang and it was John Percy who was working on the section of the *Handbook* called "The Sky Month by Month." He called to say that the data supplied by the Nautical Almanac Office in Washington, DC, appeared to be for 1990, although it was headed up 1991. This was a startling development indeed, and after confirming John's suspicion, Roy called the Almanac Office.

At first his enquiry was received with stunned silence followed moments later with, "Oh my God,

ROY L. BISHOP (1939-) teaches Physics at Acadia University in Wolfville, Nova Scotia, where his family roots go back for more than two centuries. During the '60s, he was away from Nova Scotia earning his MSc from McMaster and his PhD from the University of Manitoba, both in nuclear physics.

Roy Bishop's first contact with the RASC was with the Winnipeg Centre but, after returning to Acadia, he joined the newly re-established Halifax Centre in 1970. He contributed greatly to the Centre, speaking nearly every year at least once, serving as President in 1975-76 and as Secretary and National Council Representative. Some idea of Dr Bishop's wide interests can be seen from the titles of his talks which cover optics, the rainbow, illusions, SS433, tidal power, and many historical topics. His research on early observatories in Nova Scotia led to his election to the IAU in 1982.

At his Maktomkus Observatory at his home, using a mirror he produced himself, he has become an expert visual observer of everything from the Moon to deep-sky objects. As a teacher of introductory astronomy courses at Acadia, he has introduced a number of innovations including a set of conveniently mounted erect-image telescopes for student use. As a highly respected administrator, he has headed the department of Physics at Acadia for four three-year terms.

The Society continues to benefit from his many talents. He has been a valued member of the Constitution Committee, chaired the Historical Committee, been through the ten-year cycle of Presidential offices, and served as the Editor of Education Notes for the *Journal* since 1982 and of the *Observer's Handbook* since 1981. Roy Bishop received the Service Award in 1988.

Dr Roy Bishop tries to get his hands on a copy of the 1991 *Observer's Handbook*.

you're right!" Fortunately these events took place a few days before the *American Ephemeris and Nautical Almanac* went to press and so some faces were presumably a lighter shade of pink than they might otherwise have been. It is nice to think that, at least once, the RASC was of some small benefit to the Almanac Office. After all, the Society had been using their data for the best part of a century as the basis for most of the solar system information appearing in the *Handbook*. Presumably there had never been a problem, and certainly there was never any cost to the Society.

The year 1907, which marked the beginning of the *Journal*, also saw the publication of the first issue of the *Handbook*. The simultaneous undertaking of two such ambitious projects was only possible because of generous government financial support and the amazing enthusiasm and ability of C.A. Chant who edited both publications. In fact, events were to prove that the commitment was a bit too much for an organization of only 300 members. It appeared for a time that the Society would have to give up this endeavour and members would have to rely, as they had done in the past, on outside sources such as *Canadian Almanac* (since 1847), and *Popular Astronomy* (since 1893).

Anyway, *The Canadian Astronomical Handbook*, as it was known, was published by the Society for 1907 and 1908. Chant did the lion's share of the work, F.L. Blake of the Toronto Meteorological Office carried out much of the computation (for which he was paid $30), monthly star maps were reprinted from the *Knowledge Diary and Scientific Handbook*, and three other members made valuable contributions. Elvins distilled eighty-three years of wisdom into a few pages of hints for beginners interested in observing solar system objects and included a photograph of the Moon made by a well-known member, D.B. Marsh. J.M. Barr, who made a name for himself in the understanding of spectroscopic binaries, wrote an excellent summary of various types of variable stars, and the English amateur astronomer, W.F. Denning, a corresponding member of the Society, prepared a section on meteor showers, a topic for which he was renowned. In the preface to the 1908 edition, Chant wrote:

> The first issue of this Handbook was received with much favor, and there is reason for believing that it has decidedly increased the interest in Astronomy throughout the country.
>
> In preparing this, the second issue, the Editor had hoped to make great improvements by utilizing many valuable suggestions received from students and observers in various parts of Canada, but the limited funds available have delayed the realization of these hopes.

As it was, he reduced the number of pages in the *Handbook* from 108 to 93 by eliminating the section on occultations (which in any case had been limited to predictions for Toronto) as well as the articles by Elvins and Barr. The by-laws were also left out but a membership list was included instead. Costs of printing were reduced from $326 (for an unknown press run) in 1907 to $200 for 600 copies in 1908. Even so, it was decided that the expense was too much, and so no *Handbook* was published for 1909 or 1910. One would have thought that those years of Halley's Comet would have been ideal to capture public interest and boost sales. Presumably, a number of the 600 copies of the 1908 *Handbook* were sold individually, as there were only about 400 members and exchangees, but no actual sales figures are available. Of course the *Canadian Almanac* provided some pretty stiff competition at 30 cents per copy. It was a virtual compendium of facts on all aspects of Canadian life with about 30 of 350 pages devoted to astronomy. In any case, the Society seemed to make no effort to promote the *Handbook* and if the selling price was only 25 cents (the going rate some years later), there would have been a net loss of 8 cents on each one sold. The early *Handbook* was certainly not the money-maker that it was to become 60 years later.

Starting with the November–December issue of the Journal in 1908, and over the next two years, predictions which would normally have appeared in the Handbook were published in the Journal. One such section in September–October, 1909, included a bi-polar plot of the orbit of Halley's Comet. The idea behind such a diagram in which both the Sun and Earth are fixed is quite common now, but was certainly unusual then, and may have actually been an original concept of Chant's. On the whole, however, the experiment of putting predictions in the Journal did not work out too well, and the Council decided late in 1910 to reinstitute the Handbook for 1911 – so late that Chant could not get it ready until after the new year.

The new name that was chosen, The Observer's Handbook, might also have caused some problems. The British Meteorological Office had begun publishing a volume in 1908 under the very same title, but apparently no one objected and both Observer's Handbooks continued their existence. The British Astronomical Association nearly became a third user of this popular title when they undertook the publication of a similar volume to replace The Companion to the Observatory which ceased publication in 1920. Good will again prevailed, and the BAA agreed simply to use the name Handbook. Chant noted, rather wistfully, that twenty members assisted in the publication of the BAA Handbook.

The rejuvenated RASC Handbook had fewer pages when it recommenced in 1911, about seventy pages being the norm until 1937. Throughout this period, the annual financial statements continued to give no indication that the Handbook was being sold to nonmembers, though a note in the Journal hinted that this did happen at least in 1914 when copies were made available at public meetings in Toronto.

World War I brought some restrictions. Grants from the federal government were cut in half, and the Handbook was cut back to thirty-two pages for 1918 and 1919. This was accomplished by omitting the portions which were pretty much the same from year to year. The regular features returned in 1920 along with some improvements. Lists of double stars, nebulae and clusters suitable for small telescopes, and phenomena of Jupiter's satellites were all included now. On the other hand, accounts of the year's progress in astronomy, which had been a feature in 1912 and 1913, and descriptions of comets seen in the previous year, found in the editions of 1914–17, were now dropped. Star maps, which were always on a rather small scale, were also omitted. Instead, readers were referred to other sources, such as the University of Toronto Extension Department which sold four seasonal maps, diameter nine inches, for one cent each.

The 1920s and 1930s saw an increased demand for the Handbooks. Many Centres began to set aside five or ten minutes of their meetings when some member would explain the use of certain parts of the book or would discuss events predicted for the next month or so. Sales to nonmembers also began to become significant. Several universities and schools purchased quantities of the Handbook though the 25 cent price (20 cents for bulk orders of ten or more, starting in 1940) kept any profits down.

A complete recitation of all the yearly variations in contents would be tiresome and pointless. Suffice it to say that a table of 260 bright stars was mainly responsible for the enlargement of the Handbook to seventy-two pages in 1925. W.E. Harper prepared the first of these tables which has been an important feature ever since. The 1937 edition was increased to eighty pages by the addition of a much fuller list of stars occulted by the Moon, a list of star clusters and nebulae suitable for amateur observers, a table of double stars and four circular star maps. In 1939, F.S. Hogg's name officially appeared as Assistant Editor, and several improvements were made, including tables of miscellaneous astronomical data,

Photo courtesy of Canadian Baptist Archives

WILLIAM FINDLAY (1874-1953) grew up on a farm near Hamilton, Ontario, a region of the country where his ancestors had lived for at least three generations. After high school he went directly on to McMaster to get his BA and MA in Mathematics. Then followed three years at the University of Chicago culminating in his PhD, magna cum laude, in 1901. Findlay's teaching career began at Columbia University in New York where he tutored in mathematics from 1901 to 1905. At the same time he was Secretary of the American Mathematical Society. He then returned to McMaster where he became Professor of Mathematics and Head of the Department until retirement in 1948. Findlay was also a devout churchman and held a number of positions including the Presidency of the Baptist Convention of Ontario and Quebec in 1942-43.

Findlay originally joined the RASC in 1907 but seems to have been inactive until the 1930s. He was elected to the Council of Hamilton Centre in 1931 and served as their President in 1934-35 before becoming Vice-President and President of the Society 1936-39. He came to Toronto for every general meeting and meeting of Council during these years, but delivered only two lectures himself. He did speak to meetings of the Hamilton Centre a few times and continued on the Centre Council until 1947.

twilight times, ephemerides of Saturn's satellites and the brighter asteroids, and revisions and extensions to the sections on meteors, occultations and planetary satellites. All this was done without increasing the size of the *Handbook* by judiciously giving times of sunrise and sunset only for every second day.

World War II brought some significant changes to the *Handbook* as a result of recommendations from instructors in the Royal Canadian Air Force. In 1941, a list of stars used in air navigation was added, times of moonrise and moonset were given for each day of the year in 1942, and the range of these tables was extended to include selected latitudes from forty to fifty-four degrees North in 1943. To make room for this additional information, sections on variable stars, meteors and the nearby stars were temporarily deleted since they could still be referred to in earlier editions. In fact the Table of Contents still directed users to the appropriate location for these topics. Another noticeable effect of the war was the lower quality of paper used for the 1944 and 1945 editions.

At the suggestion of RASC President W. Findlay, advertisements for the 1939 *Handbook* were placed in the bulletin of the secondary school teachers federations and in the denominational journals of various churches, but it was really World War II which brought a surge in *Handbook* circulation and a related increase in membership. Chant noted in the preface to the 1948 edition that, "During the past decade the circulation of the *Handbook* has increased from about 1,500 to 5,500. This year for the first time a number of advertisements have [sic] been included, calling the attention of readers to various astronomical accessories." Of course, advertisers were attracted by the great increase in circulation so that by 1958 there were thirteen pages of ads. Even with this advertising revenue, the *Handbook* was priced so low, especially for bulk orders where most of the sales were made, that the Society made little profit. The introduction of provincial sales tax on books containing ads meant that there was no longer a financial benefit to the Society and so advertising was discontinued after 1975.

Editorial duties began to be shared as Chant gradually eased up on his workload. Frank Hogg, who had been doing most of the editorial work for the *Journal* since 1937, also took on more and more of the *Handbook* tasks. By the late 1940s he had complete responsibility, except in name. Following his untimely death in 1951, Ruth Northcott became assistant editor of both the *Journal* and *Handbook*, and after Chant's death in 1956, editor of both. Her presidential address in 1964, "The Inside Story of the Observer's Handbook," gave a good idea of its background and the work that went into its preparation. During her tenure, Northcott increased the press run from 6,000 to 16,000 copies and the price from 75 cents to $1.50. Her successor, John Percy, who served as editor for eleven years, 1971–81, also provided an historical account of the *Handbook* – his marking the seventy-fifth edition in 1983. Thus, up to and including the 1981 edition, the *Observer's Handbook* was always prepared by astronomers at the University of Toronto.

At least in the 1960s and 1970s, they were able to get some computational help from students hired as research assistants for the summer. Ruth Northcott also got four women members of the Montreal Centre to prepare the tables of moonrise and moonset, an arduous section to calculate in the days before electronic computers. Since 1981, Roy Bishop at Acadia University has been editor. He has further expanded the large group of contributors to the *Handbook*, but has had to manage without any personal assistant.

All these editors continued to make improvements to the *Handbook*. Northcott, during her editorship (1958–70) increased the number of pages from 85 to 100. Percy, by the end of his term in 1981, had increased the *Handbook* to 144 pages, and under Bishop the page count reached 238 by 1992.

There were four general reasons for this expansion. In the first place, many enhancements and additions reflected the ever-increasing breadth and depth of observing interests. Data on the Sun and Moon, the planets and their satellites were the main additions in the 1960s, predictions of grazing occultations and comet ephemerides came in the 1970s while a fuller treatment of eclipses and an expansion of the lists of deep-sky objects reflected greater interest by amateur astronomers in these areas in the 1980s.

Secondly, efforts have been made to make the *Observer's Handbook* more readable and instructive without sacrificing the tabular data. Users are introduced to a different type of variable star each year; suggestions are given for further reading and for visiting meteorite craters, planetaria and observatories; information about telescopes and optics, and Terence Dickinson's introduction to the Planets section are all evidence of that trend.

Thirdly, material has been expanded to make the *Observer's Handbook* more useful in other countries. Extension of occultation information to cover the United States, use of Universal Time throughout, and a star map of the southern sky are indicative of these improvements.

Finally, the tremendous advances in astronomy generally are reflected in some new or expanded sections of the *Handbook*. For example, radio astronomy first appeared in the pages of the *Handbook* in 1965. As a result of the tremendous increase in knowledge which the Voyager missions supplied on the outer planets, a greatly expanded section on their satellites appeared in the 1980s. More information on galaxies and a description and finder chart for the black hole candidate Cygnus X-1 are also included now.

All of these improvements have meant that the price of the *Handbook* could be increased without sacrificing circulation. In fact, the number of copies sold has been in the 13,000–16,000 range since

1968, but the price has gone from $1 per copy to $14.95 for the 1993 edition. During the same period. the total cost of producing and distributing the Handbook has only increased by a factor of about four, so the profit it generates is now about $50,000 per year.

It is a record of which the Society can be very proud. Not only does the Handbook go a long way to keeping the Society self-sufficient, but its large circulation ensures that the RASC is fulfilling its very important mandate, "to stimulate interest and to promote and increase knowledge in astronomy and related sciences." The Observer's Handbook has been described as "the single most useful book for naked-eye, binocular, and telescope observers." It is found in the domes of most major observatories in the Western Hemisphere and is used in many countries worldwide.

Whether the static state of Handbook circulation in recent years really indicates a saturated market is debatable, but the Society has taken steps to fill another niche. The Beginner's Observing Guide, intended to be a less intimidating introduction to the sky, first became available in 1992 at a price of $5.00. The observational and educational experience of its editor, Leo Enright, bodes well for its future success.

BULLETIN

The Bulletin, or National Newsletter as it was originally known, first appeared in 1970. As early as 1960 a committee formed to investigate co-operation between observing Centres recommended that, "a new medium for the publication of nonprofessional articles of interest to observers be established by the National Council in the form of a quarterly or semi-annual supplement to the Journal." Vern Ramsay, a member of the committee, felt that the success of a national observing program would depend heavily on such a publication. Some members of Council, however, did not see the need and wondered if it might not detract from Centre newsletters which were recently springing up. For her part, Ruth Northcott, as editor of the Journal, pointed out that the pages of the Journal were open to material of special interest to amateurs, but none appeared because none had been offered. Council consequently decided to defer any action until an observing program was developed in more detail though one or two bulletins were issued on an irregular basis by the different observing groups.

The years passed and the national observing program faded, but the need for improved communications still lingered. In 1968, the Council of the Centre français de Montréal passed a resolution requesting that they be permitted to pay only half the usual fee to the National Treasurer because they were not satisfied with the Journal or its content. The idea was quickly quashed, but Council did recognize that, on linguistic grounds, the Centre certainly had reason to be dissatisfied. Almost nothing was published in French in the Journal at that time, though the problem was not with any editorial policy. There was a clear desire to make the Journal more bilingual, not by translating English articles into French, but by having greater original French content. But again, if papers were not submitted in French, they could not be printed. The dissatisfaction with the Journal went beyond the linguistic issue however, and Montreal found moral support in other Centres.

The following year, Fil Park spoke on behalf of Ottawa Centre. Many of their members said they got little out of the Journal, and their Centre Council had suggested that many good articles in Centre newsletters would be of interest to members across the country if they could be published in a relatively inexpensive way in a separate format. Again Ruth Northcott noted that "it may well be that in many of the Centre publications there are articles which could, with a little bit of work, become suitable for inclusion in the Journal, but these are not [being] submitted."

The consistent willingness of the Journal Editor to

accommodate contributions from amateurs largely fell on deaf ears. There was, perhaps, a natural reticence on the part of many members to offer their efforts to what they perceived as an international publication with high standards to uphold. Perhaps if Ruth Northcott had had the strength to prevail, she might have brought more amateurs into the fold of the *Journal*, but fate was to give her only a few months more of life.

Against this background, the *National Newsletter* was born. A new Committee for the Co-ordination of Centre Activities (COCOCA) was formed in September, 1969, with John Percy as Chairman and six other members representative of geographic and linguistic diversity. He was very soon authorized to proceed with a Newsletter on a trial basis as an insert to the *Journal*, starting with the February, 1970 issue. And so it came out, printed in the same style of type as the *Journal* itself but on green pages numbered L1–L4 and stapled in the centre of the *Journal*. There was never any doubt that the publication would be considered a supplement to the *Journal*. Had it been mailed separately, it would not have qualified for the preferential second-class rates enjoyed by the *Journal* because of its scientific content.

In that first issue, Editor Percy optimistically noted, "The National Newsletter has the potential to lessen some of the tensions from which our Society perennially suffers. It is a forum to which all can contribute and from which all can benefit, whether they be from St John's or Victoria, amateur or professional, English speaking or French speaking, observer, instrument maker, philosopher or bookworm."

The first issue contained information on a forthcoming eclipse, a transit of Mercury, and two comets. There were some reports of interesting observations. A group of Montreal observers kept the Moon under surveillance during the Apollo XI flight and found a "general obscuration for the central region of Mare Tranquilitatis at 10:10 GMT, July 20." NASA was interested enough in these findings to relay them to Houston for transmission to the spacecraft. Another report in the premier *Newsletter* came from John Howell who had observed the grazing occultation of a binary star ZC1392A and B. Ken Chilton reported on numerous drawings and polarization observations of Venus made by four Hamilton members. He also notified readers of the formation of the International Union of Amateur Astronomers, a newly formed organization of which he was secretary.

However, the *Newsletter* was not only intended to provide information for observers; it was, after all, originally under the auspices of COCOCA, and so reports of Centre activities were definitely part of its mandate. This led eventually to a regular feature called "Across the RASC." The intention here was to summarize "reports and activities conducted by any Centre that seemed ... to be worthy of imitation or development by other Centres. ... Material is often culled from Centre publications." A nice result of this was the strengthening of ties within the Society as readers of the *Newsletter* began to recognize names of distant members whom they might meet at General Assemblies.

Non-RASC publications were also used occasionally by editors. As Ian McGregor pointed out:

> The David Dunlap Doings (newsletter of the DDO), Cassiopeia (newsletter of CASCA), and North Star (newsletter of the Planetarium Association of Canada), ... have been sources for interesting and informative items which might otherwise be seen by only a smaller, more specialized audience.

The role of the *National Newsletter* was nicely summarized by McGregor in an editorial at the end of 1987:

> [I want the Newsletter to represent] the interests and activities of the Canadian astronomical

Dr J.D. Fernie as the after-dinner speaker at the 1976 General Assembly in Calgary. In the background is Alan Batten.

J. DONALD FERNIE (1933-) was born and raised in South Africa. After receiving his MSc degree from the University of Cape Town in 1955, he and his wife moved to the United States where he earned his PhD at Indiana University. They returned to Cape Town for three years where he lectured in Physics and Astronomy before emigrating to Canada in 1961. Since that time Fernie has been on the faculty at the University of Toronto, serving as Chairman of the Astronomy Department and Director of the DDO from 1978 to '88. During his earliest years at U of T, he developed photoelectric photometry and was responsible for revitalizing the 48 cm telescope originally built by R.K. Young in the 1920s. His career continues to be very productive with well over a hundred papers mainly on photometry of variable stars, especially Classical Cepheids. He also finds the history of astronomy attractive and has published a number of papers in that field as well as an entertaining and informative book, *The Whisper and the Vision*, in which he tells some delightful adventure stories of astronomers of the past.

Don Fernie accepted the nomination of First Vice-President of the RASC in 1972, and in his subsequent Presidency, he visited most if not all the Centres of the Society. He served as Book Review Editor for the *Journal* and wrote a number of reviews himself, and was also Assistant Editor of the *National Newsletter*, graciously permitting many of his witty articles to be reprinted from the *David Dunlap Doings*.

community. Whenever possible I have included reports on amateur conventions such as the Alberta and Mount Kobau star parties, consumer related articles which since we have no paid advertising allows a more critical evaluation and dialogue on telescopes and equipment, and reports on observing activities and techniques.

I believe it is important for Canadians to be aware of the accomplishments of other Canadians. Thus the discovery of Supernova 1987A by Ian Shelton and two comets by David Levy are highlighted, comments on historical and modern observatories are printed, and reports on the techniques used by members to do their astronomy are shared across the membership. ... About 60 percent of the material submitted directly finds its way into its pages.

That the *Newsletter* enjoyed a wide audience is suggested by the response to occasional requests for information in its pages. Professor Ed Kennedy wrote about the rarity of Gregorian telescopes, and as a result received four letters from owners as far away as Natal, South Africa. When the BAA appealed for auroral observations, they received nine responses from Canadian members, and Marc Gelinas, the editor of French language contributions to the *Newsletter*, got ten observations in answer to his request for times of Venus' dichotomy. At other times various contributors have benefitted from criticisms and offers of assistance from others. The *Newsletter* has certainly been a catalyst for communication among members.

The *Newsletter* gradually became more and more independent. Beginning in 1978 it was no longer stapled into the centre of the *Journal*. The coloured paper which had been necessary to give it a distinctive identity was no longer used, and eventually the "L" prefix to the page numbers was dropped. A separate cumulative index was published for *Newsletter* contributions 1970–78, with annual indices after that. Beginning in the Annual Report of 1979, the cost of printing the *Newsletter* was reported separately from the *Journal*. Printing costs kept pace with inflation and went from $7,559 in 1979 to $13,904 in 1990.

The editors of the *National Newsletter* were:

John Percy, 1970–72 Frank Shinn, 1978–80
Norman Green, 1972–74 Ralph Chou, 1981–84
Harlan Creighton, 1974–77 Ian McGregor, 1985–90

Strictly speaking these people were recognized, and reported to Council, as chairmen of the National Newsletter Committee. The other members of the Committee were sometimes called assistant editors or editorial staff (unpaid of course). It was not until the by-Law revision of 1989, that the *National Newsletter* Editor was put on the same footing as the *Journal* Editor, as an appointed officer of the Society.

In 1991, under a new editor, Patrick Kelly, the old newsletter took on a new bilingual title, *Bulletin*, and a bright, new image with two-colour printing and a larger format. The last vestige of its origins remains in the by-line "Supplement to the *Journal* of the RASC."

Unfortunately, the advent of a national newsletter in 1970 did not stop the old complaints about the technical nature of the *Journal*. By providing a separate medium for news of Centre activities and reports of observers, the Society may have inadvertently isolated the *Journal* even more from its potential readers. The fact that revenue from the *Handbook* is nearly sufficient to pay the costs of all the Society's publications apparently does not carry much weight with some members who continue to feel that they are the ones bearing the financial burden of a *Journal* they rarely read. Others are quite content with the status quo.

The differing needs and aspirations of the members continue to pull the Society and its publications in all directions, but in the traditional Canadian spirit of compromise and tolerance, stability seems to reign and progress occurs slowly, step by step.

NORMAN GREEN (1913-) was born in England and emigrated with his parents to Hamilton. After graduating from the Technical School, Norman joined the Hamilton Centre in 1930 and became active at star nights, as a speaker and as Secretary Treasurer and Vice-President. He worked for Bell Canada for a time before pursuing theological studies and then, following ordination by the Anglican Church in 1943, he and his wife, Jean, accepted a call to the parish of Fenelon Falls. They returned to Hamilton in 1952 and Reverend Green resumed his active role in the Society. He began assisting with planetarium demonstrations at McMaster in 1958, and the same year became President of the Centre.

On the national scene, he was elected Secretary in 1964, a post he handled for a total of ten years with tactful charm and careful attention to detail. Meanwhile, Green was accepting a great number of invitations to speak on astronomy to various schools, community, church and Scout groups. For a while in 1967, he was giving an average of two talks per week. When the McLaughlin Planetarium opened in Toronto the next year, Green was appointed Assistant Director. He became its administrative head in 1971.

From 1972 to '74 he edited the Society's *National Newsletter* and continued as Assistant Editor for three more years after that. His successor as Editor, Harlan Creighton, spoke of Norman Green's "flawless sense of what is appropriate plus his dedication to hard work for the benefit of the Society, its members, and astronomy." These admirable qualities won him the Service Award in 1970. He has been the Honorary President of Hamilton Centre since 1976. In retirement at Niagara-on-the-Lake, he enjoys leading walking tours through the historic town, speaking occasionally to the Hamilton and Niagara Centres, and presenting talks on astronomy to various organizations throughout the region.

The Reverend Norman Green presents roses to RASC President Helen Hogg at the Annual Meeting in Hamilton, 1958.

Close encounters at an Edmonton Centre Starnight, Queen Elizabeth Planetarium.

CHAPTER 8

Encounters of All Kinds

A central role of the RASC has always been to advance public awareness and appreciation of astronomy. Star nights and public lectures are the most traditional means but many other ways are used to reach the people. The Society's very existence is a clear signal that it's okay for John or Jane Doe to join with thousands of others in pursuing an interest which the rest of the world might regard as unconventional or even eccentric.

MEETINGS

Most, if not all, lecture meetings held by Centres are open to visitors, though regular attenders are expected to join the Society and usually only members hear about the meetings through their Centre newsletters. These meetings will be described later, but from time to time lectures are planned and advertised specifically as events open to the public. For instance, the national Society in co-operation with CASCA has sponsored a Helen Sawyer Hogg Public Lecture annually since 1985. These are given by outstanding speakers in whatever city the RASC or CASCA happens to be meeting that year. The list includes:

Owen Gingerich: The Mysterious Nebulae 1610–1924 (Toronto),
Barry Madore: The Hubble Space Telescope (Winnipeg),
René Racine: Small is Beautiful (Vancouver),
Hubert Reeves: The Early Moments of the Universe (Victoria),
Roger Cayrel: La Construction du Télescope CFH [Canada-France-Hawaii] (Montreal),
Joseph Veverka: The Voyager Adventure (Ottawa),
Kimmo Innanen: The Prediction and Discovery of a Martian Trojan Asteroid (Toronto),
Alan Hildebrand: The Cretaceous/Tertiary Boundary Impact (or The Dinosaurs Didn't Have a Chance) (Calgary).

All of these speeches have subsequently been published in the *Journal of the* RASC.

The tradition of public lectures is as old as the Society itself. The earliest occasion on record at which a distinguished visitor addressed an open meeting of the Society took place in 1891. Mrs R.A. Proctor, widow of the famous British popularizer of astronomy, spoke to an enthusiastic Toronto audience on the Lick Observatory and its work. This, the first of the large mountain-top observatories in the United States, had just opened in 1888, and there was great interest in the attainments of its giant 36-inch (91-cm) refractor.

A list of RASC public lectures would be long indeed, spanning the years and the country. A sample could include Ernest Rutherford speaking in Toronto on "Some Cosmical Aspects of Radioactivity" in 1907, Harlow Shapley addressing audiences in Montreal and Ottawa in 1922, and J.S. Plaskett talking to a Winnipeggers in 1931 about the structure and rotation of the Galaxy. Three members of the Vancouver Centre (F.K. Bowers, J.R.H. Dempster and A.K. Goodacre) attracted over 250 people to a timely talk they gave in November, 1957, on the Russian artificial satellites. A cross-country tour by Professor Wilhelmina Iwanowska of

Michael Watson speaking to over 500 people at a public lecture on Comet Halley, 21 November, 1985.

The gallery of the University of Toronto's Medical Sciences Auditorium was packed too.

Live TV coverage: Ian McGregor, Michael Watson, Ralph Chou and program host Randy Attwood observe a lunar eclipse.

RASC members set up telescopes outside the David Dunlap Observatory for the enjoyment of visitors.

Torun, Poland, in 1973 created great interest in the 500th anniversary of the birth of Copernicus, and the 1986 apparition of Comet Halley brought out crowds to public lectures organized by the Society in Toronto and Halifax. As these last three examples illustrate, there is nothing like some scientific or technological breakthrough, an historic anniversary, or an uncommon celestial occurrence to get media attention and free publicity for related special events such as public lectures.

Miniseries were also popular in the early years of the Society. G.E. Lumsden gave a series of an elementary character on "Constellation Study" in 1891. For the benefit of new members as well as the general public, Professor A.T. DeLury gave three series of popular lectures in the spring of 1903, 1904 and 1905, the last group of eight lectures to an audience of 400 people on average. Dr C.A. Chant took up this idea and gave elementary courses of six sessions each on the "Physical Constitution of the Heavenly Bodies" in 1906 and on "Some Recent Advances in Astronomy" in 1907. He also gave short three-lecture courses in 1917, 1920 and 1927 but at least in Toronto, the Society apparently gave up the idea of such minicourses for many years after that date. Perhaps the University Extension Department began to fill the need and radio broadcasts, in which the Society played a part, also began in the 1920s.

One RASC Centre which has run a very successful course on a continuing basis is Victoria. Each summer since 1931, the Centre has arranged a series of talks which became known as "Summer Evenings with the Stars." Such a friendly, informal title is probably one reason for its success. Other factors come out in the following account for 1935:

> The titles of several of the "talks," not "lectures," were put in the form of questions – such questions as one is commonly asked when the subject of astronomy is brought up in conversation – What is a star? How big are the stars?

JOSEPH A. PEARCE (1893-1988) joined the RASC in 1916 while stationed at Rockliffe Camp, Ottawa. He had just returned, wounded, from the War. His brother, Dr Leslie Pearce of Brantford, joined the Society at the same time but died of influenza just two years later. Once the war was over, Joe Pearce returned to the University of Toronto, completing his BA in 1920 and MA in 1922 before going to the Lick Observatory in California on a fellowship.

In 1924, Pearce accepted a position at the DAO in Victoria and was appointed Astronomer the following year when R.K. Young moved to Toronto. He and Mrs Pearce both joined the Victoria Centre in 1925 where he served as Recorder, Secretary, Vice-President and President (1928-29), and Honorary President (1932 and 1981-85). He was one of the originators of the Centre's "Summer Evenings with the Stars" program. He generously donated many of his own books to establish the Centre's library and many years later, in 1961, he similarly gave over 300 books to the newly-organized Department of Astronomy at the University of Victoria.

The radial velocity work for which Pearce is best known was carried out in collaboration with J.S. Plaskett and established for the first time in a thoroughly satisfactory manner the rotation of the Galaxy and hence its mass. He was elected President of the RSC in 1949-50, and received an honorary DSc degree from the University of British Columbia in 1955. In 1988 an asteroid was named after him by the IAU. As a member of professional organizations such as the AAAS, the AAS, and the RSC, he travelled frequently to meetings and rarely missed an opportunity to address one or more Centres of the RASC en route. From 1934-40 he served the Society as Vice-President and President. The RASC presented him with the Service Award in 1962.

Dr J.A. Pearce watches as Mrs J.S. Plaskett unveils a memorial to her husband at the base of the 183 cm telescope at the DAO.

How far away are they? Following a thirty minute talk and extended discussion, participants then moved outdoors where two or three telescopes were set up. The papers and radio station gave much useful support in the form of publicity before and reports after each meeting. Of the non-members attending, 17 were elected to membership.

The Centre made a modest net profit of $3.70 that year from the nominal fees of $1 for adults and 50 cents for juniors. The next year, there were seven weekly meetings, attended by an average of sixty-seven people. Talks by two professional astronomers, four by amateurs, and a visit to the DAO made up the series. Following the lectures, several groups gathered around the telescopes (mostly made by RASC members) and looked at such sights as Jupiter, the Moon, double stars, clusters and Comet Peltier which was a naked-eye object in August, 1936. A photo of this comet, taken by Pearce and McKellar, and a set of six star maps were given out as souvenirs. Prizes were offered in some years for writing an essay on one of the summer lectures and for sketching constellations. In 1939 a notable feature to encourage careful observation by beginners, was the use of mimeographed diagrams of the stars forming the constellations. The recipients were asked to number the stars in the order of their brightness and to note their colours. Proper values were announced at the next meeting.

The Victoria summer course flourished so well that in the early 1950s attendance reached nearly 200 and the profits permitted the Centre to make a $100 donation to the Frank Hogg Scholarship Fund. Success was not achieved without a lot of work, however. In 1955, for instance, an elaborate display was set up in a prominent window of Eaton's department store and 180 notices were mailed out. At the present time, "Summer Evenings with the Stars" continues after sixty years to be one of the Society's most successful endeavours in public education. It is now held weekly in August each year.

Sometimes outside organizations ask the Society to supply a speaker, and generally some member is able to fill the bill. Even in 1893 when the Society had a pretty small membership base, Miss S.L. Taylor, a member of the Society and a teacher at Toronto's Lansdowne School, spoke to the Toronto Women Teachers' Association, Mr Pursey, the librarian, addressed an open meeting of a benevolent order, Mr Paterson, later a Society president, read a paper to a group in Deer Park, and the Society took charge of an open meeting at the Young Women's Guild.

In 1925, it was reported:

> In Toronto we have had a growing number of requests that lectures on the elementary facts of Astronomy be given in Churches, Sunday Schools, Colleges, Clubs, Schools, etc., and in order to deal with the matter satisfactorily the Council appointed a Speakers' Committee ... The idea has been to request members in various parts of the city who are equipped to give an illustrated lecture, to undertake to accept invitations to speak in their locality. This plan in operation [for a few months] has so far resulted in about twenty lectures being given, at which invitations to join the Society have been distributed.

To make such talks easier to organize, Dr Chant prepared a sixty page booklet in the form of three lectures entitled "An Introduction to Astronomy" and selected a set of 100 slides to accompany them. Duplicate sets of the slides were made available and were widely used. During 1929, it was reported that a lecture illustrated by Chant's slides was given in fourteen places in British Columbia alone.

The demand for speakers still continues. In 1989, Toronto Centre's past-president, Randy Attwood, found that their public education program needed a strategic plan to cope with the heavy demands for speakers and course instructors from schools, libraries, Cub and Guide troops, church and community groups, school boards and special educational organizations. These needs were on top of the regular public education programming in which the Centre was very active.

SHOWS AND DISPLAYS

Exhibitions and hobby shows provide an appropriate venue for the RASC to meet the public. Winnipeg Centre set up several telescopes at the Red River Exhibition in the 1960s, Halifax Centre participated in the Societies Show held at the Nova Scotia Museum in the 1970s, and Saskatoon and Victoria took part in hobby shows in the 1970s and 1980s. Toronto Centre had a very long involvement with the Canadian National Exhibition. From 1928 until 1967, RASC members were there with their telescopes, giving fair-goers a look at some bright object such as the Moon or planet. In 1949, as an example, eleven of the fourteen CNE nights were clear enough for viewing through six telescopes. A total of 7,190 people came and each received a four-page leaflet called "Sky Facts," printed and supplied by *The Toronto Star* in exchange for an advertisement on the back page.

Star parties, Astronomy Days, mall and exhibition displays are less formal than lectures and often attract a much larger audience from a much broader cross section of the public. People are free to come and go as they please, to look through telescopes, and to ask whatever questions come to mind. The Society was only a few months old when "several of the members expressed their willingness to place their telescopes at the disposal of any persons desirous of seeing Saturn and other celestial objects." Open-air meetings soon became the standard means of concluding each season in June, and in 1908 several hundred visitors came to the University of Toronto campus to view the heavens through "several large telescopes." Public interest in the 1910 apparition of Halley's Comet completely swamped the unsuspecting Toronto amateurs who had set up telescopes in front of their houses! When the comet rolled around again in 1985, there were fortunately many more RASC members with telescopes which they were smart enough to set up in well-advertised locations in parks.

Other Centres were slower to initiate public star nights. Ottawa and Victoria perhaps had less need since both had professional observatories which opened in 1905 and 1918 respectively. Both observatories invited the public one evening per week to inspect the telescopes and have a look through them if skies were clear. Toronto got its David Dunlap Observatory in 1935 and it too opened to the public every Saturday evening from April to October, but here a few RASC members also set up their telescopes on the observatory grounds. Ottawa and Victoria established a similar custom some years later. Montreal began public star nights in 1947 and by the 1960s Calgary, Edmonton, Winnipeg, Hamilton, Niagara and the Centre français de Montréal had joined in.

Bad weather, of course, can present problems whenever observing is planned. Rain dates can be scheduled, but more often than not variable cloudiness leaves everyone in doubt as to whether to proceed or not. At least one Centre has got around that difficulty by having a hot-line number with a taped message giving up-to-date information. Another excellent plan to circumvent weather set-backs on public star nights is to have some indoor displays or films as well. Co-operation with a planetarium is one way to handle this as Calgary, Edmonton and the Centre français de Montréal have done. Montreal and Edmonton set up large tents for some of their successful evenings in the 1960s. Edmonton charged a small admission fee in 1963 and made over $2,000 profit.

Even if skies are clear, it's a good idea to have something else for the public to do besides waiting for a turn at a telescope. Television monitors attached to a telescope-mounted video camera can

An RASC Information Desk

Display for Campers

Astronomy Night '64 in Calgary.

Nova East '87 in Fundy National Park.

also help to keep restless children happy with a foretaste of what's to be seen. Hamilton Centre began using this technique in 1962 when 2,000 adults and children showed up for their field night.

Centres with their own observatories often have open houses for the public on a regular basis. Calgary, Hamilton, Montreal, Quebec, Regina and Winnipeg each have welcomed hundreds of visitors annually to their facilities. Saskatoon, in 1972 alone, recorded 7,100 visitors to the university observatory.

Notable sky events usually get media attention, and curious crowds will turn out to witness an eclipse, a conjunction of planets, an opposition of Mars or a bright comet. The biggest attraction of this type in the Society's history was Comet Halley. Thousands of people stood out on cold December nights in 1985 to see the fuzzy glow of the famous phantom. Toronto Centre estimated that 15,000 saw the comet through their telescopes. Winnipeg Centre had 4,000 turn out at Steinbach and that was only one of their four Halley Skywatch locations.

City folk almost forget the spectacular sight of the Milky Way arching across a sky sprinkled with sparkling stars. Only on holidays far from the glow of city lights are they reminded that there's more to the universe than the solar system and a few bright stars. Even in 1900 when electric lighting was just coming into common use, people appreciated the view from the country. President Lumsden, on holiday in Muskoka at the turn of the century, would sometimes have fifty people wanting a look through his telescope.

Some of the Centres started to hold star nights outside the city in the 1960s. Toronto began to use the Metropolitan Regional Conservation Areas in 1964 and even further afield attracted as many as 600 to a star night in Orillia (100 km north of the city). Sibbald and Huron Provincial Parks were used as sites in 1973 and 1974. In 1980, members of the Saskatoon Centre travelled to the small town of Kindersley, about 150 km west of the city and were surprised at the large and appreciative crowds who lined up for a look through the dozen telescopes aimed at various objects and Winnipeggers incorporated a camping weekend at Riding Mountain National Park with two presentations to visitors. Kitchener-Waterloo Centre went out to conservation areas in their region in 1987.

On a more ambitious note, during the summers of 1974 and 1975, a group of Toronto Centre members presented over a hundred public education programs to more than 10,000 campers in Ontario Provincial Parks. The project was sponsored by the federal government's Opportunities for Youth Program and was carried out with the co-operation of the Ontario Ministry of Natural Resources. London Centre's Project Z was a similar case in point. In 1978, following a more modest but successful project five years earlier, federal funding was obtained under the Young Canada Works program to enable seven university and high school students to hire a van and to travel around to schools, camps, malls and parks where presentations, displays and star nights attracted a lot of interest from a different group every day during the summer.

ASTRONOMY DAY

This tremendous variety of activities just outlined was very effective in stimulating public awareness of astronomy and of the Society. Until recently, however, each Centre was on its own as far as activities and dates and there was no national co-ordination. The independence of Centres to make their own plans still continues. Some Centres open their observatories once a week to the public. Some have several star nights each year, while others are able to manage only one public event each year. But for the last few years Astronomy Day has been set nationally or internationally, usually about the time of the first quarter moon in April. Having a definite date serves as an incentive and a focus for all the Centres to get involved and tends to generate more publicity than

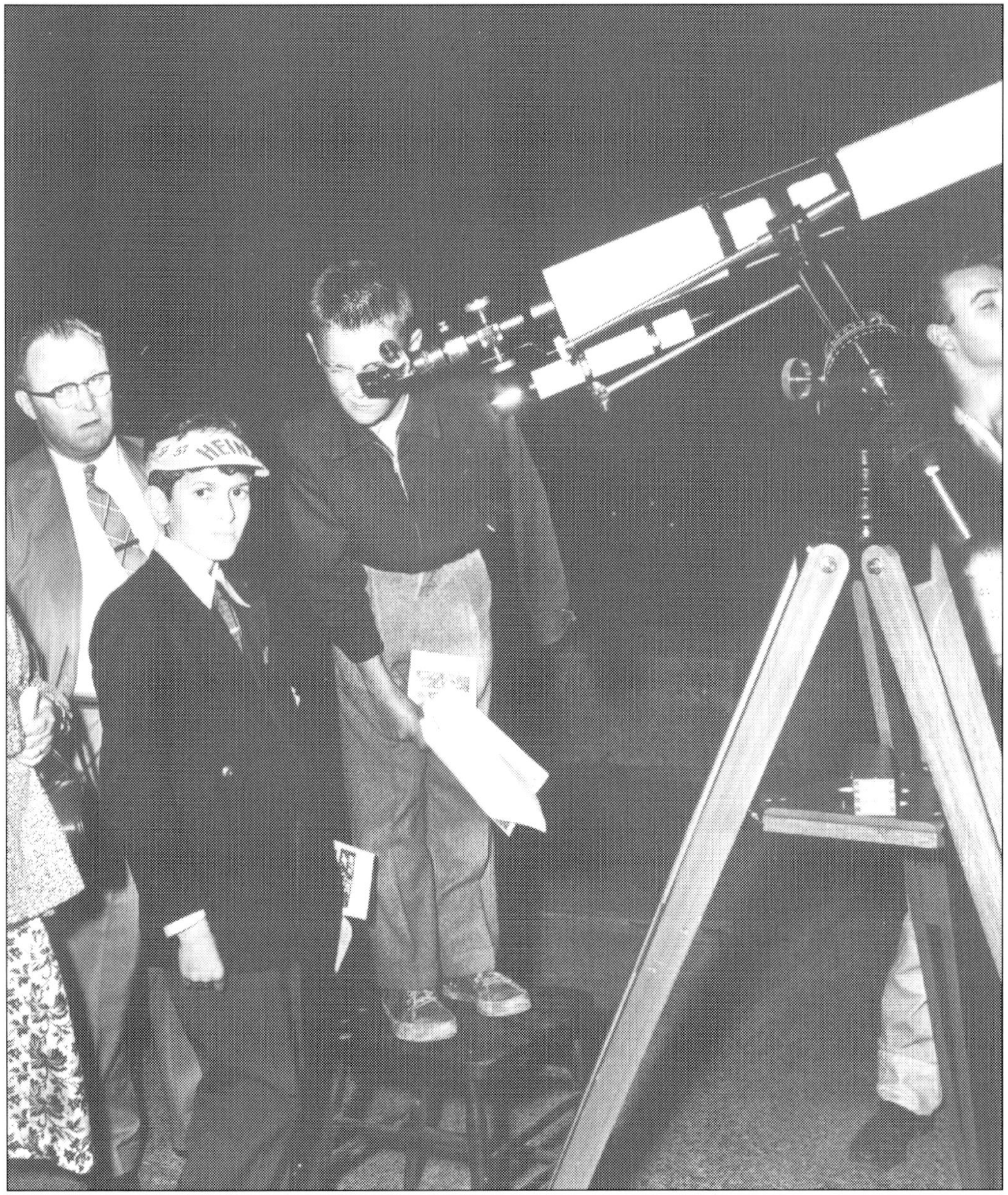

Photo courtesy of York University Archives, *Toronto Telegram* Collection, box 482, file 3191

Publicity is part of any successful star night. Two young visitors to the Canadian National Exhibition hold copies of the leaflet "With the Stars."

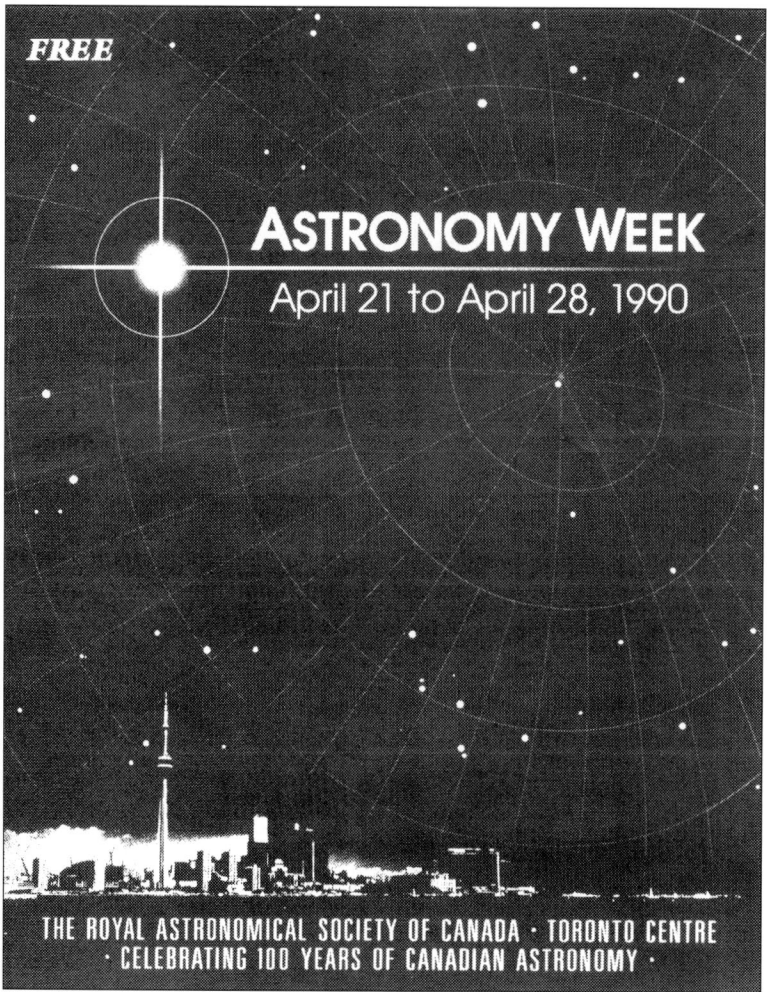

The cover of an impressive twelve-page guide put out by the Toronto Centre.

just another star night would. Astronomy Day was even announced in the House of Commons in 1987 and Astronomy Week, which marked the Society's Centennial in 1990, was proclaimed by the Metropolitan Toronto Council.

Though the idea of Astronomy Day originated in the San Francisco Bay area in 1973, Kingston Centre was the first to take up the idea in Canada in 1979, and Leo Enright, one of the moving forces behind it, soon was appointed Astronomy Day Co-ordinator by the National Council. Astronomy Day activities include public star parties, exhibits, telescope-making demonstrations, films, lectures, planetarium shows and tours of observatories. Solar viewing can be a real eye-opener for many as they see sunspots and solar prominences for the first time.

The role of the co-ordinator is to propose suitable dates and themes, to provide resource material, suggestions and posters through the national office, and to prepare reports based on Centres' activities. Steve Dodson of Sudbury's Science North took over the job with much enthusiasm in 1988. In his report for 1989 he included "Some Other Neat Things to Do":

> The Vancouver Centre displayed a Poncet mount and sold sun dial kits to children and old Sky and Telescope magazines at a bargain to raise money for their library fund. Victoria set up their mobile 20-inch scope in Beacon Hill Park. Ottawa displayed a locally-made Schiefspiegler telescope. Winnipeg had their photometric equipment on display. Montreal had a collection of antique refractors and offered tours of their Centre observatory and darkroom. Kingston had models of the solar system and its objects and of the surface of the Sun. Calgary had a display on light pollution. As expected most Centres had a variety of telescopes on display along with exhibits on astrophotography results and techniques, slide shows, posters and handouts on Centre activities, and computer displays of the sky.

Obviously these extensive and varied programs thrive only through commitment by large numbers of keen, dedicated volunteers.

Mall displays are often one aspect of Astronomy Day activities, though they are also arranged at other times in the year. Hamilton seems to have been the first Centre to "show their stuff" at a shopping centre in 1975. These displays provide a wonderful opportunity to chat with shoppers on a casual basis about many aspects of astronomy and about the Society's role. Mall programs, especially in the smaller Centres, were threatened with curtailment when owners began to require liability insurance of one or two million dollars. The problem was eased when the Society agreed to purchase insurance on a national basis.

YOUTH PROGRAMS

While most of the Society's public activities appeal to people of all ages, for youngsters especially it's hard to beat the thrill of looking through a telescope at the Moon or Saturn for the first time. No picture book can explain so clearly that those objects in the sky are really globe-shaped worlds in some ways like our own but in more ways, very different. Throughout its history the Society has done a lot specifically to interest young people in astronomy – in the schools, at science fairs, in the Boy Scout and Girl Guide organizations and through special activities within the Society itself.

Toronto schoolchildren in the 1890s had some nice surprises. Visitors from the Astronomical Society came with their telescopes so that the pupils could see the transit of Mercury in 1894. At other times members of the Society lent their telescopes to schools for use whenever possible. Ottawa and Montreal Centres both initiated clubs specifically for junior members in the 1940s and members of the St John's Centre visited all the high schools in their area in 1966. In recent years, though, the students have tended to be the ones who travel. In the 1970s,

The photograph shows Rolland Noël de Tilly on his last trip from Montreal to Quebec before retirement, January 26, 1971.

ROLLAND NOËL DE TILLY (1906 - 1983) was perhaps best known as the Secretary of the Centre français and the Société d'Astronomie de Montréal, a post which he held from 1968 until 1981. It was a large responsibility, involving several hundred members, and he carried out his duties so well and for so long that he was named "Le secrétaire perpetuel de la SAM" but he had a number of other responsibilities too. He edited the publications for the Centre from 1967-71 and was their representative on the national RASC Council from 1976- 80. As a supervisor for Canadian Pacific Railways, he was able to get free passes to travel to Toronto for these meetings. On his normal Montreal-Quebec run, he used to give astronomy "lessons" to employees and passengers.

For ten years, Noël de Tilly wrote a column for Les Cercles de Jeunes Naturalistes under the pen-name "l'Ami des Etoiles." He corresponded with several youngsters and visited numerous schools where he gave talks and slide shows. Because of his interest in young people, the AGAA decided in 1986 to establish a plaque called "Les Pleiades" to be awarded to a young amateur astronomer under 18 years of age. The attractive plaque, of enamel on copper, was designed by Rolland's daughter Pauline.

From SAM, he received the Prix Georgette Le Moyne in 1972, and l'Etoille d'Argent in 1980. The Astronomical Society of France presented him with the Medaille Camille Flammarion and the RASC honoured him with the Service Award in 1979.

Toronto Centre members organized an Astronomy Club Day each year when members of high school astronomy clubs were invited to a day of talks, tours and a star show in the Planetarium. As many as sixteen schools were represented. At the same time the Centre offered enrichment courses for students of the Toronto Board of Education in grades 6–8. In the 1980s, London and Windsor Centre members went out to present programs at children's summer camps, and Calgary had a "Youth Department" meeting twice a month on Saturdays at the Planetarium. On an ongoing basis, many Centres host dozens of visits to their observatories by school groups and other young people's organizations.

Science fairs have also been a popular way to encourage students to explore topics beyond the regular curriculum. The Youth Science Foundation was established to promote and organize Science Fairs across Canada and the RASC was a sponsoring organization right from the start in 1961. Every year until 1989 the Society arranged for someone, usually a member, to judge the astronomy exhibits and awarded one or more prizes for the top astronomy projects among the national finalists in the Canada-Wide Science Fair. In the early years, the prize was a book but latterly cash prizes of up to $100 and memberships in the Society were awarded. Some Centres, Calgary and Niagara for example, sponsor awards in local science fairs. One of the most outstanding Astronomy exhibitors at the national level was Peter Brown from Fort McMurray, Alberta. After winning first prize for his solar exhibit as a thirteen-year-old in 1984, he went on to repeat the achievement the next two years. Since then he has taken a special interest in meteors and meteorites, serving on a search team following a suspected meteorite fall and also as the North American Co-ordinator of the International Meteor Organization.

Liaison with Scouts and Guides began in the 1930s and has flourished ever since. These organizations award badges to boys and girls for achievement in all sorts of activities from art to archery, too wide a range for most leaders to supervise alone. The astronomer's badge requires recognition of some of the prominent constellations and an understanding of some basics which RASC members are usually only too happy to help with. Occasionally, Society members have involved the young scouts in some rather exciting projects. About 1939 a Scout "Cyclorama" in Toronto was attended by 100,000 visitors. A booth and display illustrating the "Starman's Badge" was planned and arranged by Toronto Centre members and this attracted great interest and several new members. E. Russell Paterson of the Montreal Centre was an inspiring leader. For a number of years he organized parties of Scouts (not Girl Guides unfortunately) at Camp Bois Franc to observe Perseid meteor showers. He taught the boys how to plot the paths on constellation maps and even added photography to the program in 1939. For the 1932 solar eclipse, he stationed over two hundred Scouts at fifty-foot intervals near the edges of the path of totality and at right angles to it in a genuine scientific experiment to determine the precise limits to the path. To interest the boys and to provide them with instructions needed to observe intelligently, prior training was given at various localities. Paterson later received a letter from Shigeo Ishii of the Tokyo Astronomical Observatory stating, "I have been making reductions of observations of the solar eclipse ... for astronomy of position and it was my pleasure to include your precious results." What a thrill that must have been for the youngsters involved and what a contrast with the total eclipse of 1979 when many schools insisted that the pupils stay indoors and watch the spectacle on television.

If there is one common thread to all the successful activities which have attracted and held the interest of the young people, it is observation. There seems to be no better way to open the minds of the younger generation to the mysteries and marvels of the universe than to let them see for themselves.

MARY W. GREY was born Mary Scribner in 1927 in Chipman, New Brunswick. After receiving her BSc in Civil Engineering from the University of NB, she worked for the Geodetic Survey of Canada where she got her first taste of practical astronomy. But, as Mary says, her "passion" for astronomy took fire after moving to the DO and after joining the RASC.

From 1964 to 1977, Mrs Grey held various offices in the Ottawa Centre including a term as President in 1975-76. Nationally, she chaired the Committee on Publications, Sales and Advertising in the '60s, wrote notes on the Dominion Observatory for the *Journal* in 1968-70, became well-known at General Assemblies for her warmth and humour, and ultimately, in 1982, embarked on the ten-year tour of Presidential Offices. Mary Grey received the Service Award in 1990 for her years of devotion to the Society.

After ten years at the DO, Mrs Grey moved to the National Museum of Science and Technology in 1974, where she became the Head of the Astronomy Division and later the Senior Curator of Physical Sciences. Her career there has been devoted to the promotion of public interest in and knowledge about astronomy through teaching and broadcasting, and writing a newspaper column as well as the Museum's quarterly, *Sky News*. Mrs Grey's interest in history and heritage was evident in her informed talks on the history of the Dominion Observatory and especially in her successful efforts in getting the 38 cm DO refractor installed at the NMST in a building of its own, where it is used for the pleasure of the public. It was dedicated as the Helen Sawyer Hogg Observatory in 1989. In recognition of her promotion of astronomy at the Museum, Mary Grey received the Civil Service Association of Canada's prestigious Merit Award in 1989.

Mary Grey standing beside the Wilson Coulee at the Calgary Centre Observatory, October 17, 1987.

Teachers are always on the lookout for resources to supplement the regular classroom routine. Those who are members of the Society undoubtedly slip some astronomy into the curriculum, perhaps by talking about their own interests, by suggesting their students attend a star night or other event, by sponsoring a school astronomy club, or in a few cases by actually teaching astronomy courses. Peter Ryback, a member of the Ottawa Centre, taught Astronomy at the Cobalt-Haileybury High School in northern Ontario during the 1980s and got a $30,000 grant from the Ministry of Northern Development to finance a fine observatory for the school and community.

A number of RASC initiatives have been aimed specifically at promoting the teaching of Astronomy in school. As early as 1928, a course in elementary Astronomy was prepared for schools in Winnipeg. Departments of Education in Saskatchewan and Ontario began to develop courses in Astronomy in 1966 and sought the help of the RASC. Ed Kennedy and John Percy became valued advisers. Saskatoon Centre was involved in a teachers' summer school in 1971, Toronto Centre held a teachers' workshop in 1974, and Halifax Centre assembled several slide sets on various aspects of astronomy to be made available to science teachers throughout the Maritimes. Quebec Centre agreed in 1980 to handle orders and distribute a 448-page series of lecture notes comprising an Astronomy course written by Dr J. Roy. At the national level, the Society together with the Astronomical Society of the Pacific sponsored a weekend workshop on "Teaching Astronomy in Grades 3–12" in Victoria in 1988.

Through the Society, professional scientists have naturally taken an active interest in Astronomy education. Besides their obvious direct participation through their own articles, lectures and courses, some have spoken out to encourage wider and better dissemination of the subject. Ernest Hodgson lectured to the Ottawa Centre in 1925 on "The Teaching of Astronomy in our Schools," R.K. Young spoke to the Ontario Educational Association in 1927, and A.V. Douglas wrote for *The Teachers Magazine* in 1932. When Chant's book, *Our Wonderful Universe* came out in 1928, R.E. DeLury thought it should find a wide use in the teaching of elementary astronomy and wondered, "Is not the time ripe for introducing such a course into our high schools?" R.J. Nicholls was involved with the Berkeley Conference for Astronomy Teachers in 1955. Since 1973 however, John Percy has been pre-eminent in promoting astronomy education. Through surveys, committee work, writing, workshops, and through the feature of the RASC *Journal* called "Education Notes" his influence has been pervasive.

MEDIA

Newspapers, radio and television have all been important media for the RASC to get its message across. Society meetings were announced and detailed reports published in a number of Toronto papers until about 1907. This exposure not only kept the Society's activities before the public but also maintained a public interest in the astronomical and physical topics which were the subject of the meetings. Other Centres also got a boost from attention given by the local press. But in more recent times, the newspapers, particularly in the larger cities, have outgrown the reporting of local club meetings. Instead, regular astronomy columns have sometimes taken their place. In the long run these have probably done more for public education than reports of RASC meetings. Paul Sykes, long-time member of the Vancouver Centre, began writing "The Starry Heavens" for the Vancouver *Province* in 1932. After the first three of his monthly columns came out, the Centre executive invited him to attend his first RASC meeting and were startled to find he was just fourteen years old! A weekly astronomy column of very long standing has run in *The Toronto Star* since 1940. Originally written by Peter Millman, it was taken over by Frank Hogg in 1941 and then by his widow, Helen Hogg, following Frank's death in 1951. When Helen retired in 1981, *The Star* at first intended to terminate the column but reversed their decision after much protest by readers. Terence Dickinson has been *The Star*'s astronomy columnist ever since. Newspapers in Niagara Falls, Saint John's and Halifax have also run regular astronomy columns at various times in the past dozen years or so. Newspaper features like these have undoubtedly been read by far more people than the Society could hope to reach directly through most of its own activities. While the authors have generally been RASC members, the Society cannot really take any credit for these columns. (The same could be said about many popular astronomy books which have been written, for example Chant's classic, *Our Wonderful Universe*, Helen Hogg's, *The Stars Belong to Everyone*, and Terence Dickinson's *Nightwatch*, to name but three well-known examples.) However, the Society did have a direct hand in a very successful program which got astronomical items into the Southam newspapers in the period 1929–31. These fortnightly columns under the editorship of W.E. Harper, have already been alluded to in chapter five.

Probably the earliest radio broadcasts anywhere with an astronomical theme were talks by Dr Oscar Lee Dustheimer over WEAR, Cleveland, in 1925. The first Canadian broadcasts of this sort came a year later. CNRT, Toronto, had a series of scientific talks which included "The Stars of Winter" on February 19, 1926. Drs C.A. Chant and E.F. Burton were heard on January 28 and February 11, 1927 over this station. Also in 1926, Winnipeg amateur D.R.P. Coats pioneered by broadcasting the positions of stars which listeners could plot on graph paper. He explained the procedure in his Radio Column in the *Grain Growers' Guide*. Harper, who edited the astronomy items for the Southam newspapers, at the same time was engaged in a series of sixty-four radio talks over CFCT, Victoria, from 1928–33. Another important group of radio talks was P.M. Millman's series over CBC called "This Universe of Space." These were also published in booklet form in 1961 and over 4,000 copies were sold by the CBC at $1 each. In recent times, Gerry Knight of Vancouver gave astronomy talks every other Friday morning on CBC's "Early Edition" and Calgary Centre member Don Hladiuk has been giving a once-a-month segment on the current sky on the local CBC program, "Eye-Opener." Terence Dickinson is heard frequently on Jay Ingram's nationally broadcast "Quirks and Quarks" and many other RASC members give interviews and provide a Canadian or local slant to astronomy news. Again, the Society cannot claim to have sponsored or arranged for these programs but it does take pride in the fact that its members over the years have shown the initiative as individuals to spread interest in astronomy to such a wide public audience.

RASC members have also been active on the television scene. Hamilton's Ken Chilton pioneered with a cable TV program called "The Sky Tonight" in 1972. This was also picked up by four cable networks in Metro Toronto. London Centre's Peter Jedicke began cable TV broadcasts in 1976. In 1979 Quebec Centre made videotapes for cable TV broadcast in which they visited various amateur observatories in the province. St John's Centre aired an astronomy series in 1980–82. "Astronomy Toronto"'s first season in 1981 comprised eight cable television programs which were shown repeatedly throughout the year. Over the next nine years, sixty programs were produced by a crew of Toronto Centre members and hosted by Randy Attwood. These were rebroadcast by several cable networks across southern Ontario. Guy Westcott of Winnipeg worked on cable TV shows in 1986.

Whether the encounters have been through the media or of a closer kind, the RASC has been an active catalyst in bringing the Universe to the people.

PETER JEDICKE (1955-), like his brother Robert, also well-known in the RASC, was raised in Niagara Falls but went to university in London, Ontario. Though he got his degree in Physics, Astronomy was among the courses he studied and he joined the London Centre in 1974. He quickly became a key figure in the Centre. During his Presidency 1977-78, he gave enormous assistance, time and energy to Project Z, the Centre's highly successful summer astronomy program. He also "single-handedly organized the 1979 General Assembly, re- wrote the London Centre's Constitution and By-Laws and obtained the Centre's non-profit status." He then became the Centre's National Council Representative and began to write reviews and articles for the *National Newsletter* and *Journal* and to speak at General Assemblies. His paper on "Neutrinos, the Sun and Canada" won the Simon Newcomb Prize in 1987. But what everyone remembers, of course, are his hilarious songs and the inimitable way they were sung at GAs by Peter and his friend David Levy.

Peter Jedicke has made tremendous contributions to public education in his area. He hosted a TV show "Telescope" on the cable network and regularly promoted astronomy at the London Regional Children's Museum, where he helped to get a small inflatable planetarium up and running. He has visited many elementary schools to give slide shows and arrange observing sessions, and has organized local science fairs. For all these activities and more, he was presented with the Society's Service Award in 1984.

Since 1985, Peter Jedicke has taught at Fanshawe College of Applied Arts and Technology. He was married in 1987 and accepted a second term as Centre President.

Peter Jedicke (l) and David Levy (r) prepare for a cable TV show in 1979.

CHAPTER 9

The Scope of Observing

Amateur astronomers come in all varieties. Many look skyward only when prompted by some special event; some are happy to bring out their telescopes for a public star night but rarely use them otherwise. More dedicated observers, those whose activities are described in this chapter, may go on a real binge once in a while, perhaps while on holidays, and a few feel cheated if they must miss any clear night of opportunity, even if the temperature plummets to -32°C. The median number of days of observing per year is twenty, according to a recent RASC survey.

Professional astronomers can be broadly divided into those whose research is mainly theoretical and those for whom making observations is a significant part of their work. Some theoreticians may even profess a perverse pride in how little they know of the constellations. Even observational astronomers may be so reliant on instrumentation that they are less able to find their way around the night sky than some amateurs. But where professionals shine is in the interpretation of the observations. Armed with an extensive background in mathematics and physics and a thorough knowledge of the literature, professionals understand, in a way few amateurs ever could, just what must be done to make the observational bricks fit into a larger structure. Except for the *Journal* which few professionals use as a vehicle for publishing their research, and some instances of amateur-professional co-operation, the Society cannot claim to have much direct influence on the work of its professional members. Therefore no attempt will be made here to summarize the accomplishments, observational or otherwise, of the Society's professional members. In any case, the importance of their work to the advancement of science is incomparably greater than that of the amateur members.

If the Society has had little impact on professional observers, the same could be said about many amateurs. There are undoubtedly many amateur astronomers who enjoy observing without any help from, or even knowledge of, the RASC. Observing the night sky can be a solitary, even a very personal activity, or it can be enjoyed in the company of other enthusiasts who add to the pleasure by sharing experiences of familiar sights seen through different eyes or possibly seeing a difficult object for the first time. For certain activities, like meteor watching or timing grazing occultations, group planning and action is almost necessary.

As long as the Society has existed there have been outside organizations and people interested in the

Len Chester (r) watches while one happy little girl gets a safe look at the filtered sun through the Society's 10-cm Cooke refractor.

> **EARL R.V. MILTON (1936-)** joined Montreal Centre as a teenager where he was active in meteor observing, the AAVSO nova search program and the Asteroid Club for Juniors. When he moved with his family to Edmonton in 1952, he brought his enthusiasm with him and soon interested a number of members of that Centre in auroral, planetary and meteor observations. While studying Chemistry at the University of Alberta, he started and headed a new observers group, and reported regularly on their activities at Centre meetings. He went on to graduate studies but managed to find the time to speak at Centre meetings, to keep up his active observing and to serve as Vice-President and President of the Centre.
>
> Earl Milton was only 23 when he received the Chant Medal for his extensive observational work, particularly in auroral and meteor programs. Subsequently he chaired the national Committee for Co-operation between Observing Centres and was the Aurora Co-ordinator for the National Committee on Observational Activities. He played a very important part in gathering eye-witness reports of the Bruderheim fireball of March 4, 1960, reports which narrowed down the search area and led to successful recovery of many fragments of the meteorite.
>
> After obtaining his PhD in Chemistry, he worked in Physics at the National Research Council in Ottawa and at the Regina campus of the University of Saskatchewan before accepting a post as Associate Professor of Physics at the newly-opened University of Lethbridge. He continued to serve the Society on the National Council, and as Aurora Co-ordinator, and spoke occasionally to meetings of the Edmonton and Calgary Centres.

The aurora at Fort Churchill, Manitoba, May, 1972

Photo by W. Morrow

co-ordination of observations among groups and individuals. In the 1890s, members sent auroral observations to W.A. Veeder in Lyons, NY, sunspot observations to the Institucion Solar in Montevideo, Uruguay, and corresponded with the BAA on lunar observations.

In the Victorian period, members had role models in some famous contemporary amateur astronomers. Wealthy gentlemen like Wilhelm Beer in Germany, the Earl of Rosse in Ireland and William Huggins in England were at the forefront of science. But as astronomy developed more and more into a profession, the hope of someone without an advanced education doing something "important" seemed to fade. In Canada, this transition was signalled by the opening of the DO and the appearance of professional papers in the *Journal*. Even after the AAVSO formed in the United States in 1911, RASC members were slow to participate in their program. It was not until Peter Millman organized meteor observations in the 1930s, that significant numbers began to participate in systematic programs. Later still, a few members with a special interest in planets, double stars, noctilucent clouds, and so on, were attracted to specialized groups which had formed in the United States, Britain and elsewhere. So the Society itself had no need to get involved in the co-ordination or collection of observations. Nevertheless, some members saw merit in co-operation among Centres which were actively engaged in observing.

In 1959 a national committee was set up to investigate these concerns and its members, Isabel Williamson, Vern Ramsay and Chairman Earl Milton identified several problems – lack of communication between observers, poor cohesion between observers and the rest of the membership, little information on observing programs, trouble getting beginners into observing groups and difficulty in interesting more than a few in active participation. The committee recommended that Centres appoint local recorders in each field of observation undertaken, and that national recorders in each of the fields be selected from these local people. Instructions and reporting forms, where needed, would be distributed to Centres from the national office, though no standard methods would be enforced. An additional RASC publication was suggested, less formal than the *Journal*, where amateurs could publish articles on the construction and use of observatories, telescopes or other equipment.

This latter recommendation, as we have seen, was not accepted at the time though in a sense it came to pass a decade later with the birth of the *National Newsletter*. On the other hand, the system of discipline co-ordinators was set up in 1961, and some bulletins were issued from the national office. In the long run, enthusiasm faded as the few members who pursued structured observing programs preferred to report to the outside organizations in their areas of interest. The Mariner space probes to Venus and Mars in 1962 and 1965 and the Apollo XI mission which landed Armstrong and Aldrin on the Moon in 1969 were so spectacular that the great majority of members who cared about planetary and lunar observing settled for recreation and public star nights rather than keeping detailed records of their work. By 1969, the Committees on Observational Activities and on Educational Activities were merged into one new Committee on the Co-ordination of Centre Activities (COCOCA). Once the *National Newsletter* was well-established, the Committee was dissolved with the intention that the *Newsletter* would take over COCOCA's role and become the effective medium for gathering and disseminating information on Centre activities in public education and observation.

To introduce newcomers to various fields of observing, at least three Centres, Ottawa, Montreal and Winnipeg, put together booklets in the early 1980s and the national Society launched its *Beginner's Observing Guide* in 1991.

STAR PARTIES

Star parties can mean gatherings arranged for the public when members of the Society point their telescopes at a few well-known sights and let their visitors enjoy the view. Other star parties are arranged for amateur astronomers themselves. The most famous of these, and the original one, is Stellafane, held each August near Springfield, Vermont, most years since 1922. Basically it is a telescope-makers convention with the judging of home-built scopes of all kinds as its centrepiece, but observing and invited talks are an important part of the program too. RASC members from Montreal began to attend in substantial numbers after World War II when it was not unusual for several carloads to drive down for the weekend. The Ottawa Centre also often sent a good-sized delegation, especially in the 1970s when their Observers Group was very strong. Tom Tothill and Fred Lossing won awards for their telescopes in 1970. In 1974, Ottawa's Dave Penchuk and Toronto Centre member, Moody Kalbfleisch, won first prizes. The year of 1978 saw an especially strong contingent of about seventy Canadians from Halifax to Kitchener. Jack Newton had organized the afternoon tent-talks that year. Another big American event with a more commercial flavour is the Riverside Telescope Makers Conference. It is held near Big Bear City, California, every Memorial Day weekend and draws thousands including a good number of RASC members.

Observing is the main focus of several annual Canadian regional star parties, each extending over three or four days. Some of these are sponsored by groups which are not connected with the RASC, though many members do support and participate in these events. Starfest was organized for August, 1982 by the North York Astronomical Association at River Place campground, north of Mount Forest, Ontario. It has been held annually in August since then and is currently the largest astronomy meeting, amateur or professional, in Canada. The Okanagan Astronomical

Very little is known about the Society's eighth President, **W. BALFOUR MUSSON**. He may have been a son of George Musson, a tea broker in Toronto since at least 1872. In any event, George Musson's address was 1 North Street, the same as W.B. Musson's when the latter name first appeared in the city directory in 1884. This address was very near Andrew Elvins', in the vicinity of what is now Bay and Grosvenor Streets, and in fact an 1873 address for George Musson was just across the street from Elvins. So it would not be hard to imagine that Elvins brought young Balfour Musson into the Society. One can trace W.B. Musson's career through the directories. He was at first a clerk with a firm of wholesale grocers, later an agent with the West Shore Railroad and the New York Central. From 1941 to '47 he was listed as a manufacturer's agent, after which his name is no longer found in the Toronto directories.

W.B. Musson was a member of the Society since at least 1892. Fifteen contributions appeared in the *Publications* and *Journal*, including three on variable stars and two on stellar classification. He successively held the offices of Librarian, Secretary, Vice-President and President during the years 1896-1909. His last known contact with the Society was in 1930 when he wrote an obituary of J.A. Paterson. His name is not in the members list for 1931.

Except for a formal photo published in the *JRASC* **11**, 62, the only known picture of Musson is the one shown.

A Victorian star party about 1900 on the lawn of D.J. Howell's home in Lambton Mills, just west of Toronto. Shown (from left to right) are Z.M. Collins (with hat on), an unidentified visitor, A. Horton, A. Elvins (in front), W.B. Musson, a lady visitor, Miss M.A. Howell, Mrs Musson, Mrs Webber, J. Webber, Miss McEachern (in white), J.E. Maybee, J.A. Paterson, D.J. Howell (side view), Miss Eva C. Howell, G.G. Pursey (blurred).

Society held the first of its star parties on Mount Kobau in 1984. The site, at an elevation of 1,861 metres, had been selected some twenty years earlier by professional astronomers as the best in Canada for a proposed new 3.8-m telescope that was never built. Travellers up the difficult road to the summit are rewarded with a 90 km vista on a clear day and a multibillion light-year view on a clear night. The event has become an annual four-day camping affair in late August, with many members from the western Centres participating. Some Calgary members, however, felt the drive was too far, and so in 1987 they started the Alberta Star Party at Chain Lakes Provincial Park, 120 km south of Calgary. About 140 attended the first of these annual events on July 24–26 that year. The Alberta Star Party is now held in parks throughout Alberta and is jointly organized by the Calgary and Edmonton Centres and by the Lethbridge Astronomical Society. Winnipeg Centre organized the Manitoba Star Convention (MASCON) at Riding Mountain National Park in 1981, with the idea of providing some viewing for the general public as well as observing and talks for registrants. This plan turned out very well with an estimated 4,000 visitors dropping by, mainly for solar viewing, over a five day period. The Halifax Centre adopted a similar arrangement for its Nova East event which was first held at Fundy National Park in 1987. Typically about thirty members and their families stay for the entire weekend acting as hosts for several hundred interested park visitors on the one evening devoted to public viewing.

Opportunities like these regional star parties are becoming increasingly necessary as year by year the night sky is getting less accessible to everyone. More and more people are living in cities whose light glow spills out into the atmosphere like a giant and permanent mushroom cloud extending far into the country. Looking for a faint nebula or comet in such skies is about as futile as trying to hear a whisper at a rock concert. True, there are nebular filters which are designed to blot out the "noise" and allow only the frequencies of the "whisper" to pass, but the results cannot compare with the real thing. The only alternatives involve travelling far from urban areas, or observing only the brighter objects like the planets, or simply forgetting the sky and becoming an armchair astronomer. Of course it must be recognized that the very urbanization which has spawned the light pollution has fostered the development of the RASC itself, as well as various commercial, technical and educational establishments which have directly or indirectly increased our ability to appreciate and understand the universe. It's just too bad that few lighting engineers yet realize that the outdoor lighting we all need can be better designed so that more goes down where it is needed rather than up where it only spoils the view of the night sky.

POLLUTION AND THE ATMOSPHERE

There is evidence that light pollution was becoming noticeable even in the late 1800s. An illustration of this is the zodiacal light, caused by sunlight scattered from microscopic particles in the plane of the planets, and seen only from dark sites. Andrew Elvins remarked in 1898 that "he had never successfully observed the zodiacal light in Toronto since the electric system of lighting had been established." Others had been more fortunate. George Lumsden saw it one clear evening in the spring of 1898 from the northeast part of the city (presumably near his home in the Bloor and Sherbourne area). Nonetheless members were definitely aware by this time that conditions were much better in the country. Of course "the country" was not that far away. G.G. Pursey paid a visit to Markham (about 27 km from City Hall) and was very much impressed with the "vast difference between the best of the atmosphere around Toronto and what is enjoyed in the country". And even in March, 1912, from a site just 18 kilometres west of downtown Toronto, C.A. Chant found the zodiacal light "decidedly brighter than the Milky

HENRI SIMARD (1922-73) grew up in Matane, Quebec, served as a parachutist in World War II, and then worked as a forester, eventually becoming director of measurement for the International Paper Company (CIP) in Montreal. It was a position which he earned through much hard work and many extra courses in mathematics, statistics and computer programming. In the course of his business trips around the province, he often spoke to local astronomy groups and encouraged many to affiliate with the Société d'Astronomie de Montréal and to join the RASC.

Originally he was a member of Quebec Centre, but his work brought him to Montreal where he joined the Centre français in 1960. He served as their President in 1963-65, edited the *Annuaire* (a French language version of the Handbook), contributed regularly to the monthly *Bulletin d'Astronomie* and received the RASC Service Award in 1967. Simard's diplomacy and stature within the Centre was important in strengthening the Society's role among francophone members in Quebec. He was elected national Second Vice-President in 1966 which led to a term as President in 1970-72, following which he was re-elected President of the Centre français.

Henri Simard was a highly skilled telescope maker. He designed and made his own unique mirror-tester and one of his products, a 15-cm reflector with tube made of wood, won him first prizes at Stellafane in 1967 in the Newtonian category and for its excellent optics. He shared his talents with others as Director of the Optical Section of the Centre, a group which he established.

His daughter Martine (Normandin) pursued astronomical studies to the post-graduate level, but followed a career at Northern Telecom in Ottawa. She has kindly supplied this photo of her father with one of his prize-winning telescopes at Stellafane.

Way, but not nearly as bright as the illumination from the city lights to the east." The thought of such observations being made at locations which are now part of the suburban sprawl is both lamentable and laughable.

Certainly the clock cannot be turned back. No one wants to live without electricity, but since the 1970s astronomers have become more active in suggesting ways to use outdoor lighting more carefully. Shielded fixtures which direct the light down and efficient low-pressure sodium lamps which confine their emission to a narrow region of the spectrum are two ways of cutting costs and at the same time helping to keep the sky dark.

The experience of the David Dunlap Observatory was typical of many suburban locations. When the DDO opened in 1935 about two kilometres south of the little town of Richmond Hill, Ontario, farmers' fields extended in all directions. By 1970 the sky over the Observatory was 1,000 times brighter than the natural, unpolluted sky and within the next twelve years shopping centres, fast food restaurants and housing developments in the vicinity caused an increase by another factor of ten. The DDO astronomers did persuade municipal authorities to adopt standards, and consequently the Observatory is still carrying on spectroscopic work with the 1.88-m telescope, disabled but not destroyed by the sea of development around it.

Within the RASC, two young Toronto Centre members, Robert Pike and Richard Berry headed a very informative study of sky brightness in Southern Ontario in 1975 in which they presented data indicating the rapid degradation of night sky conditions. The Calgary Centre's Wilson Coulee Observatory is another prime example of a fine facility that has felt the threat of urban expansion. The Centre has won the co-operation of local authorities and is now bringing its valuable experience to a national RASC committee on light pollution. The national Society has also made contact with Ontario Hydro who have promised their assistance in making municipal utilities aware of the issues, and with the Ontario Ministry of the Environment who sought public input on environmental concerns. This is definitely an area where the Society hopes the future does not look bright.

SUN, MOON AND PLANETS

Fortunately, the urban observer can still see the Sun, Moon and planets, even from downtown. As long as the Society has been in existence, members have observed and photographed the Sun, and drawn and plotted sunspots from its projected image. Andrew Elvins took great pleasure in the almost exact similarity between his drawings of sunspots and those reported by the BAA in 1894. Pursey carefully recorded sunspots and faculae every clear day, at least from 1896, and by 1901, he had accumulated 1,100 drawings. Some more extraordinary accomplishments from the turn of the century include A.F. Miller's spectroscopic observations of sunspots and of prominences and T.S. Shearmen's observations of the corona (not during a total eclipse). Some of his observations won the approbation of no less an authority than G.E. Hale. In more recent years, one could mention Ottawa member Steve Craig who made hundreds of drawings of the Sun in the 1960s, twenty-year-old Robert Pike who routinely observed the Sun with his homemade 15-cm reflector and was lucky to see a white-light flare on July 4, 1974, and another member, John Hicks, who photographed the Sun using a Hydrogen Alpha filter on a regular basis in the 1980s. Many others monitor the Sun daily.

The Moon used to be a favorite object of study, and many members enjoyed sketching various features on its surface under varying angles of illumination, and also photographing it, but except for public star nights, hardly anyone gives it any attention now. Lunar observing pretty much died out once the Moon was generally accepted as "dead" (or

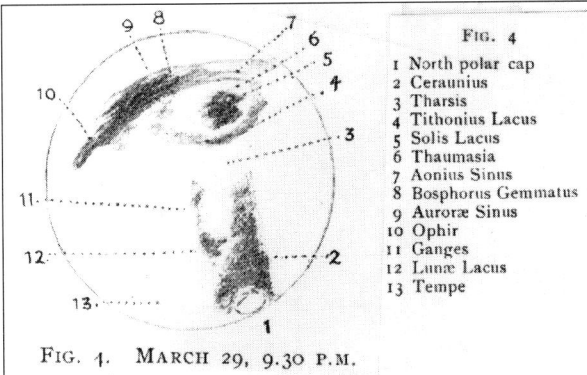

A.F. Miller made the sketch of Mars shown above in 1918 using his four-inch refractor at 300 power. From many sketches made as the planet rotates, maps of Mars like those at the bottom, can be compiled. The upper one by B Clark of Oshawa was done in 1958, the lower one by Russ Sampson of Edmonton in 1988.

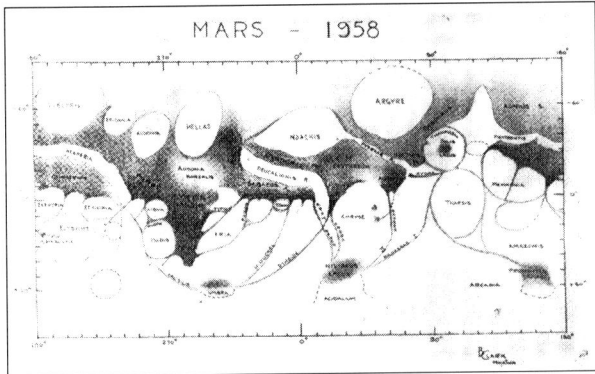

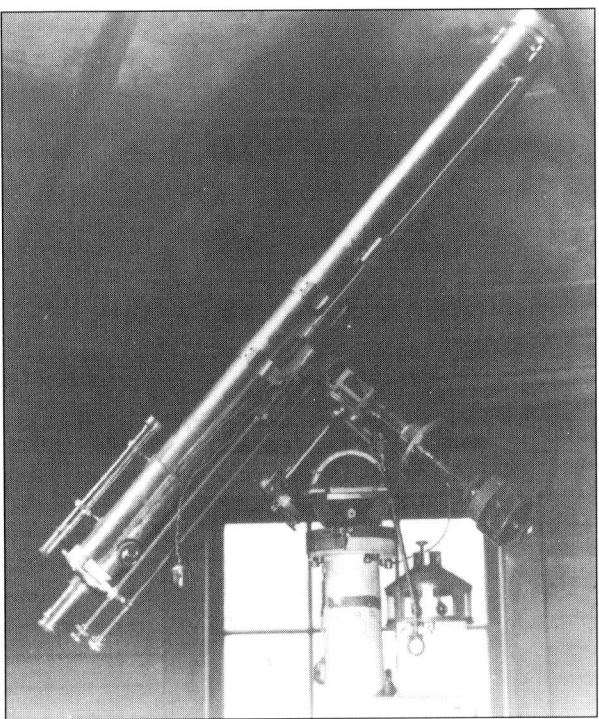

A.F. Miller's 10-cm Wray refractor in its Toronto observatory. (Photo copied for RASC courtesy DDO.)
For photographs of Miller, see page 21 and 22.

ALLAN F. MILLER (1851-1947) was born in Dublin, Ireland, the son of a sea captain. The brilliant stars of the Southern Cross, seen from the deck of his father's ship, formed a lasting impression. After his father retired, the family moved to Canada. Allan attended school in Toronto, graduated with high honours, and embarked on a career in business. In 1876 he joined the Toronto General Hospital as Assistant Secretary and worked there for 43 years, eventually becoming Secretary-Treasurer. His administrative abilities were highly regarded by the Hospital. He was proud of the fact that one of his children, Frederick, became a noted doctor at the University of Western Ontario.

In 1881, Allan Miller met Elvins and joined his coterie of astronomical enthusiasts. The following year, in time for the transit of Venus, Miller acquired a fine four-inch refractor made by Wray. This he mounted, with clock drive, in an observatory at his home at 280 Carlton Street, and equipped it with a Browning spectroscope, a filar micrometer and other accessories. (The telescope is now owned by the Victoria Centre.) Miller did some remarkable, pioneering work with this equipment. He made spectroscopic observations of Nova Aurigae (1891), photographed the solar spectrum and, in 1892, was probably the first person in Canada to study solar prominences. (In this endeavour he had the advice of G.E. Hale with whom he corresponded.) Miller also investigated the spectrum of light emitted by fireflies. His astrophysical knowledge was so impressive that he was appointed a member of the National Committee for Canada for the IAU, a body normally composed of professional astronomers. Miller spoke at meetings on many occasions from the earliest days and wrote 28 papers for the Society's publications until his eyesight deteriorated very badly. He served the Society as Librarian, Vice-President and as President in 1918-19.

J.J. Wadsworth's observatory on Talbot Street, Simcoe, Ontario.

The Moon photographed by Wadsworth in 1896.

A watercolour sketch made through Wadsworth's 12-inch telescope by Eva Brook.

nearly so), and eventually the space program provided such detailed and magnificent photos of all parts of the Moon's surface, there was no longer much point in making earth-based observations.

Studying the Moon was once so popular with members that a separate "Lunar Section" was formed in 1895. Sketches in pencil, pen and ink, crayon and watercolours were made of various features and photographs were taken of the lunar disc. Elvins, Lumsden and J.J. Wadsworth, a public school inspector in Simcoe, Ontario, were singled out by President Paterson for their systematic mapping of portions of the Moon's surface. Our Lunar Section was short lived, but its counterpart in the BAA lives on though for a time during World War II when it looked as if the BAA might have to suspend publications, the RASC *Journal* contained a couple of papers on lunar craters which might normally have appeared in the BAA *Memoirs*. (The same could be said for some papers by American planetary observers Walter Haas and Hugh Johnson from this period.) Following the setting up of RASC National Observing Committees in 1961, lunar group, under Co-ordinator Jim Low, compiled a catalogue of previously unrecorded lunar domes and confirmed or refuted those previously reported.

Responses to a recent RASC survey show that planetary viewing is by far the most popular observing interest of members at the present time. Observers are attracted to the planets because of changes that can be seen on them and because they are visible even under rather poor sky conditions. Most amateur astronomers are content to turn their sights on the planets on an occasional basis, noting the most obvious features, like the phases of the inner planets, the dark markings on Mars or the cloud belts on Jupiter or Saturn. A few "hard-core" observers, like Robert Loblaw of Calgary, report on Martian dust-storms or changes in the cloud patterns on Jupiter and Saturn to ALPO, the American-based Association of Lunar and Planetary Observers.

CHARLES M. GOOD (1904-80) received the Service Award in 1960 having been largely responsible for the establishment of Montreal Centre's Observatory, and for very ably directing their public relations activities, including the highly successful star nights in Westmount Park. In many respects his career with the Montreal Centre followed the same course as Frank DeKinder's. He first took an interest in meteors in the '40s and continued to observe meteor showers with other Montreal members at least until 1956. Though he contributed to practically every observational activity, his special interest was in timing occultations. He chaired an occultation section of the Centre starting in 1955, and the Montreal group became renowned for the extent of its contribution in this field. In the years 1967-69 when he was Chairman of the Lunar Occultation Section of the AAVSO, Montreal members contributed nearly half of the world-wide total reported to that organization. Charles Good began serving on the AAVSO Council in 1958, eventually becoming their President in 1971.

When Isabel Williamson wrote a tribute to Charles Good in *Skyward*, she noted that, "He served the Centre in many capacities, including a term as President from 1954 to 1957 and again from 1965 to 1966, but the position he cherished most was that of Librarian, a post he held from 1945 to 1965, with a brief break while President. He knew his library well, for he had read every book in it, and he used wisely the limited funds at his disposal to expand it."

As a permanent mark of their esteem, Montreal Centre established the Charles M. Good Award in 1981 which is presented to members for contributions to the Centre.

Charles Good observing sunspots by projection.

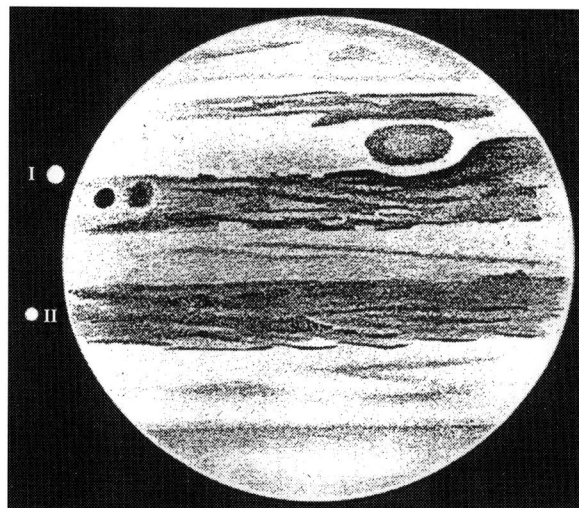

A drawing of Jupiter by G.E. Lumsden, 20 September, 1891, showing a double shadow. The sharp shadow is from Satellite I. The other may be from Satellite IV which was much further to the left at the time.

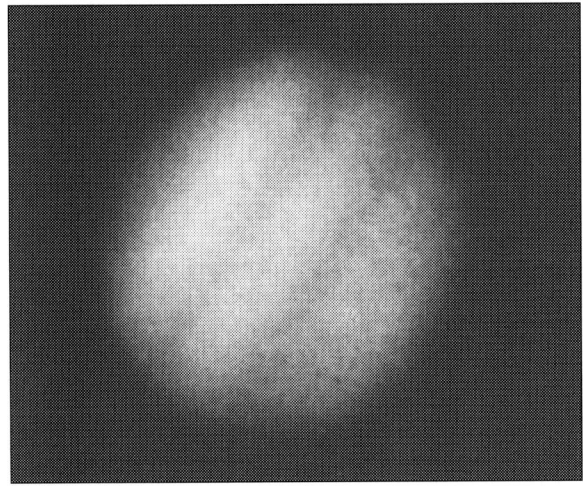

A photograph of Jupiter taken in 1981. The lack of detail is typical of what can be captured on film during a photographic exposure in comparison to what can be glimpsed visually during moments of steady seeing. Technology is just now reaching the point where electronic images made with CCDs can rival the drawings of a century ago.

The phases of the inner planets can provide some interest but looking for surface features is far more difficult. Before the space age there was more interest in seeking elusive surface details on the inner planets. One RASC member who systematically made drawings of Venus in the 1940s was Francis Morgan of Montreal.

The phases of Venus are of some interest because the times predicted geometrically for a sphere do not agree with those observed for the cloud-shrouded planet. The variation of brightness with phase is also not perfectly understood. A flurry of reports of the phases being seen with the naked eye were published in 1929–31, and in 1935. Four members of the Hamilton Centre made telescopic drawings and polarization measurements in 1968–9. Marc Gelinas of the Centre français de Montréal, appealed in 1984 for observed times of the dichotomy of Venus. He received ten replies and found the time to be 6.5 ± 3.4 days prior to the predicted time. A repeat in 1985 gave 2.17 ± 0.25 days prior to prediction. Following the thin crescent of Venus from the evening to the morning sky is a challenging project too.

Transits of Mercury across the Sun always arouse curiosity because they occur infrequently and because they were once important, along with the even rarer transits of Venus, in determining the scale of the Solar System. The transit of Mercury on November 10, 1894, gave members of the Society a wonderful opportunity to show and explain the phenomenon to schoolchildren and the public. J.R. Connon of Elora photographed the event. The May 7, 1924, transit was timed by Albert Hassard and A.F. Miller and observed by London Centre members, the one on November 11, 1940, by some Victoria Centre amateurs, and the 1953 event was timed by Torontonians J. Ketchum and Hassard. The transit of November, 1973, occurred at sunrise for Toronto observers, but eight carloads of eager members left the city at 5:30 a.m. for the Scarborough Bluffs to view it.

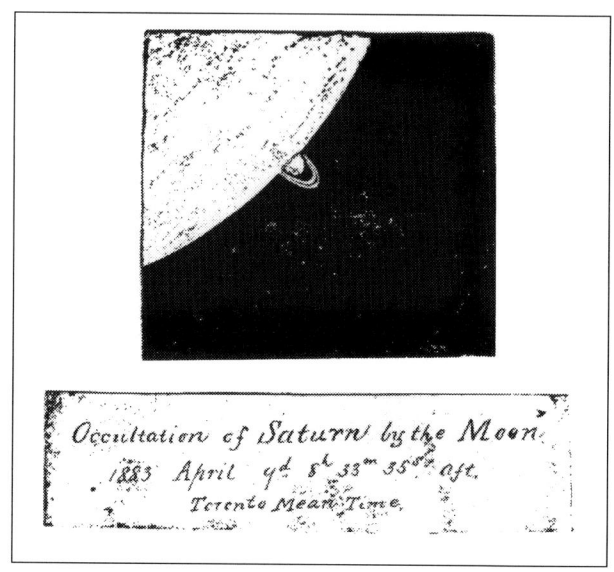

Mars is attractive because of its growing and shrinking icecaps, the dust storms and the changes in its coloration. It is only near times of opposition (about every two years) that any surface detail can be clearly seen in moderate sized telescopes. The planet probably attracts less attention now than it did in the 1920s when debates circulated in the daily papers about life on Mars, but it still is a popular object of study. Miller was a diligent observer of the "red planet" and even made a globe showing its surface features in 1919.

The two giant planets, Jupiter and Saturn, are observed carefully because of their changing cloud patterns and quasi-stationary atmospheric features. Their satellites too are quite intriguing as they move around, across and behind the disc of their parent planets. Dr J.C. Donaldson, a member in Fergus, Ontario, was particularly cited in 1896 for his "enviable reputation as a successful observer of planetary detail." The satellites of Jupiter were of intense interest to the early members, but their enthusiasm was sometimes a bit excessive. George Lumsden and Donaldson both pointed out what they thought were peculiarities in Satellite IV. Elvins staunchly maintained that he had "predicted" the fifth satellite of Jupiter which was subsequently discovered by Barnard in 1892. W.E.W. Jackson also got caught up in the excitement and thought he had seen six satellites. In fact two of the objects were stars.

Though several members, including Elvins, regularly made sketches of Saturn, there was probably never a better drawing of the ringed planet made by any Society member than that by Lumsden showing seven satellites on May 4, 1891. He used a 26-cm reflector, 208 power and classified the seeing as superb. He was also able to locate two satellites of Uranus. Miller made an interesting observation of a bright spot on the outer ring of Saturn on May 13, 1919 which was corroborated by similar descriptions in The English Mechanic. Uranus was the subject of an extensive series of magnitude estimates by Montreal Centre members in the 1950s. Neptune and Pluto don't get much scrutiny and in fact only 13 percent of RASC members surveyed in 1984 had ever seen little Pluto.

Observing of all kinds got a big boost in the early 1960s when the Society organized the activity on a national basis. Nowhere was this success more evident than in the Planetary Section under Geoffery Gaherty Jr, of Montreal. In 1961, members made 182 drawings of Jupiter's disk, recorded 1,119 transit times of various features across the planet's central meridian, and made 35 timings of satellite phenomena.

COMETS AND ASTEROIDS

Few comets were mentioned by RASC members until the national program in the 1960s. Among those earlier sightings was Rordame's Comet of 1893. This object was very bright, even at discovery, and many people, including a farmer near Elora, might have claimed priority in seeing it first. It was sketched by Elvins and photographed by Connon of Elora, and Miller observed the comet spectroscopically as well as seeing its coma (or outer envelope) occult a star.

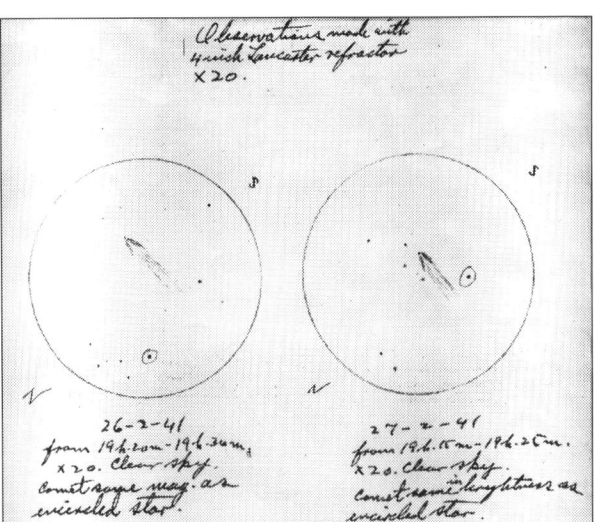

Two sketches of Comet 1941c made by Delisle Garneau

Comet Perrine, in 1902, was studied spectroscopically by Miller and observed by others. Of course, Comet 1910a, and Halley's Comet of the same year attracted widespread interest. Especially remarkable were observations from Victoria and from Birson, Saskatchewan, when Halley's Comet passed directly between the Earth and the Sun, and F. Slocum's subsequent paper on the mother-of-pearl clouds seen on the same occasion from Yerkes Observatory in Wisconsin. Photographs were taken and published of Pons-Winnecke in 1927, a two hour exposure for the picture by R.M. Motherwell at the DO serving to remind us that plates were still very slow. A bright comet in 1939 was independently discovered by a resident of Sedgewick, Alberta, Lewis V. Smith. He received the Donohoe medal from the Astronomical Society of the Pacific for his accomplishment, though he unfortunately missed out on having his name attached to the comet because of a delay in reporting it.

But again, it was the National Co-ordinating Committee which sparked an increase in observing in the 1960s. Jim Low of Montreal, as Chairman of the Comet and Nova Section, reported about 200

Comet Levy, one of the best of the century, photographed by Damien Lemay 20/21 August, 1990. An extended, diffuse object like a comet is much more attractive in a photograph since its dim light can be integrated over the duration of the exposure.

ROLF MEIER (1953-) grew up in Ottawa and received his Bachelor of Engineering degree in Electronics from Carleton University in 1977. Even at age 25, when he was nominated for the Chant medal, he had been a skillful telescope maker and observer for a long time. On nearly every clear night he could be found at Ottawa Centre's Indian River Observatory, coming early and staying late. His astrophotos were greatly admired by all who saw them but his discovery of a comet on April 26, 1978, really made headlines as it was the first telescopic discovery of a comet from a Canadian location. He discovered further comets in 1979, '80, and with the discovery of his fourth comet in '84, he became North America's greatest living comet hunter, a distinction he held until surpassed by David Levy in 1989.

In 1984, he married Linda McCrae, herself an outstanding Ottawa member. That year she won three Centre awards - for variable star observations, for the best article in *Astronotes* and as Observer of the Year. She has since served the Centre as Treasurer. Rolf has held a number of Centre Offices, including Chairman of the Observers' Group, Editor of *Astronotes* (for 12 years), National Council Rep and President. His 25 cm Newtonian reflector was installed at the Indian River Observatory in 1983 for use of members. A number of articles, displaying his wide-ranging interests, have included "Identifying a Meteorite", "Recent Aurora Activity", and "Solid State Photometer."

Rolf Meier with the 40 cm Ottawa Centre telescope.

comet observations in 1962, mainly of Comets Seki-Lines and Humason, as well as nearly 2,000 searches of sixty-seven areas of the sky, each ten degrees square. None of these searches was productive, however. Success came in a big way in 1978, when Rolf Meier of Ottawa made the first of a series of remarkable comet discoveries with the Ottawa Centre's 41-cm telescope. Normally, an experienced observer might hunt for hundreds of hours before making a discovery. Meier searched for only fifty hours before spotting his first comet. He went on to discover three more comets in the next six years, and his fellow member, Doug George, found one of his own in 1989 using the same telescope. David Levy, after years of fruitless searching while still living in Canada, found seven comets in the dark skies of Arizona between 1984 and 1991. Canadian professional astronomers, Sidney van den Bergh (in 1974) and Christine Wilson (in 1986) have also added to the list of comet discoveries, and Levy himself, working as a professional observer at Palomar Observatory, found several more comets photographically.

Asteroids, those city-sized chunks of rock sometimes glorified with the name of minor planets, were also slow to interest Canadian observers. Doug Welch, while still a young member in Ottawa, took an interest in finding asteroid positions by reducing his own observations in the mid 1970s. Quebec Centre members watched Eros occult Kappa Geminorum on January 23, 1975. Some Edmonton and Calgary Centre members observed the asteroids Kleopatra and Wratislava occult faint stars in 1980 and 1983. Kitchener member, Clifford Cunningham, did a considerable amount of photometric work, and Chris Spratt of Victoria wrote a series of interesting papers on asteroids for the *Journal*. Thanks to his initiative, Asteroid 4113 was named RASCANA by the IAU in honour of the Society's centenary.

METEORS, METEORITES, ARTIFICIAL SATELLITES, AURORAS AND ATMOSPHERIC EFFECTS

Even though comets and asteroids were of little concern to RASC observers for most of the Society's history, the debris from these bodies swept into the Earth's atmosphere as meteors and occasionally onto the Earth's surface as meteorites, has always been of intense interest. The obvious reason for this fascination is that no equipment is needed. Meteors can be plotted, counted, or just enjoyed by anyone lying out under a dark sky. The brightest of these shooting stars, the fireballs, are of course so spectacular that reports of sightings come from hundreds of miles around, leading in some cases to recovery of a meteorite that has survived its passage through the atmosphere to land on the ground.

On a number of occasions, professional astronomers have given encouragement to members of the Society to make careful meteor observations. Sir Robert Ball in 1891 and W.F. Denning in 1900 were the first to do so. Several members, including Annie Gray and Sarah Taylor, observed the Leonid meteor shower in November, 1892, and Miss McBain of Port Dover did likewise in 1898, but there were no reports for 1899, the year the Leonids were expected to peak. Interestingly, though, an observer at York Factory sent observations of shooting stars seen on November 15 and 16, 1900, some even seen in daylight. In 1915 the National Academy of Sciences in the States awarded the Leander McCormick Observatory in Virginia a small grant for the purpose of encouraging meteor research. As a result, the Director of the Observatory, Canadian-born S.A. Mitchell, wrote "Observations of Meteors Needed" for the *Journal*. The same year, Dr R.K. Young prepared a set of rules and star maps for the *Journal* and these were made available to the Leander McCormick Observatory for free distribution. It is only on the gnomonic projection which Young used for the maps that meteor paths can be plotted as straight lines. His charts were still in use by Society members in 1931. W.E.W. Jackson, in his presidential address to the Society in 1923 said he would like to see many more members making meteor observations, one reason being that the motion of the radiant points for the showers was in doubt.

No one did more for visual meteor observing than Dr P.M. Millman. The Leonid shower was again expected to peak in 1933, and he began that year to organize teams of observers. As it turned out there were fewer Leonids than in 1931 or 1932 but Millman's enthusiasm spread quickly as is clear from reports of the Perseid shower in August, 1934. As a result of his writing to Centres and individuals with offers of charts to those who would use them, 45 groups comprising 132 visual observers, mainly in Ontario, recorded 6,280 meteors. Among the observers it is wonderful to see familiar names of professionals and amateurs side by side, and to see that a number of the participants were students who later became professional astronomers. The Millman family observed from their cottage at Port Sandfield, Malcolm Thomson and Miriam Burland in Ottawa, Donald MacRae at Kincardine, Ruth Northcott at Windsor, William Findlay and T.H. Wingham at Hamilton, F.J. DeKinder at Montreal, D.R.P. Coats at Winnipeg, E. Russell Paterson with a group of Boy Scouts at Camp Tamaraconta, and of course Peter Millman himself at the DDO along with a group of observers, mostly women, including future gold medallist, Shirley Patterson. Photographic and spectrographic equipment were soon added to Millman's program. The Montreal Centre and the Regina Astronomical Society had active groups of meteor observers following the War, and naturally when Millman moved to Ottawa, that Centre also became known for its contribution in the field. The Giacobinid meteors provided quite a show in 1946, especially for Quebec Centre members who counted

8,000 in a two-and-a half-hour period on October 9. The International Geophysical Year which really ran for eighteen months in 1957–58 was another incentive to observe meteors, since they could yield information on the variation in density of the Earth's upper atmosphere. This data, of course, became vitally important in the early stages of the space age. However, radar techniques, unimpeded by weather conditions and daylight, began to surpass visual methods for studying meteors, and the last of Millman's Meteor Reports in the *Journal* appeared in 1968. Nonetheless, serious meteor observing still attracts a few devotees, while watching shooting stars just for fun is as enjoyable a group activity as ever.

Another important amateur activity which grew out of the IGY was the Moonwatch program. Networks of observers worldwide were pressed into service observing the first artificial satellites to give a firm basis for studies of their motion, particularly the effects of atmospheric drag. The Soviets gave no information about their Sputniks to the western world so almost everything had to be found out from observations, and the "Moonwatchers" were very important in this respect. Enthusiasm for satellite observing waned after the Moonwatch program came to an end about 1960, though some Montreal

An informal set-up of eight meteor observers surrounding a central recorder whose duty it is to note the time and to record the observations called out by the others. In more sophisticated arrangements, enclosed "coffins" keep the observers warm and free from dew.

Photo originally published in JRASC 57, 139.

members took an interest in the decaying orbits of the giant reflecting satellites, Echo I and Echo II in 1968–69 and Ted Molczan, a Toronto Centre member stimulated considerable interest in the late 1980s with his useful observations of satellites on the verge of re-entry into the Earth's atmosphere.

Fireballs present a unique opportunity to combine celestial observation with a terrestrial discovery. Millman was a strong influence in this area too as he devised standard forms to be used in collecting fireball information. On a few occasions, especially when there were sightings from a number of different directions, astronomers were able to predict the area of impact and actually recover a meteorite or its fragments. It is especially useful for scientists to get a "fresh" visitor from interplanetary space, one that has not had a chance to get contaminated with Earthly impurities. The key to this is assembling good visual data on which to base the calculation of the path before impact. The earliest instance of Society involvement in an event of this sort was in connection with the meteorite which fell near Shelburne, Ontario, on August 13, 1904. Two large pieces, the largest weighing about 13 kg, were displayed at the RASC meeting on October 4, 1904. Though the larger fragment ended up at Queen's University in Kingston, a plaster cast of it was presented to the Society in recognition of assistance rendered. There have been several instances since where amateur members of the Society have helped the professionals in gathering information on fireball sightings and in the actual search and retrieval expeditions. One of the most spectacular was the Bruderheim meteorite which seared its way through the atmosphere just north-east of Edmonton on 4 March, 1960. The staff of the new Queen Elizabeth Planetarium, the University of Alberta's Geology and Physics Departments, the National Research Council and the Edmonton Centre all co-operated in the recovery of more than 300 kg of blackened chondrites.

IAN HALLIDAY (1928-) was born in Lloydminster, Saskatchewan but took his BA from the University of Toronto in 1949, winning the RASC Gold Medal. Except for a year at the University of California in Berkeley, he continued at U of T for his MA and PhD degrees. Halliday joined the Positional Astronomy Division of the DO in 1952 but, after receiving his doctorate, transferred to the Stellar Physics Division, specializing in meteor spectroscopy. When the DO closed in 1970, he went to NRC as a Senior Research Officer in the Planetary Sciences Section. During his very productive career, Halliday not only continued his work on meteors and comets but investigated the diameter of Pluto, studied craters for meteoritic evidence and set up a twelve-station meteorite recovery project in western Canada (MORP). As an authority in his field, he belongs to several international societies, and has received a number of honours including the distinction of having a minor planet named for him.

Ian Halliday began speaking at RASC meetings while a graduate student at Toronto and continued to do so, on occasion, throughout his career. He served Ottawa Centre as Secretary, Vice-President and then as President in 1961. He was the author of many papers and reviews in the *Journal*, prepared a regular feature called "Advances in Astronomy" and was the Assistant Editor (1964-69) and Editor of the *Journal* (1970-75). The Society recognized its very good fortune in having someone of Halliday's calibre, willing and able to carry out these duties on top of family life and a strenuous research career, and paid tribute to his generous nature and natural courtesy when presenting the Service Award to him in 1974. His dedication to the RASC continued during his term as Vice-President, President and Past-President (1976-86) and as Honorary President (1989-93).

Dr Ian Halliday with the three largest pieces of the Inisfree meteorite fall, 1977. Their recovery was the direct result of MORP.

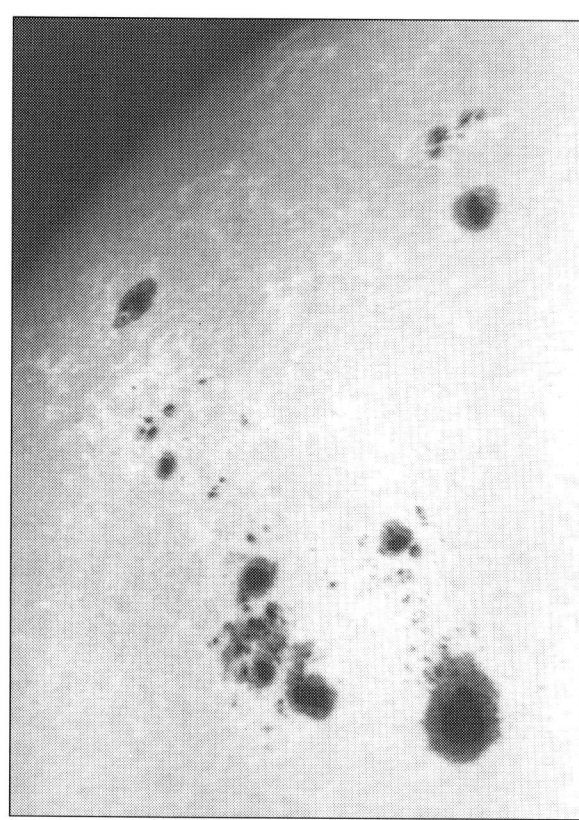

A portion of the solar disc photographed by Damien Lemay, August 22, 1990.

Accounts of auroras, which like meteors required no observing equipment, were also very prominent in Society reports, especially in the early days. Canada, of course, is ideally situated for studies of the northern lights. In the 1890s W.A. Veeder of Lyons, New York, eagerly solicited auroral observations on forms he had designed to corroborate his thesis that aurorae exhibited a 27–day periodicity equal to the Sun's period of rotation. July 16, 1893 saw one of the most magnificent auroral displays ever witnessed. The centerpiece of the show was an arch nearly 300 km high at the middle and with a breadth of nearly 4,000 km. Swishing and rustling sounds accompanied the spectacular appearance. Accounts of auroral phenomena, including daytime observations and reports of related sounds, appeared fairly regularly in the pages of the *Journal* until the 1940s and Chant especially took a strong interest in them. Another American researcher, Dr C.W. Gartlein of Cornell wrote a "Request for Auroral Observations" in the 1939 *Journal* and he got quite a large response from Montreal Centre over the next decade. Edmonton Centre made good use of their northern location with twenty-six of their members contributing over 5,000 hours of auroral observations during the years from 1953 to 1961. Earl Milton caught the essence of auroral studies when he wrote:

> For the most part, the personal work involved in participating in such an observing program is not overwhelming and the reward for the time spent is great. Besides the feeling of making a contribution to knowledge, the observer is often rewarded by the beautiful spectacle of a colourful auroral display. Words or scientific symbols cannot capture the inspiring beauty of the night sky performing its pulsating dances, splashing colour here and there on the glowing arcs, bands, draperies, rays and surfaces of the polar aurora.

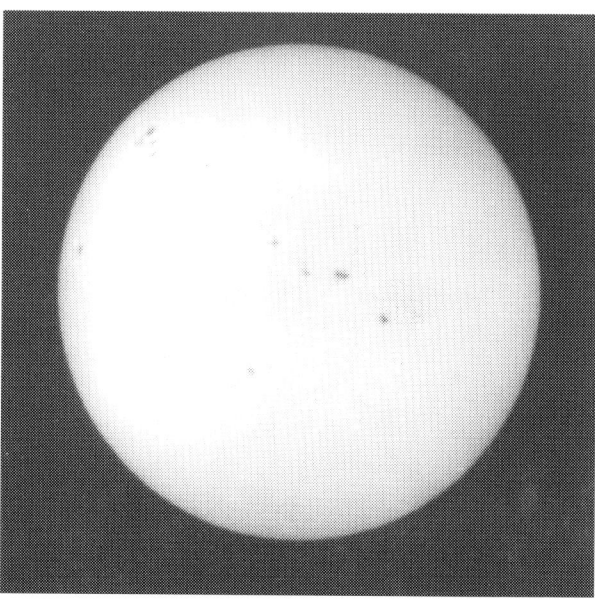

One of a series of solar photos by J.R. C[onnon], July 31, 1892.

Solar observers are sometimes able to predict good auroral displays. Dan MacLennan of Halifax and Leo Enright of Kingston began to do this successfully as solar activity picked up in the late 1980s. Leo's wife, Denise Sabatini, set up a telephone hot line to alert members to be on the lookout for increased auroral activity.

Other phenomena requiring no special observing equipment were frequently written up. Reports of halos and solar pillars, mock suns, sun dogs, rainbows, mirages, noctilucent clouds, ball lightning and even water spouts were all examples of atmospheric effects found in the *Journal*, especially from 1916 to 1940. Many of these accounts came from A.F. Hunter, but as happened fairly often, papers by one individual in the *Journal* often spawned a number of other contributions on the same subject. There was an increased interest in noctilucent clouds in 1988 and Mark Zalcik of the Edmonton Centre headed a North American network of observers of these high-level clouds.

STARS AND NEBULAE

Beyond the Solar System, the universe provides virtually limitless opportunities for astronomers. The individual stars themselves usually look no different through a telescope than they do to the naked eye and no stargazer is crazy enough to stare at points of light unless they have some redeeming feature. However, some stars which appear single to the unaided eye or in small telescopes reveal themselves as double or multiple in bigger instruments. Superficially, these doubles can be pretty to look at if the components are of different colours and can present something of a challenge for observers to try to resolve with the equipment at hand. More importantly, if motion of the components can be detected over the years, measurements of the relative positions of the stars can provide valuable information about the masses and sizes of the stars themselves. Dr Donaldson had excellent eyesight and prided himself on dividing some very close double stars with his 9-cm Cooke refractor. Miller had been monitoring the double star Pi Aquilae with his 10-cm refractor since 1883. By 1903 he was using a micrometer to measure separations and position angles for 66 Ceti and Rigel B. He also used Lumsden's 26-cm reflector to resolve the companion to Sirius in 1895 and the companion to Mira in 1900. Paul Comision, recalling his days as a student member in the 1940s, reminisced that "The big game in deep sky [observing] was the splitting of double stars. A typical conversation was "OK, but you were using a 4-inch reflector – I did it with a 2.4-inch refractor." Visual binaries are largely ignored by serious observers these days. Huge new frontiers have been opened up by larger, modern telescopes capable of showing deep-sky objects (nebulae, clusters and galaxies) so much better than the instruments of thirty years ago.

Sometimes binary star systems happen to be located so that one component seems to move in front of the other. The stars themselves may not be

The earliest stellar photograph in the RASC Archives compared with a recent ten-minute CCD image of a spiral galaxy, M51, taken by Jack Newton. The faintest stars in Miller's 45 minute exposure are nearly a million times brighter than the faintest ones in Newton's ten minute exposure. This tremendous gain is a result not only of a much larger light-gathering aperture (probably 1000 times greater in area) but to the great increase in sensitivity of modern detectors compared with emulsions of a century ago.

Globe and Mail photo, January 10, 1939

BERTRAM J. TOPHAM (1893-1962) was the first winner of the Chant Medal. He was born in England and trained there as a machinist and electrician but emigrated early in his life. He spent five years as an artillery-man with the Canadian forces in France, 1914-19, and was awarded the DCM in 1916. A shell explosion rendered him almost completely deaf for the rest of his life. Though Bert Topham knew nothing of planets or astronomy at the time, he recalled standing in the trenches and watching a bright "star" moving week by week past fainter stars. He even anticipated its return the following winter. On returning to Toronto after the war, he obtained a copy of the *Observer's Handbook* and concluded that he had been watching Jupiter. This was the start of his fascination with astronomy.

By the late 1930s he built a large observatory behind his home in what was then known as Fairbank, a suburb of Toronto. With his 16.5 cm refractor, he observed rather faint variable stars and was commended by Leon Campbell of the AAVSO for his very precise magnitude estimates. He also systematically searched for novae and comets. On moonlit nights he ground and polished telescope mirrors and welcomed visitors to his observatory. Word of mouth alone attracted thousands of people to his Castlefield Avenue home over the years. According to Dr Gartlein of Cornell University, Topham was among the most assiduous auroral observers on the continent, and Dr Millman praised him for his outstanding contributions to meteor research. Severe electrical burns sustained in 1945 sadly restricted his work thereafter.

Bert Topham's name is perpetuated in an award for outstanding observers established by the Toronto Centre in 1984.

resolved, or separated, in the telescope, but the light from the system appears to vary. These eclipsing systems, of which Algol is the most famous example, are but one type of variable star. Much more frequently, stars which seem to vary in brightness are single stars which expand and contract either in a regular or irregular fashion in periods anywhere from hours to years. With about 30,000 of these stars known, professional astronomers cannot hope to monitor them, and so estimates of their magnitudes by amateur astronomers are important to science. Most of these estimates are made visually by comparing the variable to stars of known constant magnitude. When many observations are processed and averaged (as the AAVSO does), the results are often reliable to 0.1 magnitude. For smaller variations or greater accuracy, photoelectric photometry is used and a few RASC members, including George Fortier of Montreal, Murray Kaitting of Kitchener and Ray Thompson of Toronto are involved in this work.

In the early years of the Society, some members occasionally thought they had discovered variability in certain stars. Donaldson believed, for example, that he had detected variability in Alpha Geminorum and W.B. Musson reported in 1902 that Gamma Orionis was gradually increasing in brightness. One of the outstanding amateur members in the early years of this century was J. Miller Barr of St Catharines. Using only binoculars, he made observations of variable stars and thought he had discovered changes in RU Cass, BD +30 42, and Xi Boo though it now seems unlikely that these are variables. Variable star work took a big step forward in North America with the formation of the AAVSO. One of their earliest Canadian members was Bert Petrie who was fortunate to have borrowed from the AAVSO a 15-cm reflector whose mirror was figured by E.C. Pickering, the founder of the Association and the director of the Harvard College Observatory. While Petrie was still an undergraduate at the University of

British Columbia, he wrote in the *Journal* in 1926, "There is scarcely a better way of becoming familiar with the ... constellations than by searching for variables ... It becomes an absorbing study to follow ... the light variations. Pleasure and profit [derive] therefrom, at the same time aiding science." In the course of his paper, Petrie mentioned that W.H.F. Waterfield of Nakusp, British Columbia, had discovered a new variable star in Vulpecula. Petrie became an outstanding Canadian astronomer and director of the DAO. Another important motivator was D.W. Rosebrugh, resident of Niagara Falls, Toronto Centre member, and later, after moving to the United States, an AAVSO president. He wrote two papers for the *Journal* in 1935–36 encouraging variable star observing. With the AAVSO collecting observations on a worldwide basis, there was no real need for the RASC to have a variable star program of its own. (A short-lived attempt was made in 1971.) At least since the 1940s there have always been a few RASC members who contributed heavily to the AAVSO. In 1943, for example, seven Canadians submitted 2,463 observations, led by Paul Nadeau of Quebec with 1,152. Similar figures in 1945 showed that Canadians were contributing about 7 percent of the total AAVSO variable star estimates. In 1966, Canadians contributed about 3,400 observations to AAVSO, with over half of these from Hugh Ross of Vancouver. A few steadfast RASC members continue to participate in this very worthwhile program. The most prolific is Warren Morrison of Peterborough who had sent in 69,976 estimates to AAVSO by 1992. Steven Sharpe of London is a close second with 58,518 and Chris Spratt of Victoria ranks third among Canadian members with a total so far of 33,135. Using only binoculars and a 6-cm refractor, Morrison discovered Nova Cygni in 1978 and was the first to see the 1985 outburst of the recurrent nova, RS Ophiuci. Extragalactic supernovae were photographed by Kingston Centre member Gus Johnson (living in the United States) and by Jack Newton but the most spectacular coup of this

WARREN MORRISON (1955-) grew up in Peterborough, Ontario, and obtained his BSc in Physics from Trent University in 1977. Warren was interested in astronomy even as a youngster and recalls reading every astronomy book that he could obtain and making some observations with binoculars and a small telescope. He began observing variable stars in 1971. In 1973, he joined the RASC as an unattached member, and the following year became a member of the AAVSO. He has belonged to the Kingston Centre of the RASC since 1979. Warren Morrison was awarded the Chilton prize in 1979 primarily for discovering Nova Cygni 1978. Since then he has continued to make important contributions to astronomical research, particularly in the study of variable stars. Year after year he consistently made over 3000 variable estimates annually of excellent quality, placing him in the top few AAVSO observers worldwide. In 1985, he discovered the latest outburst of the recurrent nova RS Ophiuchi, as reported by the IAU telegram system. This further work earned him the Chant Medal in 1986. His keen observing skills and deep interest in comets led to his appointment as a co-ordinator of the International Halley Watch 1983-87.

Though he works in Peterborough, Morrison's home is on a farm in nearby Cavan township, where he enjoys the country living and the dark skies.

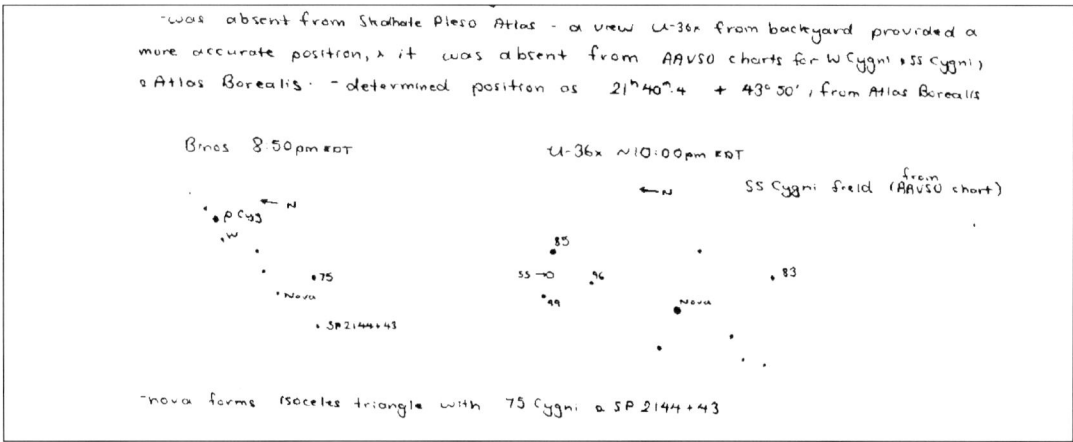

A portion of Warren Morrison's record book documenting his discovery of Nova Cygni 1978.

type was undoubtedly Supernova 1987a first seen in the Large Magellanic cloud by former Winnipeg student member, Ian Shelton, while working at the University of Toronto's southern observatory in Chile. He was subsequently presented with a life membership in the Society in recognition of his discovery.

Galactic star clusters such as the well-known Pleiades and the globular clusters, of which M13 in Hercules is probably best-known, are very popular objects for amateur stargazers. These are beautiful objects to show visitors or to capture on film. Beyond our own Milky Way are countless other galaxies, or "island universes" as they were once called. These range from the spiral galaxy in Andromeda, a mere 2.5 million light years away, to the most distant reaches of the universe. To look at these objects is to be confronted with questions about our place in the universe, about destiny and creation. Hundreds of examples fainter than the famous ones just cited can be just as alluring but more challenging to find or to photograph. In fact the challenge of locating and seeing these objects is the main attraction for some observers.

One of the oldest and best-known lists of such nebulous or "fuzzy" objects was compiled by Charles Messier in the eighteenth century. These 110 objects, now designated M1, M2, etc., are popular with amateurs because they are within reach of nearly all amateur telescopes. Locating and recording the appearance of each of them has become a sort of measure of competence among amateur astronomers, and the Society recognizes any member attaining this level of accomplishment with a "Messier Certificate" as described in chapter three. Twenty Messier objects was the median number seen by members surveyed in 1984.

OCCULTATIONS AND ECLIPSES

Finally there are the spectacular sights occurring because of the motion of our nearest neighbour – the Moon. As the Moon makes its monthly journey around the Earth, it appears to move in front of the stars, occulting them for up to an hour. Occultations of faint stars are hard to observe, especially if the event takes place against the illuminated limb of the Moon, but the brighter stars or planets can be seen, even with the naked eye, right up to the moment of disappearance. Through a telescope, these views can be quite striking, particularly with Venus whose phases mimic the Moon's, or Jupiter and its retinue of bright satellites, or Saturn and its remarkable ring. While any given occultation is visible from a wide area on the Earth's surface, it is seen from a different perspective by each observer. Therefore, by accurately timing the disappearance and reappearance of occulted stars, amateur astronomers are able to provide useful data to professionals concerned with the Moon's motion. Occasionally, when binary stars are occulted, the separation of the components can be deduced.

When the north or south "end" of the Moon barely contacts a star, a so-called grazing occultation occurs. This can be quite fascinating as individual mountains and valleys on the Moon's limb alternately cause the star to seem to blink on and off. If a series of observers stationed at intervals can make similar but different timings of the same occultation, an actual profile or cross section of the Moon's topography can be reconstructed. Of course any particular grazing occultation is visible only over a very narrow strip of the Earth's surface, and so observers anxious to time such events must be willing to travel. Grazing occultations began to attract Canadian observers in the late 1960s. Joseph Matte of the Quebec Centre made some observations in 1966 and Niagara members successfully timed four disappearances and reappearances of Eta Leonis in April, 1967. Calgary members took a special interest in grazes in the 1970s and co-operated with the University of Calgary in the hopes of learning more about the stellar diameters of Antares and Regulus. One memorable event occurred on November 19, 1982 when thirty Halifax members watched Mars come and go behind the mountains of the Moon.

While grazing occultations have attracted interest only rather recently, ordinary occultation observations were made even in the 1890s. Memorable ones included Lumsden's sighting of the occultation of Uranus on April 12, 1892, and the widely seen disappearance of Mars behind the Moon on July 11 of the same year. Donaldson noted the companion to Antares during the occultation of 1893. Elvins and Blake used the 15-cm refractor at the Toronto Observatory to watch for the daytime occultation of Jupiter on June 14, 1896. In 1898, four different observers, each with his own telescope, watched the Moon cover the 4.4 magnitude star Theta Aquarii and timed the event to the nearest second, but the problem in those days was knowing how well the clock being used kept time. Though telegraphic time signals were available to professional observatories, wireless broadcasts and short wave receivers were not common until the late 1920s. It was in 1930 that a *Scientific American* article by the noted authority on the motion of the Moon, Ernest W. Brown, was reprinted in the *Journal*. It encouraged amateurs to time lunar occultations, and C.S. Beals further advanced the cause by appealing particularly to members of the Victoria Centre. J.A. Pearce agreed to calculate the predictions and reduce the observations, and R.K. Young agreed to do the work for *The Observer's Handbook* so that predictions would appear for Ottawa, Toronto, Winnipeg and Vancouver. In the 1960s occultation work took on an urgency as the Moon's motion required refinement in anticipation of lunar landings. A big year for occultations was 1969 as the Moon passed in front of the Pleiades three times. Montreal Centre's 346 timings were reported by C.M. Good, who was AAVSO's first vice-president and chairman of their Occultation Committee at the time. The same year, Ottawa Centre reported several excellent observations to Her Majesty's Nautical Almanac Office in England. The International Occultation Timing Association (IOTA) is now the popular clearinghouse for all types of occultation observations.

When the Moon happens to pass through the Earth's shadow, a lunar eclipse results. Depending on the geometry of the situation, and to some extent also depending on atmospheric conditions on Earth, a lunar eclipse may be more or less dark. It is of interest for observers to estimate just how dark a lunar eclipse is and to time when the shadow seems to sweep over various prominent craters on the Moon's surface. The lunar eclipse of March 10, 1895, drew a small group to the grounds adjoining Toronto's *Mail and Empire* Building where, through a 8-cm Bardou refractor they timed immersions of several lunar features and occultations of some faint stars — stars that would normally be obliterated by the bright moonlight. At present, and dozens of eclipses later, members are still enjoying making similar observations but for many people, a lunar eclipse is simply a beautiful sight not to be missed.

Often, two weeks before or after a lunar eclipse, a solar eclipse occurs. This is only to be expected, as the Moon is then halfway around in its orbit and if geometric conditions are still right, the Moon will appear to pass in front of the Sun. As with the occultation of any star, the event is visible over a large part of the Earth's surface, but only from a narrow band does the eclipse appear central. Since the Sun must be totally obscured to see its delicate outer atmosphere, or corona, and to see stars in a darkened sky, a total solar eclipse is vastly more dramatic than a partial eclipse. While there may be some curiosity in seeing the Moon "take a bite out of the Sun" as Toronto members did in 1869 and 1892, the impact cannot be compared to the breathtaking spectacle of seeing totality.

A stay-at-home could wait a lifetime to see a total solar eclipse only to be clouded out on the appointed day. So travel to the path of totality is the only way for most people. Once bitten by "the bug" some avid eclipse-chasers will go to any part of the globe to see any solar eclipse.

Society members had their appetites whetted in 1896 when Miss Mary Proctor (daughter of R.A. Proctor) sent a prospectus advertising a cruise from Philadelphia to the coast of Norway where the eclipse of August 8, 1896, would be viewed. Again the following year, Miss Brown of the BAA wrote about a planned trip to India to see the solar eclipse on January 22, 1898. There is no record of any A&P Society member joining either of these expeditions, but the Toronto group decided in 1897 that it should plan a trip of its own to the southern States for the total solar eclipse on May 28, 1900, the return fare to be $25. Plans progressed and at the meeting of October 3, 1899, Thomas Lindsay and George Lumsden aroused a lot of interest but in the end there was no real expedition. Lindsay, however, took a telescope and spectroscope to Wadesboro, North Carolina, where he joined parties from Yerkes, Princeton, the Smithsonian Institution and the Royal Astronomical Society. Lumsden, by then president of the Society, went to Thomaston, Georgia, as a member of the Lick Observatory Expedition and visited the US Naval Observatory in Washington on his return. T.S. Shearmen, in charge of the Woodstock College Observatory, arranged for an eclipse group to sail on the steamer *Yanasee* out of Norfolk, Virginia.

Trying to plan an expedition to another country was one matter, but when it was realized that a total solar eclipse was predicted for Labrador in 1905, the Society was determined that something should be done. A committee was formed and subsequently, on November 11, 1904, the RASC Council passed a motion "requesting the Government of Canada to organize an expedition to observe the solar eclipse of August 30, 1905, and asking that a number of RASC members who are qualified observers be granted the privilege of accompanying the expedition at no personal expense." This was communicated to Prime Minister Laurier who approved the idea in principle and passed it on to Minister of the Interior Clifford Sifton, to arrange the details. W.F. King, the first Dominion Astronomer, was to direct the work which would entail photographing the corona as well as the flash spectrum and coronal spectrum. The six official RASC members named were Chant, A.T. DeLury, D.B. Marsh, J.R. Collins, J.E. Maybee and D.J. Howell though a number of other members went along in other capacities. Unfortunately, leaden clouds on the day of the event obscured the Sun completely. The only results obtained were the longitude, latitude, magnetic elements and the force of gravity where the party was stationed at Northwest River, Labrador.

The next solar eclipse to involve the Society occurred on June 8, 1918. The RASC Council asked Drs Chant and Klotz to observe the eclipse on behalf of the Society, and set aside $150 to cover their expenses. Klotz was clouded out in Denver, Colorado, but Chant, at nearby Matheson, was able to make observations on polarization in the corona through a break in the clouds. The following night, June 9, 1918, Chant was treated to a much rarer event: the brightest nova in over three hundred years blazed forth in the constellation Aquila!

Again in 1922, the Society showed its willingness to back a scientific eclipse expedition. There was considerable excitement over Einstein's prediction that starlight would be deflected as it passed through the gravitational field of the Sun and that this would be detectable during a solar eclipse. Though this had been attempted by English astronomers during the 1919 eclipse, their results were rather inconclusive. The 1922 eclipse provided the next chance to improve upon them and so it was that Chant, his wife and daughter, travelled to the west coast where R.K. Young of the DAO joined them aboard a ship bound for Australia. Their observations, made at Wallal, Western Australia, were successful, and Young's subsequent measurements of the plates confirmed Einstein's predictions with greater accuracy than any others had. The Society had generously contributed $500 towards expenses of the trip, but Chant, knowing the needs of the Society, returned it all to the Building Fund. An interesting footnote to this eclipse was the Fiftieth anniversary celebration in 1972 by the Goondiwindi Historical Society.

Southern Ontario's chance of the century to witness a total solar eclipse came on Saturday, January 24, 1925. Residents of St. Mary's got a wonderful view through a well-timed break in the clouds, but along the rest of the track, hopeful spectators were left shivering in the cold darkness with nothing to see. Across the border, in New York State and southern New England, conditions were much better and millions must have had the sight of their lives. Long's Corners, near Hamilton, was the main Canadian site where professionals from the DO and the University of Toronto set up their equipment along with many RASC members and throngs of others, but to no avail. There were, however, several reports of shadow bands from various places in southern Ontario, even from outside the band of totality. The snow-covered ground made these strange moving strips of light and dark especially easy to see. The only scientific experiments which could be carried out at Long's Corners were measurements of sky brightness, magnetic variations and attempts (the first ever, apparently) by personnel from the DO to determine the effect of the eclipse on radio communication. According to the *Journal*, there was an outpouring of public sympathy that most of the astronomers' plans were thwarted. One couple, a Mr and Mrs Leavey whose baby was born on eclipse day, tried to cheer Dr Chant by asking him to name their daughter in remembrance of the event. He obliged with the suggestion "Helen Corona."

Canadians got a subsequent chance to view a solar eclipse in 1932 and the Society formed a planning committee chaired by J.R. Collins. This time the

EDWIN E. BRIDGEN (? -1988) served with the Canadian Expeditionary Force in the First World War. After returning to civilian life, he was elected to membership in the RASC at a meeting of the Montreal Centre in January, 1923. He was fascinated with the lecture of the evening by C.A. Chant on the subject of his eclipse expedition to Australia. So Bridgen was quite excited, two years later, when he got the chance to travel by train to Hamilton with several members of the Montreal Centre to view the total solar eclipse. Unfortunately, they were all disappointed by snow and overcast skies, but eventually, in 1932, he was rewarded with a beautiful solar eclipse at nearby Sorel. In the '40s he was an auroral observer for the Centre, and a meteor observer in the '50s and '60s. He and his wife Lucienne were also members of the AAVSO and participated in the Nova search program in the '60s.

Between 1950 and his move to Victoria in 1963, Edwin Bridgen served the Montreal Centre as Recorder, Secretary, Treasurer, Vice-President and President. He was a leader in interpreting astronomy to the Boy Scouts, Wolf Cubs, Girl Guides and church groups, and he often gave introductory constellation talks to new members and to the public. In 1960, he prepared a revised edition of G.P. Serviss' very popular *Astronomy with an Opera Glass*. After moving west, he and Lucienne faithfully looked after the RASC booth at the Dominion Astrophysical Observatory on visitors' nights. Edwin served Victoria Centre as Librarian, National Council Representative and Honorary President; Lucienne was 2nd Vice-President in 1974.

E.E. Bridgen receiving the Service Award from Ruth Northcott in 1963.

path of totality started in Quebec, swept through Maine, and then out into the Atlantic. The RASC tried for a $1,500 grant from the federal Department of the Interior to be used for equipment and its transportation, as well as for the preparation of a suitable site but, as might have been expected during the Depression, the application was turned down. Nonetheless, plans went forward and over fifty members travelled from Ontario to two locations in Quebec – Louiseville and Acton Vale. The big day, August 31, turned out to be bothered by variable cloudiness but most of the RASC crowd were at least favoured with some breaks in the clouds at the moment of totality. Professional astronomers came from many parts of the world as the triennial IAU General Assembly was scheduled to follow on September 2–9 at Cambridge, Massachusetts. Some managed to obtain photographs of the flash spectrum just as totality began.

In retrospect this 1932 eclipse seems to have been something of a watershed. Subsequently, as specialization developed among professional astronomers, fewer and fewer bothered to travel halfway around the world for something which was not their field. On the other hand technological improvements of many types made eclipses more and more accessible to amateur astronomers and to the general public. Hundreds of thousands poured into the 160-km-wide band of totality in 1932; trains and steamers were packed, and cars were everywhere. A couple of Toronto members did what later became an obvious maneuver – when conditions looked bad they simply jumped in their car and drove to a clear spot. There was even talk of using television to enable those outside the shadow to see the event, and of filming the eclipse from an airplane if the day were cloudy. This was the eclipse, discussed in chapter eight, at which E.R. Paterson of Montreal trained 254 Boy Scouts to make observations which accurately delimited the path of totality. But aside from that remarkable experiment, and a couple of members who timed the contacts and the onset of shadow bands, most of the

The solar corona photographed on 31 August, 1932 by D.B. Marsh at Acton Vale, Quebec.

Second contact of solar eclipse of 10 July, 1972 photographed by D.C. Ellis at Coppermine, NWT.

Members of the eclipse party on board ship returning from Labrador, August, 1905.

Setting up beside the chartered train near Wivenhoe, Manitoba, 20 July, 1963.

Probably the first RASC eclipse expedition outside the country was organized by Winnipeg Centre in 1970.

Some Edmonton members awaiting the eclipse of 26 February, 1979.

amateur astronomers were content to just watch the spectacle or to photograph it. In fact, this seems to be the earliest eclipse at which RASC amateurs used their own equipment to successfully photograph the corona and to take moving pictures as the drama unfolded. Fifteen members from the Hamilton Centre under the leadership of D.B. Marsh were quite successful at Acton Vale, 80 km east of Montreal. They had shipped three refractors and a clock drive pedestal ahead by truck and then came themselves in cars, stopping overnight enroute. The refractors, with equivalent focal lengths of 6.7, 4.3 and 3 metres were equipped with cameras using 13x18-cm and 10x13-cm glass plates. Orange, yellow and blue filters were used and four exposures were taken with each camera.

Western Canada had its turn on July 9, 1945. The war had just ended, making advance planning difficult. The eclipse was observed from a number of locations but the only RASC group trip seems to have involved seven Winnipeg members and a few from elsewhere who travelled to Wolseley, Saskatchewan, where the hospitable townsfolk accommodated the visitors, and the mayor and his wife fed them. The day began cloudy but the sky cleared and everyone at Wolseley had a beautiful view of 33 seconds of totality. Squadron Leaders Millman and Heard, who had been navigation instructors during the war, flew from their base at Rivers, Manitoba, and obtained moderately good photographs of the corona and flash spectrum from high above the clouds. Ironically, one of their photographs was taken right over Bredenbury, Saskatchewan, where the ground-based professional astronomers saw nothing of the eclipse.

Heard was not so lucky in 1954, at the next Canadian eclipse. He and a number of professional colleagues went to Mattice, in northern Ontario, taking with them the same long focal length astrographic lens that Chant and Young had used in Australia thirty-two years earlier. Their efforts were frustrated by heavy overcast. Millman, on the other hand, led his second successful airborne expedition. Some RASC members from Montreal and Ottawa chartered a railway sleeping car and travelled to Mattice and another group of ten from the London Centre also made their way to the same location but of course they all shared the same fate as Heard. Nine Quebec Centre members had equally bad weather in the wilds of Ungava, 1,000 km from Quebec City. The cost of their expedition, approximately $3,000, was borne in part by grants and donations from Hollinger Mines, Université Laval, various Quebec government ministries and the Centre itself. There are no reports of any amateur members seeing totality in 1954, though Edmonton Centre members did take photographs and colour movies of the partial phases from the roof of the MacDonald Hotel.

The eclipse of July 20, 1963, was a different matter. The path of totality ran right across the country, from the Yukon and the northern part of the western provinces and Ontario down through the populated areas of Quebec, New Brunswick and Nova Scotia. Isabel Williamson of Montreal was named as the Society's co-ordinator for the event, and she immediately got down to work. By December, 1962, she had already contacted Dr J.L. Locke at the DO for specific advice on suitable observing programs for amateurs, got from the Province of Quebec a supply of maps and pamphlets giving camping sites and accommodations and discussed suitable sites with the Shawinigan Water and Power Company. As she stated in her report:

> The programme will offer something for everyone, and all members will be invited to participate. They can undertake all or any part of the programme (1) as individuals, (2) by organizing their own observing parties, or (3) by joining one of the highly organized field stations established by the national committee. Series of circulars ... will keep the Centres informed; standard report forms will be made available for

each phase of the programme. Centres which have formulated their own plans are not expected to conform to the national programme but will be asked to send reports on their observations to the national committee.

By the time it was all over, Miss Williamson reported that the committee had issued a series of eight bulletins. She had personally written over 200 letters in connection with the arrangements for observing the eclipse and had organized the observing station at Plessisville. The program was highly successful in bringing Centres into closer co-operation and many reports and data were received. Calgary and Edmonton Centre members together with the University of Alberta formed an expedition to Fort Providence, NWT, where they suffered only a slight haze. Winnipeg Centre, with the co-operation of Canadian National Railways, organized a train excursion for about 120 members and friends to Wivenhoe in northern Manitoba but they were clouded out. Hamilton Centre mustered four different teams which all successfully photographed the eclipse from different locations. Toronto Centre members also spread out to a number of sites, where some had good weather and some not. About fifty Ottawa Centre members travelled to Grand'Mère, Quebec, where the Centre had rented a field. They obtained many fine photographs as the clouds parted just in time for the eclipse. The Province of Quebec was of course the most popular destination being convenient to the large population centres of Canada and the United States, and having reasonably good weather prospects. Quebec Centre estimated that 50,000 visitors came to the province to see the eclipse. The weather there was patchy, so some went away happy, others not, but the Quebec and Montreal Centres did everything in their power to help.

Two more times the Moon's shadow swept across Canada. In fact the path of totality for the July 10, 1972, eclipse was entirely within Canada, and so it was dubbed the "AllCan" eclipse. The Annual report for 1972 supplied a good summary:

> A special group of members from the Calgary, Edmonton, Vancouver and Saskatoon Centres flew to Tuktoyaktuk in the Northwest Territories and were rewarded with a perfect sky. Large parties from Toronto, Hamilton and Montreal chose points in the Gaspé region of Quebec but were disappointed and some experienced rain. Members from Ottawa and other Centres as well many visitors from the United States chose Prince Edward Island and in general experienced good viewing conditions. Winnipeg members journeyed to Baker Lake. When ground conditions appeared unfavourable, the group hurriedly boarded their chartered plane and had a breath-taking view from above the clouds. A party from St John's Centre had an excellent view from Antigonish in Nova Scotia.

Undoubtedly a big factor in the good fortune of so many was the careful preparation made by the co-ordinating committee under Ken Chilton. They selected three prime sites and eleven alternate locations in eastern Canada, and Ken personally visited most of them during his summer vacation in 1971 to get the lay of the land and to check for accommodation. Reprints from a *Sky and Telescope* article on the eclipse, supplemented with a meteorological bulletin and the committee's own information, were sent out to over 600 people who made enquiries through the Society.

The last "Canadian" total solar eclipse of the twentieth century arrived on February 26, 1979. Southeastern Saskatchewan and southern Manitoba were the only populous parts of the country to be within the path of totality and Winnipeg was right in the middle of it. Because of this there was no national committee and Winnipeg Centre had almost sole responsibility for planning for the big day. They

Damien Lemay in his original observatory built of snow in his backyard at Rimouski in 1973 and his first observatory structure, built in 1976.

DAMIEN LEMAY (1943-) is a Physics graduate, presently Assistant Director of Transmissions for Quebec Telephone. RASC members know him better as President (1990-92), author and observer par excellence. As a student, Damien became interested in astronomy through the newspaper columns of Paul-H Nadeau, and during summers as an assistant to Nadeau at the Quebec Observatory.

Damien Lemay served as the French Editor of the *National Newsletter* from 1977-89, and contributed many articles himself during this period. As well he wrote frequently for the newsletter of Quebec Centre, for *Québec Astronomique* and was the author of two booklets, *Guide Practique de l'Astronomie Amateur* and *La Photographie Astronomique*. Since 1984 he has served on the National Research Council's Associate Committee on Meteorites. The AGAA recognized Damien Lemay's outstanding contribution to Quebec astronomy by presenting him with their Trophée Méritas in 1982. A regular observer of variable stars and sunspots, he won many awards at RASC General Assemblies for displays of his work. He was probably the first Canadian (and perhaps North American) to see Nova Cygni on August 30.06 UT, 1975. But his photographic atlas of the sky was his magnum opus. Taken with his 14 cm Schmidt camera over a period of five years, the 1182 photographs which comprise the entire sky of the northern hemisphere show the meticulous care he took in the exposure, developing and printing. For this magnificent project, Damien Lemay was awarded the Chant Medal in 1987.

prepared the public for the event through an information centre set up at the Manitoba Planetarium and by eclipse seminars and lectures at schools in Winnipeg and outlying towns. They hosted the National Council which met the previous day, and they held a banquet for 300 people just hours after the eclipse. Everything, the weather especially, was superb, and many beautiful photos were taken. Both Winnipeg and Edmonton put together slide sets which turned out to be quite good fundraisers for the Centres. The biggest out-of-town contingent naturally came from Toronto. Organized by Michael Watson, two aircraft flew 106 members and friends to the former airbase at Gimli, Manitoba, where viewing conditions were ideal. It was a great finale. Canada would not see another total solar eclipse until 2,008.

Many members were not willing to be that patient, and RASC eclipse expeditions to all parts of the world became almost an annual occurrence. The Toronto Centre especially, with its large membership base, was able to arrange trips to Kenya in 1980, Siberia in 1981, Indonesia in 1983, Virginia for the 1984 annular eclipse, Papua-New Guinea in 1985, the Philippines in 1988 and Finland in 1990. Hundreds of members from various Centres travelled to Mexico in 1991 to view the longest total solar eclipse until the year 2,132.

Space launches from Cape Canaveral and the lure of good views of Comet Halley from the Southern Hemisphere provided other good "excuses" for expeditions.

EQUIPMENT AND OBSERVATORIES

There are plenty of members of the Society who enjoy looking skyward without any optical aid. Like naturalists who take pleasure in identifying plants and insects, these people have noted atmospheric effects, meteors, aurorae, conjunctions, bright comets and eclipses. Then there are those with only the most modest equipment – binoculars or very small telescopes. Indeed, in the first decade of the Society, there was an "Opera Glass Section." Viewing for pleasure, and even some scientifically useful observing, can still be done with binoculars.

For most of the Society's history, standard equipment for a serious observer was a 8-cm or 10-cm refractor or a 15-cm reflector. Purchasing a 10-cm refractor would have required at least a month's wages for most people in the 1890s so donated telescopes were highly prized and shared amongst members. George Lumsden's 26-cm reflector and J.J. Wadsworth's 30-cm were exceptionally large instruments for amateurs to possess at that time. In the 1890s, telescopes were rarely homemade as reflectors still normally used speculum metal, and lenses for refractors usually required an optician's skill to grind and polish. Zoro Collins experimented with silver on glass mirrors in 1894 and thought they "might be very successfully introduced among Canadian observers" and he and his brother tried a number of innovative designs of both reflectors and refractors. Toronto lawyer, A.R. Hassard, made his own 24-cm reflector in 1908 and later (by 1921) made a 38-cm. The *English Mechanic* and personal correspondence, mainly with J. Mellish and D. Prahl of Wisconsin, were his main sources of information. Constructing one's own instrument became a great deal easier after 1928. That year *Scientific American*'s associate editor, Albert G. Ingalls, began a regular feature in the magazine and edited the first volume of *Amateur Telescope Making*, a guide which through many editions became the virtual bible on the subject. The 1937 volume of *Advanced Telescope Making* contained chapters on the 48-cm reflector made by R.K. Young and on Meteor Photography by P.M. Millman.

For about fifty years, 1930–80, telescope making was a very common activity among amateur members. Many of these people were really attracted by the challenge of designing or scrounging parts for the mounting or by the mechanical art of grinding, polishing and testing mirrors. Some used their tele-

JESSE KETCHUM (1885-1973) was a great-grandson of the well-known Toronto pioneer of the same name. Raised on a farm near Orangeville, he went west as a young man, establishing a homestead in Saskatchewan in 1907 and later working for the railroad as station agent and telegrapher. His love of astronomy eventually brought him into contact with the Saskatoon Centre of the RASC though following retirement, he and his wife moved back to Ontario and transferred their membership to the Toronto Centre in 1951.

For nearly twenty years he was known to virtually every member of the Toronto Centre as the director of the telescope-makers group. Countless members, young and old, learned the tricks of the trade from Jesse Ketchum. His enthusiasm for his craft was well-captured in a story told by Fred Troyer:

> Often asked by a novice — usually a high-school student — "What's the best kind of telescope to buy?" Jesse would scowl, "Buy? Such nonsense. Build your own. It's cheaper and it's a lot of fun."

He could always be counted on to be on hand with one or more telescopes for public viewing at the Canadian National Exhibition or on Saturday evenings at the DDO. He spoke on occasion to school classes and Scout groups about his hobby. He was honoured by the Society with the Service Award in 1962, by his country with the Centennial Medal in 1967, and posthumously by the Toronto Centre in the naming of the Jesse Ketchum Award in 1984.

This photo, supplied by Mme Huberte Palardy, was taken in the École Technique de Québec and shows (l to r) P.-H. Nadeau, Professor M.-L. Carrier and A. Boivin, founder of the Cercle Léon Foucault.

scopes very little before moving on to something else. There must be hundreds of these instruments in various states of completion and repair gathering dust in garages and basements across the country. One problem was that the reflective coating had to be regularly replaced and it was frequently difficult to find someone who could do the job. In the old days the silver that was used tarnished from exposure to the air after about a year and even the aluminum which came into use in the 1950s oxidized after a few years at best. Once mass-produced catadioptric telescopes with their sealed optics became easily available at affordable prices in the late 1970s, few people saw the point in investing weeks or months of labour in an undertaking which might not turn out satisfactorily. The claims of commercial manufacturers, however, are not always matched by reality and so telescope-making as a hobby has not died out. Prizes are still eagerly contested each year at gatherings such as Stellafane. Large Dobsonian telescopes with easily constructed mountings and with apertures exceeding 50-cm have recently been completed by a number of amateurs. While some members have their own observatories (5 percent of those surveyed in 1984), most opt for the mobility of a compact telescope which can be easily transported to a dark site and quickly set up, possibly with the drive connected to the car battery.

Some Centres, of course, have accumulated a number of instruments through donations or purchases and some lend these out for a modest fee to their members. Observatories owned by Centres will be discussed in later chapters but it is worth stating here that even a modest observatory is a major undertaking and responsibility for any group. Finding a site which is both accessible and dark enough to promise years of use, and the work and cost involved in construction are only the beginning. Users must be trained to be careful operators of the equipment, the site must be cleaned, maintained, and secured against floods, theft and vandalism, all of which have been felt by RASC observatories. Technology sometimes becomes the tail that wags the dog as members have been known to devote their energies to a continual round of modifications and improvements to the detriment of actual observing.

The national Society, too, has owned instruments and at times aspired to have an observatory. It fell heir to a number of telescopes over the years, many of which are now missing. These include, among others, Sir Adam Wilson's 15-cm reflector and Larratt Smith's 8-cm refractor, originally owned by J.G. Howard, who left High Park to the City of Toronto. James Todhunter's 6-cm refractor, used for a while by Junior members, and a couple of other 5-cm telescopes were lent to the 4th Canadian Mounted Rifles Battalion during World War I. Their commander, Col H.D.L. Gordon, wrote that the instruments were in constant daily use from 1916 until the end of hostilities. "They were very useful in distinguishing and locating snipers, surveying the ground held by the Germans, and directing operations. The telescopes were of especial value in preparing for the attack on Vimy Ridge."

Three telescopes which went out on permanent loan were the "Carpmael Telescope," lent to the Halifax Centre in 1962 and since reported as missing, F.T. Stanford's 10-cm refractor, to the Toronto Centre in 1965, and Ruth Northcott's 15-cm reflector, delivered to St. John's Centre for their use in 1971, the latter two still in existence. The only telescope that the Society ever bought was a rather cumbersome 10-cm Cooke refractor which was purchased in 1901 for $195. The Society spent $95 in 1949 to have it repaired and to have slow-motion controls added; it is still used by Toronto Centre at public star nights.

AN RASC OBSERVATORY?

The Toronto Observatory, under the federal government's Department of the Interior, had acquired its fine 15-cm Cooke refractor just in time for the 1882 transit of Venus. Clouds obscured the event itself, and little is known about other astronomical projects at the Observatory other than a long series of sunspot observations. Once the A&P Society became incorporated in 1890, the Society's first president and director of the Observatory, Charles Carpmael, permitted members of the Society to use the telescope occasionally but unfortunately there was not greater co-operation. The Observatory seems to have been rarely used at night and yet many saw a genuine need for a public observatory which could also be used by Society members in furthering their studies. Those who thought about the future of Canadian astronomy were tantalized by examples of American observatories endowed by wealthy citizens. A deputation in 1892 approached the well-known industrialist and benefactor, Hart Massey, for support. Plans for Toronto's Massey Hall were by that time nearing completion and Mr Massey was therefore unable to undertake other projects. He did hold out some hope that the Society might have a meeting room in Massey Hall, but that never materialized. Throughout the decade efforts to encourage financial support continued. Speeches made to the Society were reported in the papers, Lumsden spoke to the Canadian Institute in 1896 and prepared a prospectus for "The People's Observatory," committees were appointed, university authorities were consulted, members visited and reported on exemplary observatories which had been established south of the border, but all to no avail. The Society did successfully oppose the proposed removal of the meteorological observatory to Ottawa in the early 1900s and Vice-President Stupart (Carpmael's successor as director of the Observatory) did generously offer the use of the 15-cm refractor whenever required. For a number of reasons, however, the Society rarely used it but persisted in attempts to get an astronomical observatory for Toronto. Obtaining a site was never the main problem. Trinity University considered providing a location on its King Street property in 1903. The City

LOUIS B. STEWART (1861-1937) was born in Port Hope, Ontario, and trained as a land surveyor. His father, George, was a Civil Engineer, and Louis worked with him on the survey of Banff National Park, prior to serving with the 92nd Regiment in the North-West Rebellion of 1885. He joined the faculty of the School of Practical Science, one of the original three members of that institution which later became the University of Toronto's Engineering Faculty. Though most of L.B. Stewart's career was spent as a professor of engineering at the U of T, he was involved in surveying in the Yukon during the gold rush of 1898, and in Hudson Bay, the Gulf of St Lawrence and the Maritimes. After the Meteorological Service gave up the Toronto Observatory on the U of T campus, Stewart used it for teaching purposes. The building is now named the L B Stewart Observatory (see photo), though the telescope has long since been moved and is now at the National Museum of Science and Technology in Ottawa.

The natural link between Stewart's main interests of astronomy, geodesy and surveying resulted in a number of papers in publications such as *Canadian Engineer* as well as the *Journal* of the RASC. He lectured at meetings of the Society a few times and served as Vice-President and President (1912-13).

Parks Commissioner was willing to offer a site on Bathurst Street on the north side of the Cedarvale ravine in 1914. The war then interfered, but in 1919 the project was revived:

> A large and representative deputation met the Board of Control of the City on June 11. ... [The group] met with a favourable reception, and the plans outlined by it were highly commended by the city press. ... An eminent firm of architects prepared sketch plans of the building, the terms of an agreement were drawn up between the city, which would provide the land, the University which would maintain the observatory, and the Society which would place its library and telescopes there. All that was needed

was a person of means, possibly someone who wished to honour the name of one who served in the forces or gave his life for his country.

All this and E.E. Barnard's stimulating public lectures the following year seemed to lead nowhere. However, towards the end of May, 1921, Chant gave a public lecture on the subject of Comet Winnecke. As Chant himself recalled:

> At the end of the lecture, [I] took ten minutes to make a plea for an astronomical observatory for the University and the city. [I] projected on the screen some slides which had been prepared some years before when it had been proposed that the city, the RASC and the University would unite in providing an observatory. After the lecture a gentleman came up to me and said he was interested in my project. He said his name was David Dunlap. In the months which followed I had some correspondence with Mr Dunlap regarding astronomical matters and he used to come to meetings of the Astronomical Society. I cherished the hope that he might provide the observatory but he passed away.

The happy ending to the story is that Dunlap's widow agreed to fund a magnificent observatory for the University of Toronto in honour of her husband. The planning and eventual opening of the country's largest observatory in 1935 is a fascinating story on its own account but has little relevance in the context of the Society's history. What must not be forgotten is that this dream which finally became reality occurred only after years of preliminary effort by Chant and many others in the RASC.

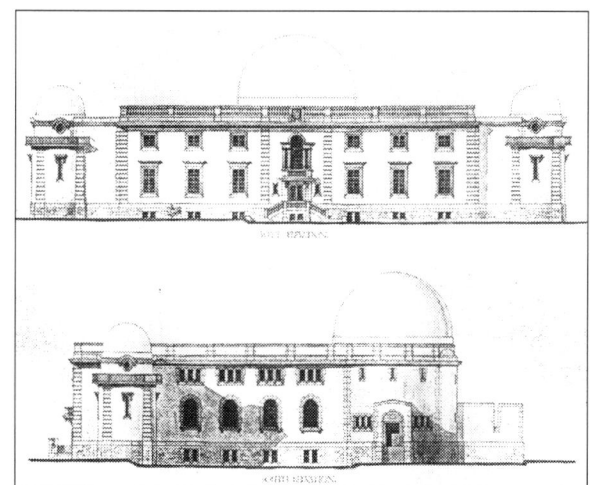

Architect's drawings for the proposed Royal Astronomical Observatory, Toronto and the site as it looked in 1915. The view looks north on Bathurst Street, from about Lonsmount Drive. The observatory would have been located atop the distant hill.

(Photo on the right is from Metro Toronto Reference Library, T33468)

CHAPTER 10

Clusters

The meetings of the Society can be divided into two categories – those which are organized by the Centres for themselves, and the annual meetings and General Assemblies which are usually arranged by a Centre but which are attended by members from all over. Technically, all the meetings in Toronto up to 1927 were organized by the General Council and were considered to be meetings of the Society as a whole.

CENTRE MEETINGS

It is probably no exaggeration to say that there have been ten thousand regular meetings of Centres throughout the Society's long life. These include "members nights" where those conducting photographic, observing and telescope-making projects get a chance to "show and tell," film nights, planetarium shows, observing sessions, courses for beginners, tours, field trips, debates, panel discussions and, of course, talks by both amateur and professional astronomers. Some general trends over time, and some highlights are about all that can be attempted here. What the average member may not realize is that these meetings don't just happen. A great deal of thought, effort and planning goes into a successful program and typically a Centre council or Executive needs to meet nearly as often as the Centre membership itself. As Alan Dyer pointed out when he was president of the Edmonton Centre in

Photo by Dan Falk

Two prominent star clusters, the Pleiades and the Hyades in the dark sky above a small cluster of RASC members at Mount Uniacke, NS, 16 November, 1985.

1978, arranging the program of regular meetings is "a question of balance ... between meetings that will please both the armchair astronomer and the active observer, ... between education and entertainment, formality and informality, ... between introductory and advanced topics, ... between the principal topic and the inevitable business announcements." What a challenge to the organizers! The large Centres have the luxury of being able to plan events for special interest groups with the knowledge that some members will support one activity while others will back something else. The smaller Centres, however, have to hope that their membership as a whole will be behind whatever program is planned. This usually means that the smaller Centres try to include a variety of activities in each meeting. In some respects the new Centres of today face problems similar to those met by the Society itself when it was in its infancy.

When the Society was small and young a good proportion of members actively participated in meetings. They took turns reading papers which they found interesting in science journals, usually the popular ones such as *The Sidereal Messenger* or *Knowledge*, and also presented results of their own research. Though there were a couple of members attached to the Meteorological Service who, as part of their work, made seismographic, magnetic and solar observations, there were really no professional astronomers until the Dominion Observatory opened in 1905. Because astronomy lectures by experts were nonexistent, any communications with professionals from the States or abroad were highly valued and were read and discussed at meetings. University of Toronto professors did lecture to the Society, but their topics were on physics and chemistry. Even C.A. Chant was purely a physicist at this time. One of his memorable lectures, held on November 26, 1895, included what may have been the first demonstration of radio waves in Canada.

Hardly a meeting passed without reports from members of their observations; anything was acceptable from simple descriptions of an aurora to micrometric observations of double stars. Some evenings would be designated for observing when several telescopes belonging to members or to the Society would be set up for viewing of celestial objects, the planets being especially popular. And of course as every meeting was a meeting of the Society, election of new members, correspondence, reports and any business of general interest would also take place.

While it is impossible to find a "typical" quote, the following verbatim report of the meeting of September 5, 1893, gives a glimpse of the sort of involvement which enlivened the meetings of the time:

> Mr. G.G. Pursey, the Librarian, reported that he had received several volumes from kindred Societies, and announced that an Index of the subject-matter of exchanges was in preparation, and would shortly be completed. Letters were received from Dr. J.J. Wadsworth, of Simcoe, and from Mr. J.A. Copland. The latter reported at some length the observations of auroral and meteoric displays made by him in North Wellington. Mr. A.F. Miller, in reporting his observations of the solar surface, referred to the difficulty the amateur sometimes has in identifying a sun-spot from day to day, which, however, was not surprising, as it had been necessary during the past year at Greenwich, to employ extra assistance in compiling a complete record of the appearance of the Sun. Mr. A. Elvins presented several sketches of spots on the solar disc, some of which had been visible to the unaided eye. Mr. R. Dewar reported having heard the swishing sound of the aurora during the display of July 23rd. An interesting discussion arose in connection with photographic work on the heavens and Mr. Miller described some of the difficulties in the way of the amateur, one of which is the procuring of a suitable lens, the ordinary photographic lens not being quite adapted for the purpose. Mr. Arthur

Harvey thought that the selection of plates was of even greater importance, and that orthochromatic plates solved the difficulty. Mr. Mungo Turnbull gave some practical illustrations of the method of using the celestial globe recently constructed and patented by him, using for the purpose the globe presented to the Society by Lady Wilson, in accordance with the wish of the late Hon. Sir Adam Wilson, Q.C.

Mr Harvey read the following ... "I am permitted to bring for your inspection, from the Canadian Institute, the Third Annual Report of the Vatican Observatory, and I think the beautiful plates will interest you very much. Especially instructive is the plate of the star-cluster in Cancer, Praesepe or the Beehive."

He went on to describe how such photographs taken many years apart would enable proper motion to be measured, and invited members to inspect the photograph with a magnifying glass.

Though photographs of brighter objects like the Sun, Moon, and planets were no longer a novelty, photographs of stars still were. It should be remembered that astronomers were skeptical of the practicality of stellar photography until 1882 when plates of the comet of that year clearly showed hundreds of stars. Books rarely had astronomical photographs until after the turn of the century, so one can understand that the publication brought by Mr Harvey aroused great interest.

Some similarities can be found to the meetings one hundred years later, especially those in smaller Centres and those usually called members nights. Members still present their observations, though usually in the form of coloured slides rather than as sketches or verbal reports, and members still discuss the suitability of various equipment and detectors. However, correspondence plays almost no role anymore, and publications, though received, are rarely discussed.

Amateur members of the Society are generally more willing to share in a meeting by making a short presentation along with others rather than to take on an entire hour-long meeting. Nonetheless, many amateurs have assumed the responsibility with enthusiasm and success and have entertained audiences with beautifully illustrated talks on eclipse expeditions, space launchings and astrophotography, instructed them in instrument design and telescope construction, and have made presentations of a truly professional calibre on a wide range of observational and research topics. Amateurs are sometimes attracted to topics well in advance of their time. Space exploration surfaced as early as 1933, when Clinton Constantinescu of Edmonton lectured on "The Rocket." R.A. Storch spoke to the Winnipeg Centre in 1944 about "Problems Involved in Travelling Through Outer Space," and Norman Green of Hamilton took "Synthetic Satellites" as his subject in 1955 two years before the first Sputnik was launched.

Separating the contributions of amateurs and professionals is sometimes impossible; there is really a multidimensional continuum of expertise. Professional astronomers, for instance, may sometimes address historical topics for which they have no formal training, and frequently amateur astronomers will bring professional expertise in a related area which could be anything from ophthalmology to philosophy.

Nonetheless, the lecture meeting when an invited speaker addresses the audience on his or her particular area of specialization, is an important part of many Centre programs providing a window on current research which members find stimulating and challenging. The Society does provide financial assistance to Centres wishing to make travel arrangements for visiting speakers but the smaller Centres with no university community to draw on usually have to rely mainly on their own amateur members to speak at meetings. In fact some Centres do not hold lecture meetings at all, but devote their energies to observing sessions, star parties and public education.

Naturally, the topics presented have always depended on the expertise of the available speakers. As already noted, in the absence of professional astronomers many of the lectures in the early years in Toronto were on physics; the name of the Society itself was a reflection of that. The tendency of professional lectures to be connected with other sciences was evident for example in Vancouver in the 1930s and Windsor in the 1940s, where there were universities but no astronomy departments. In Ottawa, on the other hand, where the establishment of the Centre was motivated by the opening of the Dominion Observatory in 1905, the usual fare there in the early years included geodesy, surveying, and positional astronomy, as well as reports of scientific meetings.

Tracing the subjects which have been presented over the years gives a fascinating glimpse of the evolution of astronomical knowledge, and provides clear evidence for just how current many of these lecture topics were. One could almost write a history of twentieth century astronomy from the reports of the RASC meetings. The Ottawa Centre was kept up-to-date with lectures by the astronomers at the DO. In 1911 R.M. Stewart spoke on "The Principle of Relativity." W.E. Harper, in his 1916 speech on "Nebulae," concluded "spiral nebulae are outside of the galactic system, and probably form separate stellar universes." (This was four years before the famous Shapley-Curtis debate took place.) And in 1927, at a time when Eddington and Fowler were just coming to grips with the true nature of white dwarf stars, F.C. Henroteau addressed the Centre on the subject, "The Strange Companion of Sirius." Astronomers from the DAO in Victoria played an important part in explaining the latest developments. In 1953, P.E. Argyle spoke to the Victoria Centre on "Photoelectric Astronomy," and Jean McDonald told Toronto Centre members "How Canada's Electronic Brain is Aiding Astronomy." Other examples of astronomers from Canadian universities speaking on "hot" topics include R. Butler of Queen's University who spoke to the Montreal Centre on "Radio and Optical Observations of ... Quasars" in 1966, J.F.R. Gower of the University of British Columbia speaking on Pulsars in 1968, G. Mitchell of St. Mary's (Halifax) in 1970, "Where are the Solar Neutrinos?" and W.Y. Chau, to the Kingston Centre in 1973 on the subject of "Black Holes." R. Roeder of Toronto spoke to the Halifax Centre in 1980 about "Double Quasars, Gravitational Lenses" and M. Duncan of Queen's addressed the Toronto Centre on "Chaos and Stability in the Solar System" in 1990. There are many, many other instances of speeches which have been just as current and equally fascinating, but perhaps less flashy because they were based on fields of evolving research rather than brand new topics.

Audio-visual technology has had an important impact on Society meetings. In the 1890s lantern slides came into use. This was a wonderful advance allowing an audience to feel a part of the scene as it was "thrown on the screen." D.J. Howell was appointed the Society's photographer, and one of his responsibilities was to make lantern slides from appropriate prints for use at meetings. In this way and through purchase, the Society accumulated a very good collection of transparencies. Nearly five hundred slides were catalogued by 1904. These were not only useful at Society meetings but were often lent to members who had been invited to speak to various community groups, and also to affiliated Societies.

Slides, of course, have changed their format to 35 mm and since the 1960s, amateur astronomers have been taking their own celestial scenes in colour. The great increase in the sensitivity of films, the larger telescopes available to amateurs, and easier transportation to remote dark sites, have meant that the slides of today are gorgeous celestial portraits. As technology has advanced, interest has accelerated and the astronomical slide show is now a very popular feature of many Society meetings. As well, since about 1960, illustrated talks about astronomy in other countries have become increasingly common. Air travel and affluence have combined with progress in photographic technology to make trips to the southern hemisphere, to famous observatories or to solar eclipses appealing subjects for many members to record and present at meetings.

Films have also been the centrepiece of many a meeting. The earliest screening took place at the Ottawa Centre's At-Home on January 29, 1914, when movies of the Yukon surveys were shown. For the most part, the films that have been used were commercially produced for educational purposes. Some with astronomical themes were shown as early as 1922, including such titles as "Worlds in the Making," "The Mystery of Space" and "The Earth and Moon." Over the years, the professional producers and artists became more and more skilled and their presentations increasingly sophisticated and realistic. The best of its time was "Universe," produced for the National Film Board by Roman Kroitor and Colin Low in 1960, and the winner of awards at festivals at Vancouver, Edinburgh and Cannes. This half-hour film began and ended with a night of observing at the DDO and included some excellent animation depicting solar prominences, the motion of gases in a comet's tail, nova explosions and the interaction of galaxies. Even Stanley Kubrick found the special effects so impressive that he asked Low to work on 2001; *A Space Odyssey*. The space age brought wonderful NASA films of Ranger to the Moon, Mariner to Mars and Voyager to Jupiter and beyond. It is only in the last few years that movies have given way to videos, and though some videos have been used at Centre meetings, the typical television screen is not ideal for large groups, and big screens are not very common. Anyway, public broadcasting has reached out to virtually everyone's living room with some excellent educational series such as "Nova" and Carl Sagan's "Cosmos".

Members are beginning to make videos of meetings and sky events and are producing digitized images from charge-coupled devices (CCDs). These aspects of electronic technology are likely to see increasing use. Software which portrays aspects of the sky at any time and any location, computer simulation of everything from orbital motion to evolving stars, and encyclopedic data bases will no doubt have a great impact on the meetings of the future.

ANNUAL MEETINGS OF THE SOCIETY

The first Annual Meeting was held on February 24, 1891. Correspondence, reports of officers and elections were on the agenda of that meeting, just as they still are at annual meetings. A big part of the meeting was the presidential report on the progress of Astronomy generally during the past year. This was the format for the first nine years, though the date dictated by the by-laws became the second Tuesday of January from 1892 on.

In 1900, the Annual Meeting became strictly a business meeting, with an event called the Annual Conversazione held two weeks later. This was a formal social occasion to which dignitaries and members of the public were invited. Following the presidential address, members and guests had the opportunity to view lunar photographs from Meudon Observatory donated to the Society by the Government of France and to enjoy some of the Society's lantern slides shown with the Society's new electric projecting lantern before partaking of some light refreshments. The intention of this sort of meeting was apparently to give the Society a chance to show its best to the public, and as it was held in the Society's premises in the Canadian Institute Building, it became known as the Annual At-Home.

The new premises of the Canadian Institute on College Street to which the Society moved in 1908, proved too small for a large public gathering and so the Annual Meeting and Conversazione were held together on January 12, 1909, at Victoria College, in the chapel and adjoining rooms. Lachlan Gilchrist operated an x-ray apparatus, and A.F. Miller gave demonstrations with a spectroscope. Refreshments and a social hour concluded the evening.

The business of an Annual Meeting was not really consistent with the more social flavour of the conversazione, and so the two events were again separated in 1910, and continued to be held on two separate evenings until 1952. During most of this long period, both meetings were held in January in the Physics building of the University of Toronto, and though not literally accurate, the second evening at which demonstrations and refreshments followed the presidential address, was consistently known as the Society's At-Home. Presidents from out of town could hardly be expected to attend two meetings only a week or two apart and so, quite frequently, the Annual Meeting was conducted without the president in attendance. Though this was not seen as a problem for many years, it ultimately was a factor in the demise of the At-Home.

Another important consideration was the growing number and strength of Centres outside Toronto. Reports from Centres were a part of the Annual Meeting at least since 1910 and these were published in the Society's annual reports. In fact, until 1926 it was largely on the basis of these Centre reports that the Council decided on the grants to be remitted to the Centres. Also, there was a growing awareness that Centres should have a more active role in the Annual Meeting. A very successful AAAS meeting in Toronto in 1921, prompted the wish that RASC meetings could be arranged each year at which papers could be read by members from all the Centres. It was an idea that took over thirty years to become reality due to the great expense of time and money involved in travelling across Canada.

A change in timing of the Annual Meeting was also becoming necessary. Not surprisingly, the treasurer and auditor found it very difficult to have financial statements ready by mid-January. The executive secretary found the year-end distribution of *Handbooks* made preparations for the Annual Meeting difficult, and Centres also were being squeezed, as they had to have their own annual meetings prior to the Society's. Consequently, the Annual Meeting moved to February in 1957 and to March, starting in 1958. The 1958 meeting was held in Hamilton, the first ever outside Toronto. The published annual reports now began to appear in the June issue of the *Journal* rather than the March issue where they had traditionally been.

In 1958, the Montreal Centre, which had a very strong observational program, urged the extension of the Annual Meeting to a second day to allow time for members of the Society to present papers. Their proposal was adopted and so, in 1959, the Annual Meeting of the Society was held in the Debates Room of Hart House at the University of Toronto on Friday evening, March 13 and the next day, Saturday, was set aside for the presentation of ten papers. Eleven of the fourteen Centres were represented, and thirty out-of-town members including thirteen from Montreal were among the seventy-five in attendance.

Buoyed by the success of their proposal, Montreal Centre acted as hosts in 1960 for a two-day General Assembly on April 8–9 which included the presidential address. The Annual Meeting, however, had to be held in March, according to the Society's by-laws, and had to be held in Ontario, the province of incorporation. So the Annual Meeting was held following a regular Toronto Centre meeting on March 11. Neither the president (from Victoria) nor the first vice-president (from Ottawa) was there, and the national secretary was ill! However all went well, and the rather perfunctory meeting was over in half an hour.

The following year, 1961, the Annual Meeting and General Assembly were back in Toronto, held on the Friday evening and Saturday of March 17–18, but for the next seven years the two meetings were held separately. This was still necessary whenever the General Assembly was hosted by a Centre outside Ontario (as it was in 1962, 1966, 1967, and 1968), but besides that, the decision had been made to hold General Assemblies on the holiday weekend in May when travel was more pleasant, though the by-laws still required the Annual Meeting to be held no later than March 15. It was not until 1969, after the new Constitution was adopted making the Society a federally incorporated body and allowing annual meetings as late as May 31 that the Annual Meeting and the General Assembly permanently came together. Plans for a joint meeting with the AAVSO in 1974 necessitated moving the date for that year to the Dominion Day weekend and a by-law amendment was passed allowing the date of the Annual Meeting to be as late as July 31. Since that time, Annual Meetings and General Assemblies have occurred either on the holiday weekend in May or at the beginning of July.

Little needs to be said about the Annual Meetings themselves. The agenda has changed very little over the years. Elections have been by mail-in ballot for most of the Society's history with the results usually a foregone conclusion. Nominations close well in advance of the Annual Meeting (presently sixty days ahead) and generally the only nominees are the ones chosen by the Nominating Committee. There were, however, actual elections for vice-presidents and presidents in 1898 and 1976, and for second vice-president in 1984. The other offices have apparently never been contested; perhaps they involve too much work and too little honour! Years ago, when there were elective members on Council, there were sometimes more candidates than positions and write-in votes were also permitted so that real elections were sometimes required for councillors. While it is a healthy sign to have people vying for office, the sad truth is that the unsuccessful candidates, at least in recent years, have become less active in Society affairs after their defeat.

Most of the motions at Annual Meetings deal with the acceptance of reports, the appointment of auditors and other routine matters. One exception was a resolution passed on January 14, 1913, strongly urging the Government to supply a larger reflector as had been requested by the director of the Dominion Observatory. Though this was but one of several avenues initiated by J.S. Plaskett to lobby the Government, success was achieved and plans were approved later that year for the facility that became known as the Dominion Astrophysical Observatory. Another motion, adopted January 12, 1951, urged the Canadian government to designate the recently explored crater in Ungava as a National Park, or to take other appropriate steps to preserve this crater as an object of scientific interest. In this case, the Cratère du nouveau Québec is still on a provincial government list of proposed future reserves. Still another story with a happy ending was the motion passed at the Annual Meeting in 1974 petitioning the appropriate authorities to erect a plaque to W.F. King, and expressing concern about the proposed removal of the 38-cm telescope from the old Dominion Observatory. The telescope found a happy home at the National Museum of Science and Technology, and the plaque was erected the following year. Other concerns beyond the Society's own immediate affairs which were acted on by the Council were already dealt with in chapter one.

GENERAL ASSEMBLIES

Each year, either on the Victoria Day or Canada Day weekend, the Society holds a national convention to which all RASC members are cordially invited. The hundred or so who attend generally return home exhilarated but exhausted from nonstop activities. Formally the reason for the "General Assembly" is to hold meetings of committees and National Council as well as the Society's Annual Meeting, and to give

Delegates at the Hossack Memorial Lecture in Toronto's Education Centre, 1963.

The first General Assembly in Montreal, 1960

Part of a GA display in 1964.

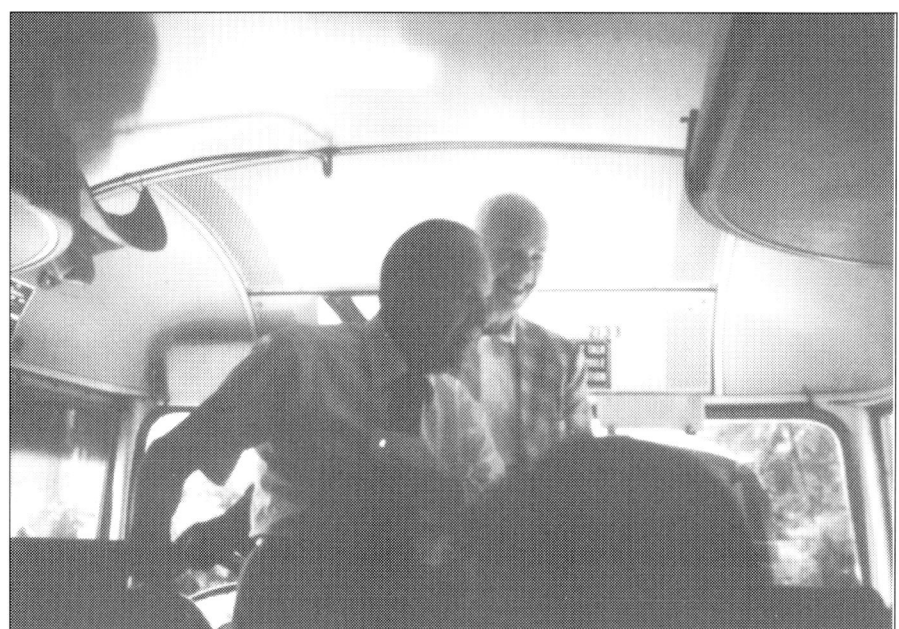
Peter Millman and Ian Halliday on a tour bus in 1973.

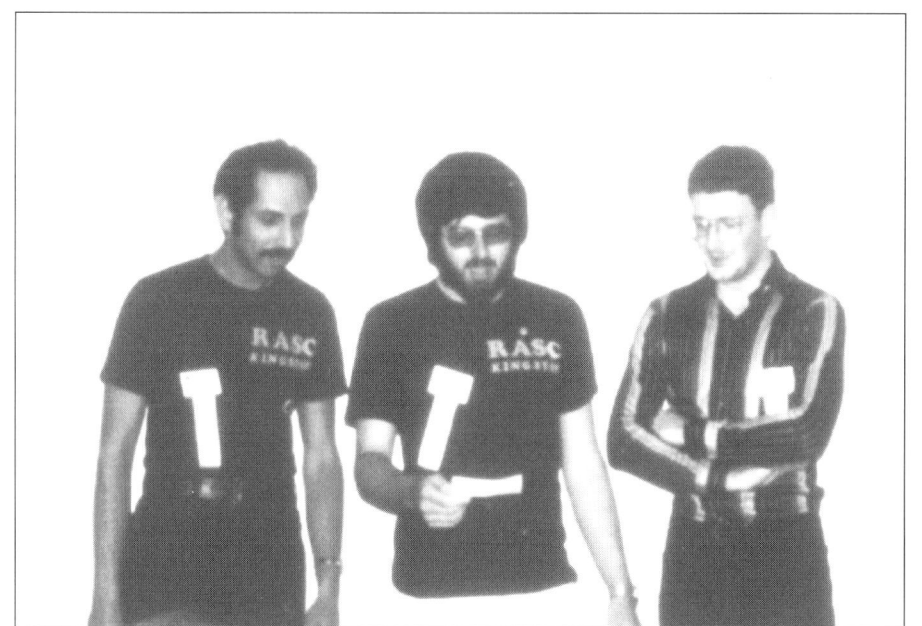
The "Kingston Trio," 1979.

Members from across the country get together.

members a chance to display their work or to present a ten-minute paper on a research or observational topic. For those with special talents, or conspicuous lack of same, there are slide shows (including the pathetic Murphy slides) and song contests for those daring enough to wrap their vocal chords around some original lyrics. Athletic types enjoy the east-west baseball game and the human pyramid. Observing is always planned for those who can stay awake day and night, and everyone enjoys lots of friendly talk at breakfasts, barbecues, banquets and beer parties. The atmosphere is more like a family reunion as members living thousands of kilometres apart but united by common interests greet each other after an absence of a year.

As long-time member Jim Low points out, General Assemblies provide a great opportunity to learn more about our country and its people as the host city changes from year to year. The format, however, has altered very little. The "GA"s have consistently been held in educational facilities. With the events always taking place in May, June or early July, university residences, cafeterias, and lecture theatres have generally been available as a convenient package at affordable rates.

Besides the paper sessions, displays and tours to observatories or other points of local interest have been standard features. A banquet, with a speech either by the president or a special guest, and award presentations have been a consistent highlight.

Though a special session at which members could present papers was held in conjunction with the Annual Meeting in Toronto in 1959, the first time that the term "General Assembly" was used was in Montreal in 1960. The following summary is intended to indicate some of the extraordinary events:

1960 Montreal Centre hosted the meetings, and the Centre français provided a tea at St Helen's Island.
1961 the Toronto Centre was the host, but tours included a trip to Hamilton's McMaster University.

Ottawa members in a song competition, 1979

Preparing a display for the '79 GA.

A human pyramid, 1979.

Do-si-do at the '89 GA.

1962 The new Queen Elizabeth II Planetarium attracted 120 to Edmonton. Calgary helped out with a tour to Sulphur Mountain.

1963 The Hossack Memorial Lecture was given by W. Petrie, on "Science and our Future" at Toronto's Education Centre.

1964 Nearly 200 attended the GA and enjoyed tulip time in the nation's capital. Commercial firms set up displays.

1965 The seventy-fifth anniversary of the Society's incorporation was marked by combining the GA with AAVSO meetings in Toronto.

1966 Winnipeg's dykes were still up after the spring floods. Lt-Gov. Bowles received delegates at Government House.

1967 Expo was in Montreal, and so was the GA, hosted by the Centre français. Helen Hogg gave the Petrie Memorial Lecture.

1968 The official opening of the PZT observatory in Priddis, Alberta, was held in conjunction with the Calgary GA.

1969 marked the centennial of organized astronomy in Toronto. Two shows were held at the new McLaughlin Planetarium.

1970 An alternating east-west tradition was building. This GA in Edmonton included the first Annual Meeting outside Ontario.

1971 Hamilton and Niagara Centres jointly hosted this GA which included a unique RASC flag-raising at City Hall.

1972 saw another joint venture, this time by the Vancouver and Victoria Centres.

1973 All eighteen Centres were represented at Ottawa, where awards for displays were given for the first time.

1974 The City of Winnipeg's Centennial made this GA a special occasion. The AAVSO joined the Society for joint meetings.

1975 Delegates to this Halifax GA all became members of The Order of Good Cheer as guests at the Province's banquet.

1976 Among fourteen categories of awards, Calgary included an innovative observing competition (won by Damien Lemay).

1977 A Wintario Grant of $1,250 helped with convention expenses for this Toronto GA.

1978 Visitors enjoyed a frontier meal of buffalo meat, sourdough bread and cranberry punch at Fort Edmonton.

1979 The Canada-Wide Science Fair coincided with this London GA. Gerald O'Neill of Princeton spoke on colonizing space.

1980 The Bluenose GA in Halifax brought the professional astronomers of CAS and RASC members happily together.

1981 Victoria Centre scheduled informal sessions for slides and songs. The Planetarium Association met right after the GA.

1982 Though numbers were smaller at this Saskatoon GA, fun and camaraderie were never higher.

1983 Two cultures and three societies (RASC, AGAA and AAVSO) met together in Quebec City aided by simultaneous translation.

1984 The seventy-fifth Anniversary of the Hamilton Centre was the occasion for this GA jointly hosted with the Niagara Centre.

1985 The RASCals were again in Edmonton, this time enjoying the spectacular new Space Sciences Centre and IMAX theatre.

1986 Excellent lectures by Barry Madore and Roy Bishop and a magnificent auroral display made Winnipeg's GA special.

1987 was the year of Supernova Shelton! Appropriately, Robert Garrison spoke about the latest development at this Toronto GA.

1988 The ASP had its hundredth Annual meeting in Victoria along with our GA and the Western Amateur Astronomers' meeting.

1989 Unattached members in Cape Breton under Raymond Auclair hosted a wonderful meeting at the Sydney Coast Guard College.

1990 The hundredth Annual meeting was celebrated with a Centenary Symposium and Canada Day fireworks on Parliament Hill.

1991 Tours of TRIUMF, the Southam Observatory and the MacMillan Planetarium were part of this Maytime GA in Vancouver.

1992 Delegates enjoyed the wonderful campus facilities of the University of Calgary, the site of the Winter Olympics of 1988.

1993 All delighted in the maritime scenery around Halifax and in hearing firsthand from David Levy about the remarkable encounters of Comet 1993e with Jupiter.

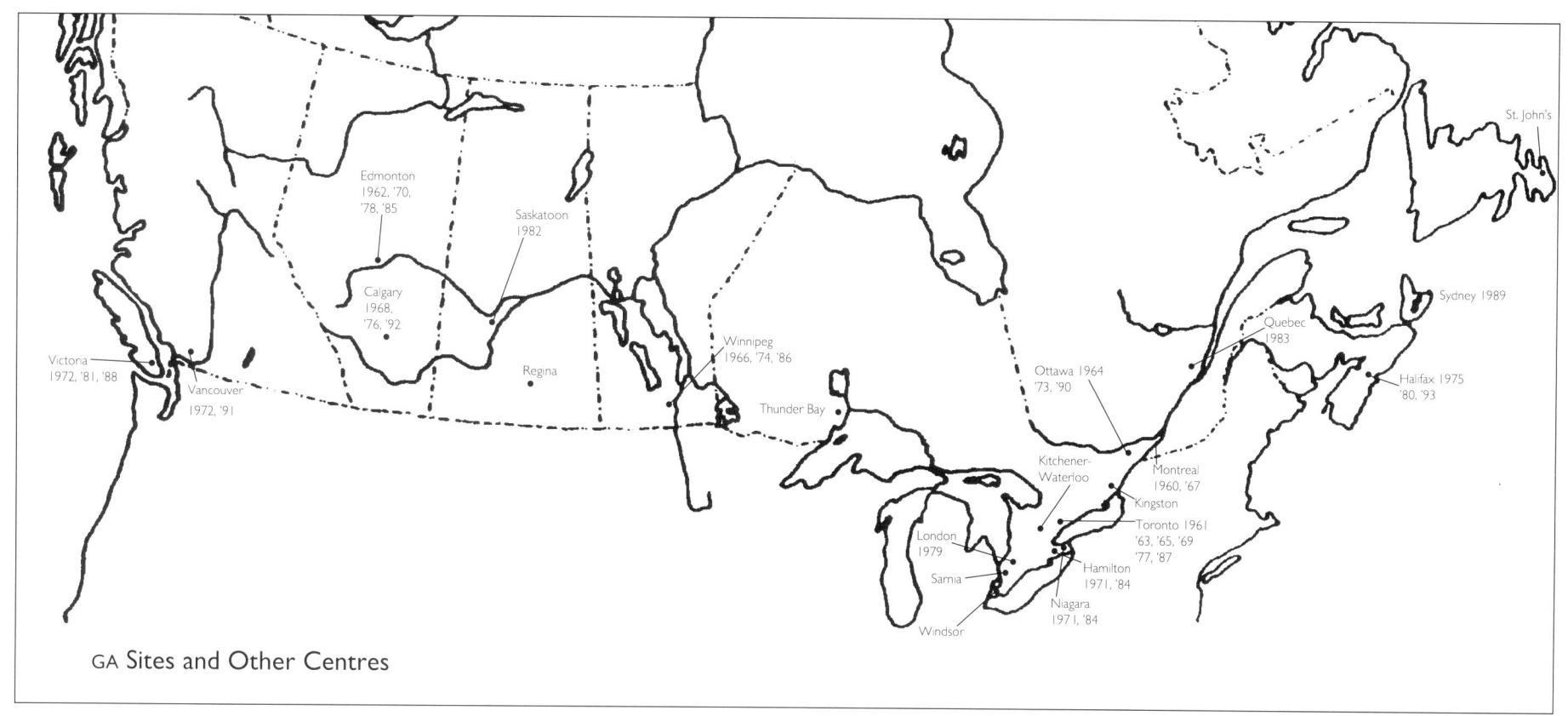

GA Sites and Other Centres

Interested spectators take a close look at a replica of *Alouette*, Canada's first satellite, at a Calgary Centre Astronomy Night, 1964

CHAPTER 11

Satellites and the First Steps Beyond

Off and on from 1890 until 1906, there were local astronomical societies in a number of Ontario towns. Meaford, Tavistock, Orillia, Simcoe and Owen Sound all were affiliated with the Toronto Society for short periods of time. Enquiries were also received from groups in Uxbridge, Wiarton, Seaforth and Port Dalhousie though no association with the RASC developed in those places. In Hamilton, an Astronomical Society operated successfully as an independent organization from 1901 until its members decided to become an RASC Centre in 1908.

Survival as affiliates must have been difficult prior to 1906 as there was no provision for the local clubs to retain any portion of the Society's membership fees for their own use. In fact, each affiliated society was expected to pay $5 per annum for the privilege of having a report of their activities published in the Society's *Transactions*. Whatever resources they required, the affiliates had to provide for themselves.

Whatever success the local clubs had can usually be attributed to the enthusiasm of one leader. In Meaford, a community of only 2,000 in 1891, it was the Reverend D.J. Caswell. He kept a group of about twenty going on a regular basis by having members read a chapter from an adopted textbook. The reading would be followed by discussion. Interspersed were other topics or perhaps an observing night, a slide show, or musical entertainment. The Toronto Society helped out to some extent by sending slides and the occasional lecturer, including the recorder and editor, Thomas Lindsay, in 1899. Caswell, for his part, enjoyed visiting Toronto where he observed with Miller, Harvey, Pursey and also with Blake at the Toronto Observatory. Caswell subsequently received several meteorological instruments from the Observatory for the purpose of establishing a weather station at Meaford. After illness forced him to retire in 1901, the Meaford Astronomical Society quietly disappeared.

The guiding light in Simcoe was Dr Wadsworth. "Much encouraged by the lively interest taken by the young people of his town in scientific matters," he and twenty-four other members established the Simcoe Astronomical Society in 1901. That year they met twenty times with attendance varying from six to eighteen:

> Personal observations were made by all members, readings were made from popular books and journals such as *Popular Astronomy, English Mechanic, Scientific American,* and *Knowledge,* original papers were presented and lively discussion ensued. The summer meetings were generally held at the Observatory of the President, Dr Wadsworth, whose 12" reflector driven by home-made clock-work enabled parties of a dozen or more to view the heavens at leisure. Other telescopes and opera glasses were used on the lawn.

Though lip service was paid to the desirability of branch societies, no formal steps were taken to encourage them financially or through representation until after 1903 when the Society's name change signalled its broader aspirations. Chant declared in his presidential address for 1905 that the RASC proposed "to multiply its membership in every part of our land and thus to make it truly national in character." The following year, by-law amendments were introduced increasing elected representation on Council and providing remission of half the fees to any Centre where members had organized a section of the Society. These steps made the difference between life and death to the Centres. Of the affiliated Societies existing before 1906, all disappeared while of those Centres formed after 1906 all but Guelph and Peterborough survived (some admittedly with temporary failures). The new arrangements evidently proved very attractive, for by 1908, Ottawa, Peterborough and Hamilton were all Centres of the RASC.

From the national Society's point of view, returning $1 per member, or half of the fees, to the Centres was a generous act of support which could never have been contemplated without the backing of significant government grants. A Centre member in 1908 received the *Journal* and *Handbook* which alone cost the Society over $3 per member to publish, plus $1 worth of services from his local Centre, all for the $2 annual membership fee. What a bargain!

For many years, members submitted their fees directly to the national office. The National Council then granted to each Centre an amount based on reports submitted by the Centre's secretary and treasurer. While Centres were thus encouraged to send in their reports promptly, they often failed to purge their lists of overdue members and so the amount of the grant submitted by the national office did not often tally with the Centre's membership roll. The practice now is for Centres to collect their own fees, withholding their portion from the amount remitted. The old problem of trying to get Centres to send in up-to-date membership lists on schedule still causes difficulties, especially for the publications which are mailed from Toronto.

The local portion of $1 per member stayed at that rate for many years and some Centres began to find the amount inadequate. The Montreal Centre had a tradition of bringing in guest speakers but realized in 1926 that they might have to levy a special fee if they were to continue. The National Council conceded that in demonstrated cases of hardship, extra amounts might be advanced to Centres and also agreed to encourage unattached members to join Centres, but even after the regular fees went from $2 to $3 per annum in 1948, the Centres still only got $1 back. They were thus getting only one-third of the fees at that point. Though that fraction was maintained following the next fee increase in 1957, Centres now got 1/3 of $5 or $1.67 for each regular member, a long overdue raise of 67 percent. It was only in 1968 that Centres got their share of the regular fees increased to 40 percent, and not until 1980 that they received the same proportion for their life members. At the present time, about half the Centres require extra revenue beyond their 40 percent share of the fees, and so charge their members a variety of surcharges. In addition to the traditional costs of paying meeting expenses, most Centres issue newsletters on a regular basis and about half have permanent observatories to maintain.

The national Society provides a number of programs to offer assistance to Centres. Travel expenses are now shared with Centres to enable their representatives to attend some National Council meetings. Special Project grants became available after the sale of the building at 252 College Street and grants have also been provided for the exchange of speakers between Centres ever since the Edmonton Centre made the suggestion in 1979. Though there were examples of unfunded exchanges in earlier years, the availability of travel funding from the national Society has been a boon to communication between Centres. The maintenance of public liability insurance was mentioned already in chapter eight.

The national Society has rarely taken the initiative to urge groups of unattached members to become Centres. Quite aside from the fact that it would not be financially prudent to do so, the feeling has always been that an organization is much more likely to be permanent when the demand for it comes from the members themselves. Even with this arm's length approach, RASC Centres now exist in twenty-two Canadian cities, with interest recently expressed by groups in Saint John, Moncton, Cape Breton Island and Sudbury. From time to time, groups outside the country — in Bermuda, the West Indies and various American localities — requested Centre status, but they were always turned down on the grounds that the Society's mandate was national not international. The RASC, of course, continues to welcome unattached members and to offer any advice or support in its power to help these extra-territorial groups. In 1957, the Council went so far as to suggest that such organizations with ten or more members of the Society might consider themselves affiliated with the RASC but apparently nothing was formalized.

The remainder of this chapter consists of short accounts of each of the first four Ontario Centres, including the two now defunct.

OTTAWA

The Society had little influence on the establishment of astronomy in Ottawa. In fact, in the 1890s only one communication from an Ottawa member, the Reverend W.J. Murphy, was reported and that dealt with plans for timing the occultation of Mars on July 11, 1892, at "the Observatory." The reference was apparently to the small government building on Booth Street which housed an 22-cm reflector and a transit instrument installed for the regulation of surveyors' chronometers in 1890. (See also p. 197) The only other Ottawa resident to be associated with the Society in the 1890s was corresponding member, Dr W.F. King, Chief Astronomer for the Government of Canada. Though King's work was entirely connected with surveying, he recognized the wider implications of astronomy and the broad public support it enjoyed. As early as 1898, he began to urge the establishment of a national observatory as a centre for scientific investigation, an idea which won government support in 1901 and which came to completion with the opening of the Dominion Observatory in 1905. The members of the Toronto Astronomical Society who had tried in vain for many years to get an astronomical observatory for their city may have felt a touch of envy. The only mention that appeared in their *Transactions* for 1901 was the following short announcement:

> Letters were read from Dr J.A. Brashear of Allegheny, Pennsylvania, and from Mr W.F. King, of Ottawa, Government Astronomer, describing the instruments to be supplied to the Dominion Government Observatory at Ottawa. The Society has learned with much interest and pleasure of the intention of the Government to establish at the Capital an astronomical observatory of the highest order.

Ottawa could be considered the first Centre of the RASC. The organizing meeting was held in the Carnegie Library on December 20, 1906, where a unanimous resolution was passed: "That an Astronomical Society be formed in Ottawa as a Section of the Royal Astronomical Society of Canada." The following officers were elected and instructed to draft a constitution and by-laws: President — Dr W.F. King, Vice-President — Dr Otto Klotz, Secretary — J.S. Plaskett, and Treasurer — R.M. Stewart. They worked quickly, and by December 31 the constitution and by-laws were adopted, and the program laid out for the first half of 1907. As explained in chapter one, some negotiating was required before approval of a new national Constitution could be reached in 1908 which gave Centres a certain degree of autonomy as Ottawa required.

The early meetings in Ottawa bore the clear stamp of the professional origins of the Centre. The speakers were nearly all connected with the DO and naturally they talked about their professional interests — mathematics, instrumentation, time and latitude, spectra, surveying, geodesy and geophysics. Generally there were one or two distinguished visitors each year who came to call at the Observatory and also favoured the Centre with a lecture. The amateur members rarely participated, except through questions submitted to the experts. Excluding the summer months, meetings occurred every two weeks, alternating between technical talks held in the afternoons at the Observatory, and evening lectures of a more popular nature held at the Carnegie Library, the Normal School or the YMCA. For many years, reports of the meetings in *The Ottawa Citizen* did much to sustain public interest, and membership stayed close to one hundred.

Though the astronomers pretty much ran the show, they were consistently generous towards all members of the Centre. An annual At-Home at the Observatory was always a popular event as this account from 1911 shows:

> After being received by Mrs Plaskett, Mrs Stewart, Mrs Smith and Mrs Klotz, over one hundred guests spent a couple of hours viewing the exhibits which had been carefully arranged so that everyone was able to get a good idea of the numerous branches of work successfully carried on at the Observatory.
>
> Transportation to and from the Observatory was much facilitated by the courtesy of the Ottawa Electric Railway Company in supplying a ten minute service in the evening. The music, which was supplied by Graziadei's orchestra and the refreshments which were tastefully arranged and served added to the enjoyment of the evening.

Though the public was always welcome at the Observatory on clear Saturday evenings, an annual open-air meeting of the Society gave members a special chance to enjoy the view through the 38-cm refractor.

The location for evening lecture meetings of the Centre was switched to the Victoria Memorial Museum in 1922 and the change seemed beneficial for a while. Attendance picked up, especially for Shapley's lecture on Star Clusters which attracted an audience of nearly 300, but a downturn in membership which had occurred during the war was not reversed. Six meetings per year was now the norm instead of the fourteen which were usual in the Centre's first decade.

In the 1930s, Peter Millman's enthusiasm got many interested in meteor observing, and even though there were hardly any lecture meetings during those years, the turnout was always close to a hundred whenever he came to Ottawa to speak on his favourite topic. From 1934 on, twenty or more people, mainly girls and boys of highschool age, would gather under the stars each August and November to plot Perseids and Leonids under the direction of Malcolm Thomson and Miriam Burland. Soon observing interests broadened; telescopes were built and brief newsletters went out containing items on timely topics like the opposition of Mars and Finsler's Comet.

By 1940, the Centre was reaching out to the community with Dominion Observatory staff in the forefront:

> On Friday January 19, 1940, at 7:30 pm, the Ottawa Centre, at the invitation of the Public School Board, held an open meeting. ... About 150 people were present, nearly half of whom were young people of school age. Mr C.C. Smith spoke on "Stars and Clocks". Following this, three telescopes were set up outside under the care of Mr J.P. Henderson, M.M. Thomson and Miss Miriam Burland. In spite of the near zero

(Fahrenheit) weather, it was nearly an hour before the last person turned reluctantly away from the views of the Moon and the three planets, Mars, Jupiter and Saturn.

A month later a joint meeting was held with the Ottawa Field Naturalists Club, at which Thomson spoke on Meteors, and telescopes were again set up. Later that year, a weekly column called "Star Facts" was started in The Ottawa Citizen. It ran for seven years with all the articles being written by RASC members, and actually made some money for the Centre. Judging by the many enquiries it brought forth, the column was widely read. It found favour in schools and RCAF stations and generated interest in the Society, especially in The Observer's Handbook. In 1944, the Centre sold 140 copies to nonmembers, and made a few more dollars that way, though with only a nickel profit on each book, it wasn't much. They persisted in their efforts however, and by 1957, with the Handbook priced at a more realistic level, they cleared $60.

During the years of World War II, the Centre made really heroic efforts to thrive even though some of their most active members were in the Forces and those who were not away were very busy with extra work and duties. Arrangements were made for visitors to give public lectures. Talks on Calendar Reform by Elisabeth Achelis and on Eddington's Life by Vibert Douglas, were two which proved very popular. Joint meetings with the Society of Chemical Industry and the Canadian Institute of Mining and Metallurgy helped to swell the audience. With the help of two teachers from the Ottawa Technical High School, Malcolm Thomson organized a Junior Club in the fall of 1941, but when he went on active service in 1942, Miriam Burland assumed responsibility. She maintained the program quite successfully for about five years, with a group of about twenty boys meeting every two weeks at the Observatory for instruction, a contest and some

MIRIAM S. BURLAND (1902-) was a student of A V Douglas at McGill University before joining the Astrophysics Division of the Dominion Observatory in 1927 - the first woman astronomer on staff. Her main responsibility was photoelectric photometry of Cepheid variables using the 38 cm refractor until she was transferred to the Seismology Division during the War. Eventually she returned to astronomy and was a member of the National Committee for Canada of the IAU in the 1960s.

Just when Miriam Burland joined the Ottawa Centre is not known but by 1930 she had become Secretary, a position she occupied until 1933. Then from 1935 until 1941 she was successively Vice-President, President and Honorary President of the Centre. In 1934, Mim (as her friends call her) began to take part in meteor work along with Malcolm Thomson and others in the Ottawa Centre. Encouraged by Peter Millman, she was soon organizing other observers and reducing the data from their observations. It was an interest she maintained for over twenty years.

Miriam Burland's name was well-known to RASC members in connection with two regular features of the Journal. For many years, until her retirement in 1967, she produced regular reports from the Dominion Observatory, and following retirement, until 1977 she wrote "About our Authors" for the Journal. As the Society's "Technical Correspondent" for several years, Miss Burland answered numerous enquiries received at the national office from members and the general public about all aspects of astronomy. Having a sincere interest in others, especially young people, and with her experience as the Public Relations Officer for the DO, she was the ideal person for the job.

The Society presented her with the Service Award in 1963.

Miriam Burland as a member of the National Committee for Canada of the IAU in 1960. Seated beside her is J.F. Heard, and standing (from l to r) are R.W. Tanner, D.A. MacRae, G.A. Harrower and P.M. Millman. The full group photo by DEWAR was published in JRASC, **55**, 94.

Fred Lossing holding the controls for the 41 cm telescope at the Indian River Observatory.

FREDERICK P. LOSSING (1915-) grew up in Norwich, Ontario, halfway between London and Hamilton. He attended the University of Western Ontario for his BA and MA and proceeded to McGill for his PhD in Chemistry which he received in 1942. Nearly all of his career was with the National Research Council in Ottawa where he specialized in properties of gaseous ions and radicals as revealed by mass spectrometry. Since retiring from NRC, he continues to do research as a guest worker at Ottawa University and is now approaching 170 published papers.

The first record of Dr Lossing's association with the RASC seems to date from 1952 when, at an annual meeting of the Ottawa Centre, he played the 'cello in a string quartet. Music is only one of his many talents. He became a founding member of the Ottawa Observers' Group in 1954, a group which he chaired in 1960 and '69. He personally built over a dozen telescopes, two of which were prize winners at Stellafane. He aluminized countless mirrors over a period of 30 years, and with friendly advice helped many members to produce their own instruments. His designs for an electronic drive system and an inexpensive photometer were widely used by many amateurs. As laudable as all this was, it was Fred Lossing's role in Ottawa's North Mountain Observatory which ultimately won him the Service Award in 1973. He chaired three committees responsible for the design and construction, the site and the management of the Observatory, built the electronics himself and always did more than his share of excavating, painting and mirror grinding. Once the work was complete, he was among the Observatory's most frequent and effective users. From 1975-78, Lossing served as Vice-President and President of the Centre and since 1990 has been Honorary President.

observing with the 10-cm equatorial mounted in the Observatory's small dome. In spite of these valiant endeavours, the membership figures reached an all-time low of forty-five by the end of the War. In the long run, however, a base had been laid and the numbers bounced back to eighty-six by 1950.

In the postwar years there was a healthy continuance of participation by amateur members in meteor observation and members nights, but there was also a revival of interest in public lectures by professionals with about four scheduled each year. Except for a year or two at Carleton College, these meetings continued to be held at the National Museum until 1955, when they moved to the Library-Assembly Hall of the DO's new Geophysical Laboratory. Still later, when government astronomy was transferred to the National Research Council, the Centre followed too in 1974, and it has enjoyed the facilities at Sussex Drive until very recently.

Nineteen forty-eight saw the first of many Annual Meetings to be held in the cafeteria at the Experimental Farm when 140 members and friends enjoyed an after-dinner astronomical sing-song led by Peter Millman and a talk by C.S. Beals about his recent trip to Europe to attend the IAU. In what seems to have been typical Ottawa style, even these annual banquets were arranged so that the Centre consistently made a profit.

An event of lasting importance in the history of the Ottawa Centre occurred in the fall of 1954. An observers group, under the chairmanship of Art Covington, began to meet once a month. Before long they were arranging public star nights and giving brief talks based on material in *The Observer's Handbook* preceding the main lecture at public meetings. Their central purpose was always to learn from one another the hows and whats of good observing. Only a few highlights can be mentioned. The group in 1963–64, nearly all keen students, was granted permission to use the instruments in the small dome at the DO. During the year they took about 200 photographs on seventy-six different occasions and presented their results at the 1965 General Assembly. Another group in 1969 made an excellent series of occultation timings which brought an extensive reply from the Nautical Almanac Office in Britain. Variable stars got a lot of attention with thirteen observers making a total of 2,216 estimates in 1970. The same year 17 people recorded 8,740 sightings of 6,608 meteors on 60 nights.

Telescope making took a giant leap forward with work commencing on a 41-cm mirror for the Centre in 1969. The project was the subject of much admiration at the GA and of a talk at Stellafane that year. The telescope itself, completely fabricated by Ottawa members, won a prize the following year at Stellafane, and in 1971 was proudly installed in the Centre's new North Mountain Observatory about 50 km south of Ottawa. The tradition of excellence in telescope-making continued with Dave Penchuk, Fred Lossing, Steve Dodson and Max Stuart being among the winners at Stellafane over the next few years.

Rolf Meier described the North Mountain Observatory this way:

> The new sixteen-inch [41-cm] telescope provided marvellous viewing. The rotating head, large Erfle eyepiece, and rolloff roof which exposed the whole sky made for comfortable observing with the feeling of being closer to the stars than ever before possible. ... In general, few restrictions were placed on keyholders' eligibility, in order that as many members as possible could enjoy the observatory. ... [Its main purpose was] to provide Ottawa Centre members with the opportunity to see the sky with all the beauty and detail which only a large telescope can reveal.

The North Mountain Observatory certainly achieved that purpose. One remarkable project involved several observers who co-operated on a search for suspected variability in the 14th magnitude quasar, OJ 278. Though the results were inconclusive, it was a useful and valuable scientific project for those involved.

Within six years of the inauguration of the North Mountain Observatory, urban light pollution became a serious problem there and so the search for a new site was started. Financed in part by the sale of colour slides and T-shirts, the relocation took place in 1977 to a secluded spot about 50 km west of Ottawa, near the village of Almonte. The reputation of Ottawa Centre observers flourished at the new Indian River Observatory; the discovery of four comets by Meier and one by Doug George were their crowning achievements. At IRO is a building of two rooms — one for the 41-cm telescope with a rolloff roof and the other serving as a clubhouse. A separate metal shed houses a 25-cm f/4.5 Newtonian built by Rolf Meier. As well, the property is the site of what has been called the world's largest amateur radio telescope.

The Indian River Observatory. In the lower view, on the left hand side can be seen some of the parabolic radio antennae.

SATELLITES AND THE FIRST STEPS BEYOND

STANLEY A. MOTT began observing meteors in the '30s. It was an interest which he would maintain for thirty years not only as an observer but as an organizer and co-ordinator on the local and national scene. He was also a generous donor to the Ottawa Centre's Observatory Fund and an enthusiastic participant in many eclipse expeditions from 1952-79.

He joined the Ottawa Centre Council as Treasurer in 1947 and served in that capacity for over a decade. He then became the Centre Librarian, a post which he continues to hold 33 years later. The citation for Stan Mott's Service Award in 1980 pointed out that he "built up an outstanding library of some 300 books, often by contributing his own money to the Library Fund. He is present at all lecture meetings and opens the library at every meeting of the Observers' Group." In 1984, the Centre honoured him by mounting a plaque naming their library the "Stan Mott Library".

Ottawa was the first Centre where amateur members took an active part in radio astronomy. This interest dates back to 1966, when the Defence Research Board kindly let members have the use of what was called "The Quiet Site," including a six-metre radio telescope and an antenna interferometer. Not much was done with the radio telescopes at that time but the site got a lot of good use, especially for summer star nights and meteor observing. By 1972, however, a few keen members were gathering interesting data on variable radio sources. Anxious to learn more about electronics and the construction of radio telescopes, they decided to build their own phase switched interferometer and in 1978, Art Covington officially opened the radio telescope at IRO, consisting of two 14 x 5-metre antennas 185 metres apart. A change in frequency from 176 to 238 MHz and other improvements in 1981 and '82 led to excellent chart recordings of the Sun and the strong sources, Cassiopeia A and Cygnus A with Virgo A and Hercules A also showing up. Changes continued to be made every year. The installation of low-noise amplifiers and a rebuilding of the dipole array in 1986–87 increased sensitivity to the point where detection of over forty discrete sources down to 10 janskys was a matter of routine.

With a few observers advanced to the point of being almost unpaid professionals, there might have been a danger that the Centre would revert to its status of seventy or eighty years before when the average member had little to do. Fortunately, the Council never let that happen. Beginners' courses were held and the Observers' Group under chairman Brian Burke, put together a very helpful 77-page manual in 1981, intended to introduce astronomy to new members.

Public awareness, maintained through media coverage, mall displays, star nights and Astronomy Day events, not only brought about a growing membership but required more and more help from members to make it successful.

Recently, members have taken a renewed interest in meteor observing, using a new set of meteor "coffins" at the home of Rolf Meier. Many members are enjoying marvellous views of deep sky objects through Rob Dick's 60-cm f/4 reflector, located at his observatory far out of the city at Big Rideau Lake. Some are taking an interest in optical work under Peter Ceravolo's direction and some are exploring new possibilities with CCD cameras.

Life in the Ottawa Centre bears little resemblance to what it was like in 1907. Public lectures, some given by government astronomers, are still a popular and important feature of the annual program. But the DO itself is closed and the telescopes have been moved away to the National Museum of Science and Technology. If the charter members could drop by for a visit, what would they think of a Centre observatory completely built by amateur members and its telescope whose aperture exceeds that which was the pride of their profession? Whatever they could make of the incredible advances in technology, they would surely respect and be exceedingly proud of the confidence and accomplishments of the Centre they established so long ago.

PETERBOROUGH

In the early years of the twentieth century, no one was a more ardent proponent of astronomy in general, and the Society in particular, than the Reverend Dr Daniel B. Marsh. Wherever his pastoral duties took him, an astronomical group was soon thriving under his leadership. While he was the minister at Springville, Ontario, he persuaded the Young Mens' Guild to sponsor a meeting at the Charlotte Street Church in nearby Peterborough on February 26, 1907. Mayor R.F. McWilliams was chairman, and Dr Marsh spoke on "A Night in the Skies" to a large and enthusiastic audience. Following the lecture, steps were taken to form a local organization, intended to be a section of the RASC, with Marsh as president. Peterborough adopted the constitution of the Ottawa Section, the National Council approved, and the new group was off to a flourishing start with forty-seven members on the roll by May, 1907.

Dr Marsh knew how to treat people. He invited the members out to his home in Springville for social evenings and a look through his telescope. On one of these occasions, a photograph was taken of the Moon and prints were later mailed out. Besides the regular monthly lectures, he arranged parlor meetings in members' homes where beginners felt more comfortable about raising questions and expressing opinions. One delightful story tells of Dr Marsh conducting church service on Sunday, June 28, 1908, while his son James was left in charge of the telescope and camera to take photos of the partial solar eclipse. Following the service, the congregation saw the eclipse through Marsh's 13-cm telescope on the lawn of the Springville manse.

A number of other telescopes were soon in use in Peterborough. R.F. Stupart, director of the Toronto Meteorological Observatory, and a past-president of the RASC arranged for the loan of a government telescope to the Centre. Homer Fiske fitted up a very complete observatory at his home and there determined time and the longitude of Peterborough within one-fifth of a second. Herbert Collier, vice-president of the Centre, became the proud custodian of a fine 8-cm telescope made by Marsh and his son for the use of Centre members, and a prominent member, the Honorable J.R. Stratton, M.P., (who had been Lumsden's successor as Provincial Secretary before being elected to the House of Commons in 1908) presented an 8-cm refractor to the Centre in 1910.

As the following report shows, that's not all that Stratton did for the Society:

> On the evening of June 22 [1909] Hon. J.R. Stratton, M.P., and Mrs. Stratton entertained the members of the Peterborough Centre of the

D.B. Marsh and his 13-cm refractor.

Astronomical Society of Canada, at their residence "Strathormond." The grounds were beautifully illuminated by suspended electric lights, while convenient seats and rugs were distributed in different parts of the lawn.

Three telescopes were placed on the grounds for the use of those in attendance, and much pleasure was derived from them.

Among the guests in attendance from outside points were the Society's Honorary President, Dr. King, C.M.G., of Ottawa, and Dr. C.M. Stratton, of Napanee.

A very pleasant evening was spent in viewing the heavenly bodies through the telescopes, and in social intercourse. In addition the 57th Regimental Band provided music, and refreshments were served from a marquee on the lawn, Coleman Bros. catering in a creditable manner.

Dr C.M. Stratton of Napanee was also a member. In 1908, he gave an interesting and instructive lecture to the Peterborough Centre on the subject "The Eye – the Medium of Vision" and apparently he made arrangements for Marsh to speak in Napanee at a meeting which, according to reports, was attended by one thousand persons.

The proximity of Peterborough to Toronto was an advantage to the new Centre. Dr Chant came most years to give a lecture. The 1909 program included visits from J.A. Paterson and R.F. Stupart from Toronto, G.P. Jenkins of Hamilton, and Captain J.E. Bernier, recently back from his Arctic explorations. A very special visitor from the United States, RASC honorary member Frank Schlesinger, was entertained at the Oriental Hotel before speaking to an audience of 400 in 1913.

In fact, Dr Marsh reported, 1913 was the best year yet for the Centre:

> There is a hearty interest among the members and several of our young men are looking toward an Astronomical Course in the University of Toronto — two intending to devote their lives to the study. The Board of Education has generously supplied us with rooms, including light and heat, without charge. The telescopes ... [are] in constant use.

At the Annual Meeting, on December 10, 1913, membership was reported at forty-four, with much interest being taken in the work of the Centre. Sadly for Peterborough, the meeting was followed by a luncheon in honour of Dr Marsh who was moving to Holstein, Ontario.

Just how important Dr Marsh's leadership was can be seen by the rapid decline after he left, though the war probably took its toll too. By 1915, membership in the Centre had dwindled to eleven, and no meetings were reported. There still was interest, evidently, for over 200 turned out to hear Dr W.E. Harper speak about Nebulae in 1916, but only a very brief report was submitted that year and nothing in 1917 or subsequently.

HAMILTON

The Astronomical Society of Hamilton was founded December 20, 1901, with the Reverend D.B. Marsh as its presiding officer. Early the next year, it became a section of The Hamilton Association for the Advancement of Literature, Science and Art which had originally been established in 1857. Marsh had also joined the Toronto Astronomical Society in 1901 and was elected to the RASC Council for 1904, thus establishing cordial relations between the two organizations. He was renowned for his observatory and the extensive work he did with it. The observatory was a wooden structure with a revolving dome which housed a 13-cm achromatic refractor made by Brashear with fittings and setting circles made by Marsh himself. Supplementary equipment included several eyepieces, a specially constructed camera, a star prism, a Herschel prism for solar work and a solar spectroscope, also by Brashear, which enabled

Marsh to study prominences.

Marsh, as already noted, moved on to Springville where he started the Peterborough Centre in 1907 - an example that apparently inspired the Hamilton Astronomical Society to consider joining the RASC. Marsh was invited back to a meeting in Hamilton on December 15, 1908, to explain the organization and objects of the RASC and he strongly urged the members to "abandon their charter and merge themselves into the national organization." Evidently his arguments were persuasive for thirty-five people signed up that night. The president of the group was G. Parry Jenkins who, before emigrating to Canada, had founded the Astronomical Society of Wales and had been a friend of the well-known amateur astronomer, Canon Webb, in Hereford, England. Jenkins accompanied Marsh to the RASC Council meeting the following week where their request to form a Centre was enthusiastically adopted, and the first meeting of the Hamilton Centre took place on January 15, 1909.

In fact, the entire first year was highly successful. Five hundred circulars explaining the new developments were sent out and the local papers gave the new Centre good publicity. There were many new applicants for membership including Lt-Gov. Gibson. Thirteen meetings of the Centre were held in the Recital Hall of the Hamilton Conservatory or in the Art Gallery of the Art School but even the large gallery was not sufficient to accommodate the

William Bruce and his Elmwood Observatory, Hamilton.

audience which gathered to hear Marsh speak on Stellar Evolution. Lecturers from Toronto were also quite willing to make the short train trip, and so Hamilton received past-presidents Chant, J.A. Paterson and Musson and the current president of the RASC, A.T. DeLury, as speakers. Professor S.A. Mitchell, then of Princeton University, visited both cities, and was introduced to his Hamilton audience by Mayor McLaren.

The Centre's first vice-president, William Bruce, invited members to visit his Elmwood Observatory on the Mountain at any time. Situated in the middle of an eight acre field (known today as Bruce Park), it was a splendid installation. To the dome formerly owned by Marsh, Bruce had added "a reception and reading room, having a flat roof, and outside stairs leading up thereto, for viewing star clusters with mounted field glasses." He had 5-cm and 8-cm telescopes made by Marsh with Brashear lenses, a 10-cm Grubb refractor and a 22-cm reflector. Bruce became president of the Centre, spoke at meetings, served as the RASC's vice-president, and continued to welcome visitors to Elmwood for many observation meetings and public star nights over the next decade, but there was a limit to what an octogenarian could do. By 1916, he was eighty-three, the war was taking everyone's attention, and interest in the Centre faded. That year there were only nineteen members and only three meetings were held.

The membership continued to dwindle down to eight by 1922. No meetings had been reported since 1917, and the Centre officially disbanded in 1926 when the bank balance of $33.37 was returned to the Society. One of the original members, H.B. Witton, formerly a member of Parliament for Hamilton, kept up his interest in the Society and maintained an excellent personal library containing works by such famous lunar cartographers as Hevelius, Beer and Madler. He had spoken to the Hamilton Association on selenography as early as 1881, had generously donated Rutherfurd photos from the 1850s to the Toronto Society, and had produced his own lunar map 30 inches in diameter. Chant enjoyed a visit to his home in 1921, where he found Witton, then ninety years of age, at work on an alphabetical index of lunar features.

A meeting to reorganize the Hamilton Centre was held in the Public Lecture Hall on April 8, 1930. About 200 people turned out to hear Dr Marsh, now living to the west of Hamilton in Norwich, talk on "Our Sun," and sixty-eight people joined the Centre that evening. The executive included as Honorary President – the Reverend Dr D.B. Marsh, President – J. A. Marsh (D.B.'s son), and Secretary – Miss Beatrice Marsh (D.B.'s daughter). Mrs D.B. Marsh joined in her own right in 1932. No wonder the Hamilton Centre decided in 1980 to name their observatory after their ubiquitous founder! For Dr Marsh, Honorary President was not merely a title of distinction. He continued to be a regular speaker at Centre meetings and led the successful solar eclipse expedition to Acton Vale in August, 1932. Even though it was during the Depression, the Centre's expenses of that trip were partly recouped by charging admission to a special public presentation in October given by Marsh and others in the eclipse party.

The General Council of the Society welcomed Hamilton back into the fold and several speakers, including Chant, J.R. Collins, H.R Kingston and R.K. Young went there to give lectures in 1930. For their part, the Marshes, father and son, reciprocated by speaking to the Toronto Centre.

Another prominent Hamilton Centre member at the time was the vice-president in 1930, W.T. Goddard. Each June, at least until 1944, he and his family welcomed hundreds of visitors to their home on Paradise Road for a "field night," the event in 1937 being almost too successful:

Twelve hundred visitors turned out to see the Moon, Jupiter, Mars and more distant objects through several telescopes erected for the occasion by the members of the Hamilton Centre. Alderman John Marsh, M.P., was in charge of Mr Goddard's 5-inch [13-cm] clock-driven refractor. The other telescopes ranged from 8-inch reflectors to 2 1/2-inch refractors. ... Due to a combination of circumstances, including a perfect night, splendid publicity with the co-operation of The Hamilton Spectator and Mr Goddard's inviting estate, this was a red-letter day in the annals of the Hamilton Centre. ... The Goddard family were busy until well after midnight supplying their guests with coffee, cakes and cookies.

Since it was generally acknowledged that many of the 1,200 had more interest in the refreshments than in astronomy, the Council made a wise decision to charge a ten-cent admission fee to "the Goddarium" of 1938, with proceeds going to a local charity.

From 1930 to 1935, the Centre met in the Public Library and after that in the Art Gallery. A regular feature before the main lecture was the so-called "curtain raiser" at which two members briefly reported on the latest astronomical news and what to look for in the current sky. This feature was replaced in 1940 by opportunities for prepared answers to be given to written questions. Norman Broadhead, the secretary at the time, reported:

> Perhaps the most elementary question asked during the year was why Orion rises on his side and sets on his feet; while the one which occasioned the most difficulty was why a planet that reduces its orbital velocity would not reduce the size of its orbit by "falling" nearer to the sun. ... It took two or three meetings to straighten this matter out, as each time the question was revamped and again submitted for further treatment. This incident is mentioned as evidence of the patience of those answering the questions

This clipping from the *Hamilton Spectator* for 28 May, 1936, gives an idea of the extent of the arrangements made for star nights on his property.

WALTER T. GODDARD was an electrical engineer by profession and owned a small porcelain factory which produced insulators for high voltage power lines. When the Hamilton Centre reorganized in 1930, he was their Vice-President, and had a great deal to do with the successful revival of the Centre and of astronomy in Hamilton. Though he never was President of the Centre, he became Honorary President in 1939, a post which he held for thirty years. His devoted service was recognized in the presentation to him of the Service Award in 1964.

Alfred E. Johns, his wife Myrtle, and three sons, Martin, Harold and Paul.

ALFRED E. JOHNS (1884-1959) grew up in rural Ontario and went to the University of Toronto for his BA and MA, which he earned in 1908. He was a scholarship student in both modern languages and mathematics and led the University football team to a championship. He decided to enter the ministry and was ordained in the Methodist church in 1910. For the next 15 years, he and his wife lived in China where he worked as Professor of Mathematics at West China Union University in Chengtu and she served the mission as teacher and nurse. The Johnses and their five children returned to Canada. He taught at Brandon College, Manitoba, 1927-31, and then joined the faculty of McMaster University in Hamilton. Continuing his studies and research, he earned his PhD from U of T in 1935, his thesis being on the bilinear transformation in the real plane. After he retired from McMaster in 1951, he served St Giles Church in Hamilton as Assistant Minister.

Reverend Johns joined the RASC in 1929 but it was only after moving to Hamilton that he became active in the Society. He was on the Council of Hamilton Centre for many years and was their Vice-President and President in 1934-37. He attracted several outstanding speakers to the Centre and gave many popular lectures himself over the years, not only in Hamilton but also to Toronto, London and Windsor Centres. From 1942-46 he was national Vice-President and President. Two of his sons became physicists and RASC members. Harold was noted for his research into radiation treatment of cancer, and spoke on at least two occasions to Edmonton Centre. Martin was on the Council of the Hamilton Centre in the '50s including two years as Vice-President, and gave planetarium demonstrations and lectures to the Centre a number of times.

rather than the persistence of the members asking them. However, if both attributes are virtues, this Centre is indeed blessed.

Mr Broadhead was credited with being something of "an innovation expert" as meetings during his subsequent presidency included such novel features as games of questions and answers, a dramatization of a dialogue on relativity, an entertainment night and a meeting completely presented by the ladies. Over the years, local members began to take more of the responsibility for the Centre meetings, with just an occasional distinguished visitor like Dr Pearce from Victoria adding a special sparkle to the program.

A few members also began to show an interest in more serious observing. In 1936, for example, Dr W. Findlay and Mr T.H. Wingham observed Perseid meteors and methodically plotted sunspots; J.A. Marsh photographed Peltier's comet and Jupiter's moons.

In the postwar years, a friendly relationship developed with McMaster University. A number of professors were active in the Society, meetings were held in Hamilton Hall and field nights began to be held on the campus in 1947. Proceeds from these popular public events could exceed $100 on a good night, so the Council decided to save up their money towards the $1,100 purchase price of a small planetarium projector. The plan was that the Centre would use it at their meetings and the Mathematics Department would find it helpful in their Astronomy courses. With the understanding that the university would build a dome and maintain the planetarium, the Hamilton Centre donated the equipment to McMaster in 1949. The formal presentation was attended by over 700 people and included a lecture and demonstration by the designer, Mr Armand Spitz, of the Franklin Institute, Philadelphia. Field nights after that were even more popular as they included not only the traditional views through the telescopes, but movies in Hamilton Hall and demonstrations of the planetarium, all for the price of admission — 35 cents.

At the time of the planetarium's tenth anniversary, it was noted that:

> It has been housed in three domes (including a parachute) and has travelled on at least 25 occasions to surrounding towns for the presentation of planetarium demonstrations to the general public. In the ten-year period the planetarium has averaged 200–250 demonstrations per year [meaning that] somewhere in the neighbourhood of 150,000 people have had the opportunity of viewing the stars and hearing about the stars, people who probably would have had no contact whatsoever with astronomy.

McMaster University found that the planetarium was a valuable teaching aid for astronomy classes and also a very excellent public relations instrument to make the university better known to the community.

Some special characteristics marked the 1960s. A separate Junior Section with its own president and secretary ran quite successful meetings during the half-hour or so just prior to the regular Centre meetings. Usually about fifteen members in the eleven to sixteen age range met for a bit of business and a fifteen minute talk given either by an adult or by one of their own. The telescope makers were another very popular group, having started up in 1958. About twenty members met regularly every two weeks and many completed telescopes with mirrors between 15 and 30 cm in diameter. An observers group and a discussion group which met monthly in members' homes, were a natural outgrowth of the telescope makers. Public education also spread its wings to include field nights, talks and displays at the Burlington Fair, and at schools and for Cubs and Scouts in town and at camps. There were twenty-three school presentations alone in 1968. All in all,

WILLIAM J. McCALLION (1918 -) received his BA degree in Mathematics from McMaster University in 1943 and was immediately appointed to the staff as a Sessional Lecturer in the Naval Training Program under the aegis of both the Physics and Mathematics Departments. After the War he earned his MA and continued his academic career at McMaster - a career which included several years as Director of University Extension and Director of Educational Services. He retired in 1978 as Dean of the School of Adult Education but continued teaching as Professor of Mathematics until his final retirement in 1987. Bill McCallion joined the RASC in 1943. He frequently was the speaker at meetings of the Hamilton Centre, served on its Council for several years and was Centre President from 1950-52. He was concerned that many Centre lectures were too difficult for the audience and that this was a factor in the large turnover of members. Consequently, during his years as a national Councillor, he made a proposal which led to the formation (in 1962) of the Adult Education Committee to investigate the possibility of using films, slides and tape-recordings.

He was instrumental in getting the Spitz planetarium for Hamilton in 1949 and as director of public viewings played a major role in personally showing the beauties of the heavens to over 150 000 persons. Appropriately, McMaster named its newly equipped planetarium in his honour in 1992.

The photo, taken at the dedication ceremony on 9 February, 1993, shows (l to r) Dr Douglas Welch, Professor McCallion and McMaster President, Geraldine Kenney-Wallace.

WILFRED S. MALLORY (1893-1969) was among those joining the Hamilton Centre when it revived in 1930. He was a teacher of Mathematics, Physics and Music at the city's Delta Collegiate where he organized the first school band. He was also a member of the Scottish Rite and the Hamilton Camera Club. Wilfred Mallory served on the Centre Council for 25 continuous years, including a term as President in 1932-33. He spoke frequently at Centre meetings on a variety of topics from Meteorology to Gravitation. When he received the Service Award in 1965, his guidance and encouragement of student members and young people was especially noted.

ERIC ORR (1904-90) received the Service Award in 1988 for his enthusiastic, dedicated support of amateur astronomy. He and his wife Mary were well known throughout the Society for their regular attendance at nearly every General Assembly since joining the Hamilton Centre in 1970. The citation for his award stated that he served on the Centre's Board of Directors for 15 years, ably managing various positions of responsibility, including that of National Representative. He generously gave material and financial assistance for the construction and development of the Centre's observatory, and for many years handled the distribution of the Centre newsletter, *Orbit*. In addition to being an active observer with a wide variety of interests, Eric Orr contributed several articles to *Orbit*, and assisted with the public education program at the observatory. The citation concluded: "Always on hand to greet the public at mall displays and observing sessions in his enthusiastic yet courteous manner, Eric [was] the "ambassador of astronomy" to many a potential new member."

Mary and Eric Orr (foreground) at the 1987 GA.

high spirits were sustained by banquets, picnics and the group meetings in members' homes. All these events were publicized in *Orbit*, the Centre's newsletter, which started in 1968.

Also about this time, the Hamilton and Niagara Centres established contact with a number of associations in upper New York State to propose a joint astronomy weekend. That initiative proved very beneficial as over the years the connection with the Niagara Frontier Council of Amateur Astronomical Associations (NFCAAA) provided exchanges of speakers, periodicals and ideas. The influence of the Centre spread even further as a result of Ken Chilton's active involvement with the International Union of Amateur Astronomers. In 1975 IUAA representatives from eight countries met in Hamilton for their triennial assembly, at which time they elected Chilton as their president and Peter Ashenhurst, also a Hamilton Centre member, as secretary.

Plans for a Centre observatory began to develop when an optical firm offered to contribute a 76-cm telescope provided that the Centre would build and maintain an adequate observatory. By 1973, they had acquired a ninety-nine-year lease on a six acre site a few miles north of Hamilton in East Flamborough Township and had prepared detailed plans which included domes for smaller instruments, a darkroom, a library and adequate space for several members to observe and work at one time. It was a very ambitious project with government support needed and a goal of $25,000 in a fund-raising campaign. In the end, the funds were very slow in coming with the result that the promise of the 76-cm telescope fell through but the observatory project progressed slowly in modified form. After a great deal of work under the direction of Peter Ashenhurst and Les Powis, a Special Projects Grant from the Society, and generous donations of time, money and materials from many members, a Centre observatory finally opened in 1979. At first it housed a 32-cm telescope but plans were in place for some members to build a 44-cm equatorially mounted instrument. In the meanwhile the Centre acquired the wonderful old 13-cm refractor of D.B. Marsh, installed it in the dome, and gradually refurbished it to a better than new condition. What was the Centre to do with all their fine telescopes but only one dome? The obvious answer was another building. So once again the members got busy and erected a new structure, 5 metres square with roll-off roof for the newly completed 44-cm telescope. On completion, the new facility was named the Chilton building, with the original observatory housing the refractor known as the Marsh building. The site as a whole was dedicated as the Leslie V. Powis Observatory on Astronomy Day 1987. The same year, a 13-cm antique brass refractor, formerly belonging to Dr Bell of Paris, Ontario, was donated by his widow. After cleaning and mounting in the Chilton Building, it soon became much admired for its spectacular planetary views. Electricity was brought into the new building, a telephone with a taped message for the public was connected, and washrooms were installed at the site. A substantial library has been built up and named in honour of the late Ian Stuart and the catalogue can now be accessed through the observatory computer. The 1990 Annual Report lists the telescopes belonging to the Centre as two 13-cm f/16 refractors, a 40-cm f/4.5 and a 25-cm f/5 equatorially mounted Newtonian, a C8, 25-cm f/6 and 15-cm f/8 Dobsonian portables and two 60-mm refractors. With a full set of eyepieces, solar filters and photometer, the observatory has become arguably the best equipped amateur observatory in the country.

All the years that went into the development of the Powis Observatory and the continuing efforts and expense to maintain it produced some wonderful benefits. Increased numbers of members attended the monthly observers group meetings and workshops, some took part in a long-term project to draw and photograph Mars, Jupiter and Saturn,

FRANK SCHNEIDER (1900-84) developed an interest in astronomy as a young boy in Yugoslavia. He emigrated in 1926, married and had two daughters, and took employment as an eyeglass specialist at the Steel Company of Canada, where he worked for 40 years. He applied his talent in optics to the making of telescopes. His first was an 20-cm reflector made soon after he joined the Hamilton Centre in 1932.

The citation for his Service Award presented in 1972 stated, "He has served on both the local and national Councils of the Society and has been a frequent lecturer to meetings of the Hamilton Centre as well as to other local groups. Frank Schneider successfully guided many Hamiltonians through the difficulties encountered in the construction of their telescopes and then set an example for them to follow as a tireless observer. Variable stars, meteors and solar eclipses were prominent among his observing interests." He also won wide acclaim for his photographic skill at the telescope and as a naturalist.

JAMES A. WINGER (1922-) was an electronics technician at Westinghouse. He joined the Centre in 1952, was elected to the Council in '56 and served as President in 1961, '66 and again from '83 to '85. He and some others organized a visual meteor watch program in Hamilton for the International Geophysical Year in 1957-58 which developed into the Centre's Observers Group. He also established a group of amateur telescope makers who held their meetings at his home; for many years at field nights, he manned the group display. In the early '60s, his two daughters were keen participants in the Centre's Junior Section. Jim Winger received the Service Award in 1984.

From left to right in this 11 July, 1956 *Hamilton Spectator* photo, are Frank Schneider, Gord Craig, Ron Daiton, Jim Winger, Les Powis, Stewart Buntain.

The dome and RASC flag being lifted into place atop the Hamilton Centre Observatory.

LESLIE V. POWIS (? -1985) joined the Hamilton Centre in 1957. He already had a well-developed interest in astronomy by that time for he reportedly was the first person in North America to see Comet Mrkos (on 11 August, 1957). The following year he joined the Council of the Hamilton Centre, serving in 1962 and again in 1971 as their President. He was elected Recorder of the Society in 1964 and held that position for eight years. Les Powis received the national Service Award in 1984.

He frequently spoke at meetings of the Centre on a great variety of topics. His interests included telescope making, ham radio and radio astronomy. He encouraged a group of local youngsters to form an astronomy club and his son, David, became its President. This led to the formation of Hamilton Centre's Junior Section. As Treasurer of the Centre, he played a big role in arranging the financing of the Centre observatory, but besides that he was responsible for finding a suitable site and took a very active part in the construction and maintenance of the building. Because of his outstanding contribution to this project, the Centre named it the Leslie V. Powis Observatory following his death.

A photo of L.V. Powis is on the previous page.

others found solar observing or deep sky photography to their liking. The Observatory saw excellent use in public education. Twice each month, groups of Cubs, Scouts and Brownies, as well as the general public visited the Observatory for a twenty-minute talk by one the members, a movie, a tour of the building and a look through one or more of the telescopes. Several hundred people generally visit the Observatory each year.

Beyond the Observatory, the Centre continues its program of star nights at parks and campsites, and mall displays feature a fine new set-up. Though membership has always had its ups and downs between 50 and 100, the Centre's excellent facilities and programs will surely be the basis for healthy growth in the years ahead.

The McMaster planetarium is also revitalized. The old projector had worn out and replacement parts were no longer available but with the arrival of a more modern projector from the Ontario Science Centre, Hamiltonians are once again enjoying star shows under the dome. The renovated facility is now appropriately known as the W.J. McCallion Planetarium.

GUELPH

In 1911, Guelph joined the ranks as a Centre of the RASC. Seven meetings were held that year with an average attendance of seventy-five. Amongst the members were Alderman H. Westoby (the Centre's president), Lt-Col McRae (father of John McRae of Flanders Fields fame), and H.E.S. Asbury. Mr Asbury had formed an observing group in Seaforth in 1898, was now active in Guelph where he made a standing invitation to all members to visit his home to use the Centre's 15-cm telescope, and later became one of those principally responsible for the founding of the Montreal Centre. During 1913, as part of its program, the Centre enjoyed visits from Dr Chant and Dr Schlesinger, who was on his RASC lecture tour. Unfortunately membership steadily declined during

CHAPTER 12

New Frontiers

The Society's first expansion outside Ontario was westward with Regina and Winnipeg joining the RASC family in 1910, the year of Halley's Comet.

Winnipeg's importance as the gateway to the West was very evident a hundred years ago with railways and surveys branching out to the frontier. As early as 1888, Otto Klotz had set up a reflector of 22-cm aperture along with a 8-cm transit instrument to establish an astronomical reference point for surveyors. The A&P Society had one enquiry about membership from Winnipeg, from a Mr H.C. Howard in 1895, but the first real evidence of any widespread scientific interest came in 1909. In September of that year the British Association for the Advancement of Science held one of their far-flung meetings in Winnipeg under the chairmanship of their President Sir Joseph J. Thomson, the famed discoverer of the electron.

It was a great novelty for Winnipeggers to have such distinguished scientists in their midst and the BA meeting aroused considerable enthusiasm among the citizens. That, of course, was just what the Association hoped would happen, and was the primary reason for their policy of peripatetic conclaves. The very same month that the BA met, the Astronomical Association of Western Canada was organized at the suggestion of J.P. Hughes. Two months later, in November 1909, Neil Bruce MacLean, Professor of Mathematics and Astronomy at the University of Manitoba, agreed to accept the Presidency. He was able to arrange for the Association to hold their meetings in the University Building on Broadway Avenue where a projector, library and the Department of Astronomy's 13-cm equatorial telescope were all made available. MacLean had graduated with W.E. Harper from the University of Toronto in 1906, and had worked with Harper for a time at the Dominion Observatory before joining the faculty at the University of Manitoba. He later worked for Sun Life as an actuary and became a professor at McGill. The AAWC remained independent for only a year, becoming the Winnipeg Centre of the RASC on November 22, 1910. Thirty-one members joined of whom ten were women.

Though Winnipeg was then Canada's third largest city, it still was pretty isolated from the world of science and academia. Finding outside speakers was very difficult. The RASC president, A.T. DeLury, was able to pay a visit in 1911, and Dr Curvin H. Gingrich, associate editor of *Popular Astronomy*, travelled in 1913 from Carleton College in Northfield, Minnesota, to address the Centre on "Modern Methods of Stellar Photometry." But as his expenses amounted to $35, nearly the entire annual budget of the Centre, such visits had to be rare indeed. World War I nearly finished the Centre but the RASC Council provided support beyond the customary $1 per member per year. For instance in 1915, Winnipeg's grant was $50 though only twenty members were on the roll, with only twelve paid-up and of these, eight were Officers or Councillors. This backing helped to keep the little band going and they had five meetings that year.

Darby Coats is shown here holding the Service Award which he received in 1968.

DOUGLAS RICHARD PROCTOR (DARBY) COATS (1892 - 1973) was a prominent member of three RASC Centres during his lifetime. His interest in astronomy was probably inspired by the achievements of his famous great-uncle, R.A. Proctor. Coats joined Montreal Centre as a charter member in 1918, but went to Winnipeg in the '20s where he lived for nearly 40 years before moving to Calgary. Wherever he went, his enthusiasm for astronomy was infectious.

Darby Coats was one of Canada's first radio broadcasters, starting with Montreal's experimental station, XWA. For many years, his name was a household word across the Prairies as the popular "Uncle Peter" host of a children's radio program; his Dickens' "Christmas Carol" was an annual tradition.

Coats was President of the Winnipeg Centre in 1934-5, 1947-8 and 1952-4. His speeches, on a great variety of topics, show that he was frequently a pioneer - "Teaching the Stars by Radio" (1929), "Atmosphere and Radio" (1931) and "The Sun and Radio Reception" (1933). In 1930 he played phonograph records of H.H. Turner and Oliver Lodge for the Centre. He was a skilled amateur telescope maker in the early 1930s, and was still speaking on this subject in 1965. Coats was a ship's radio operator in the First World War, surviving two sinkings, and was a Flying Officer in the Second World War. He observed the total solar eclipse of 1954 from the air.

He donated some books and charts by R.A. Proctor to the national Library in 1972 and wrote an unpublished history of radio communication in Canada.

Without the leadership of a few university men and the use of the university facilities, the Centre would surely have foundered. Professors MacLean, H.R. Kingston and L.A.H. Warren all spoke frequently and held various offices. Observing nights and short talks, each dealing with one constellation, planet or star enhanced the regular meetings, especially for members of the public who had little familiarity with the sky, and membership climbed to about fifty by 1920.

From this larger base, amateur members such as H.B. Allan, D.R.P. Coats, A.W. Meggett and Mrs E.L. Taylor were able to take charge of more of the activities themselves throughout the 1920s and 1930s. Observing through the university's telescope became a regular feature following lectures, and Meggett scheduled an additional evening each month for observation. At a typical observation meeting in the 1930s, members enjoyed views of the Moon, planets and double stars through telescopes set up on the university grounds. Apparently the public was admitted only by invitation, three targeted groups in 1935 being the Board of Trade, the Medical Association and the Boy Scouts.

One professional astronomer who did speak to the Centre a number of times in the 1920s and 1930s was J.S. Plaskett with the result that he was adopted as honorary president. Meetings or business occasionally called him east from Victoria, and he would make a stopover in Winnipeg whenever he could. One of the happiest of these events was a luncheon at the Fort Garry Hotel hosted by the Centre in 1930 when Dr and Mrs Plaskett were returning from England where he had just received the Gold Medal of the RAS.

Other special visitors came in August, 1932, as an extension of joint Solar Eclipse/IAU meetings in the eastern United States and Canada. On this gala occasion, the Centre held a dinner at the Royal Alexandra Hotel at which astronomers from Britain, France, Belgium, Italy and Switzerland were the guests of honour. Among the speakers that evening were the presidents of the RAS and the BAA. Later in the month, Sir Frank Dyson, Astronomer Royal, gave an illustrated address on the Eclipse of the Sun.

Though membership declined during World War II, navigation was one topic which helped to sustain the Centre. Speakers came from the Wireless School at Tuxedo, and from the Air Force Navigation school at Rivers, Manitoba, and a number of RCAF men flew in to hear Winnipeg President, Mr V.C. Jones, explain and demonstrate his own planetarium. Nonetheless, with membership down to twenty-eight in 1944, something had to be done. Centre President L.T.S. Norris-Elye, Director of the Manitoba Museum, was determined to reverse the trend, and took steps which led to astonishing results. In 1945 he had copies of the Centre's program of six lectures and two observation nights sent out to all high schools, Boy Scout leaders and Girl Guide Captains. Five hundred youngsters and adults showed up for the first lecture of the season, and almost as many for the next two. Attendance at observation nights was also excellent with six or seven telescopes in use. Capitalizing on these developments, the format of the regular meetings was altered and began with a twenty minute talk of special interest to junior members. Membership jumped to sixty that year, but unfortunately the momentum could not be sustained and numbers fell back to twenty-seven again by 1950.

About this time, the University of Manitoba moved from the city out to its present location, ten kilometres south. The feeling was that the new location would not be convenient for Centre meetings and consequently for a year or so, the Grain Exchange Building was used. It was a difficult time for the Centre with various members looking after the telescopes, books and lantern slides belonging to the Centre. There was talk of an observatory but the Winnipeg School Board rejected the idea of having it installed in the new Technical School on Notre Dame Avenue. Even the meetings seemed to degenerate with four out of six being used to present films and two of these had the unlikely titles of "Springtime in Holland" and "Meet the Ducks."

Fortunately Darby Coats took hold and got the University involved again. Meetings did move to the Fort Garry Campus south of the city. A reorganized series of Summer Observation Nights was instituted at the same location, away from city lights, which worked well with over a thousand people attending in 1953. Co-operation with the university's Evening Institute course in Astronomy gave those enrolled the enjoyment of using telescopes belonging to Winnipeg Centre members. The Perseid and Orionid meteor showers were carefully observed, the results being sent to Dr Millman, and the Telescope Makers Group was reactivated. With all these signs of vitality, membership again more than doubled, to sixty-four, by 1953. Though the Centre still uses the facilities of the University of Manitoba for its regular monthly meetings at the present time, it tried a variety of other locations over the years, including the Shinn Conservatory of Music, The Manitoba Museum of Science and Man, and the University of Winnipeg.

Over the years the membership pattern in Winnipeg has been a cycle of gains and slumps but with a general upward trend. One factor in this numerical progress has been public outreach achieved through a variety of events in addition to traditional star nights. During 1955, for example, an astronomical exhibit at the Winnipeg Auditorium included books, instruments and a Junior Spitz Planetarium and drew an estimated 3,000 people in the course of the ten day event. Participation in the Red River Exhibition began in 1958 and continued for many years. More recently, in 1978 for instance, members put together an impressive display of photographs, literature, maps and instruments for an event at Eaton's department store called "Space Odyssey." In 1980, the Centre organized several

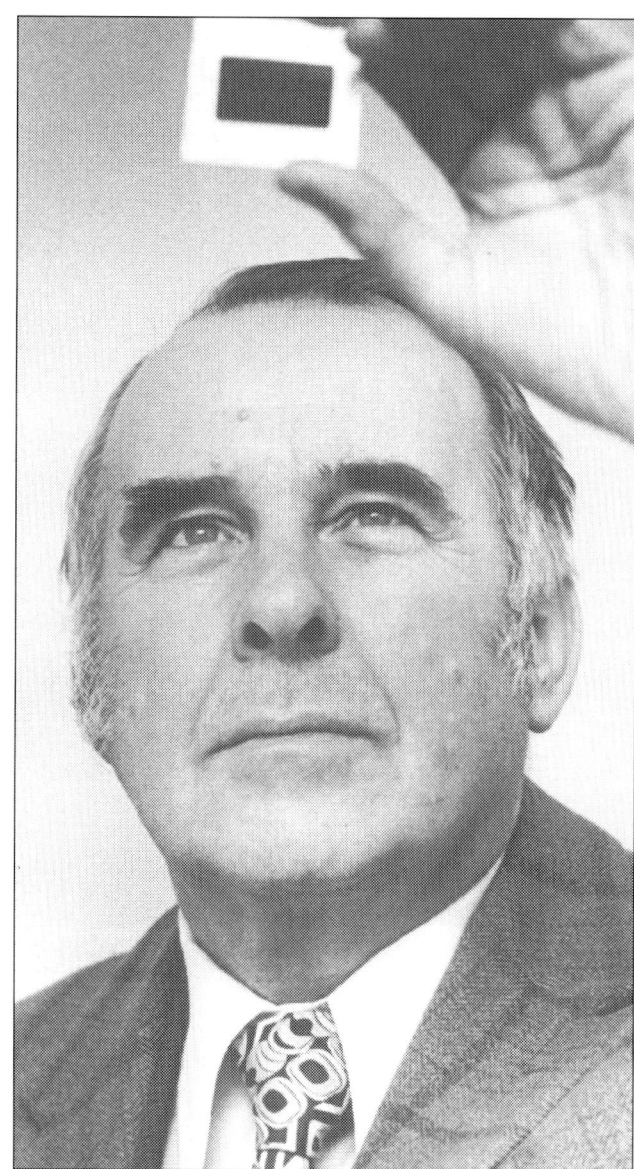

John Scatliff examines a slide he took of Comet Kohoutek in January, 1974, with the temperature at -30°. As he said, "I stood out there until my hands nearly froze to the telescope, but it was worth it."

JOHN N.R. SCATLIFF (1914-) had a distinguished medical career in Winnipeg. Having earned his MB, BS degree from the University of London in 1941, and his Diploma in Public Health from the University of Toronto in 1951, he became Medical Director of the Misericordia Hospital in Winnipeg and the Director of Venereal Disease Control for Manitoba.

Dr Scatliff played a very active part in the Winnipeg Centre and in recognition of this, received the Service Award in 1971. He had been on the Council since 1955 including a term as President of the Centre in 1958-60. He spoke frequently at meetings, instructed and tested Girl Guides and Boy Scouts for their astronomy badges, taught telescope-making in schools, gave a ten-week Extension course in astronomy at the University of Manitoba and was part of the University of the Air astronomy series in 1966.

Scatliff took part in a variety of observing projects. From 1957-60, he was an active meteor observer for the IGY Project. He built his own 15-cm reflector, and housed it in an observatory originally used by Father Rivard at St. Adolphe, Manitoba. In 1966, Scatliff joined the AAVSO and took up solar observing using a telescope with an unsilvered mirror and a Herschel wedge. Later, sketching the planets became a special interest. As chairman of three Winnipeg Eclipse Committees, he chartered a train to Wivenhoe in 1963, arranged a trip for five members to Mexico in 1970, and chartered a DC-3 aircraft from which the 1972 Eclipse was seen over the Northwest Territories.

He owned a number of historical instruments which he donated to the Manitoba Museum of Man and Nature. His observatory (originally Father Rivard's) was given to the Winnipeg Centre 1964.

events. On National Museum Day they set up displays of photographs, telescopes and other paraphernalia. During the summer, school children were treated to a special Junior Star Party at Birds Hill Provincial Park and on the Labour Day weekend, five members packed their tents and telescopes and travelled to Riding Mountain National Park, where they ran a three day astronomical weekend for the public. This was the origin of MASCON, now an annual summer event. In 1981 the Centre had a display at the Fiftieth Anniversary celebration of Bristol Aerospace.

Meanwhile, more traditional programs of public education continued. As early as 1960, Dr J.N.R. Scatliff and others spoke to school groups as well as to Girl Guides and Boy Scouts. These contacts had the wonderful result of bringing many youngsters into the Centre as junior members, and in fact one meeting each year was handled completely by these budding astronomers. Unfortunately there were some who were more like thorns than buds with the result that the Council, after serious consideration, amended the Centre's constitution making sixteen the minimum age and eliminating any lower fees for junior members. This move caused a reduction in the membership in 1974 but it presumably gave the Centre a healthy core of more mature members genuinely interested in Astronomy.

Star nights in parks proliferated with nineteen public programs presented in 1976 and in 1977. Some of these events involved out-of-town trips of 500 km in one day. Since that time, with the opening of the Centre's observatory at Glenlea, about 30 km south of the city, most of the programs have been held there. Every Saturday evening throughout the summer, the observatory is open to the general public.

The observatory, of course, was the most important development in the Centre's history but came only after years of waiting. Back in the 1950s, meteor observing which required no optical aid, was the most popular outdoor activity of the Centre, with satellite watching taking hold soon after the Sputnik launch in 1957. The 1960s saw a great increase in activity through the national observing program and a parallel surge in telescope making. Scatliff's city observatory was a real drawing card, especially for solar observing, though after he donated the building to the Centre and it was moved to St Norbert it was not so popular. The inconvenience of getting out to this dark site was probably a factor but in any case most members now had telescopes of their own. A survey found that members owned thirty-three telescopes, many of which were home-built. Other members were able to borrow Centre instruments. One of the last but most significant projects undertaken before mirror grinding lost its lustre, was the production by a group of Winnipeg members of a solar telescope for the Manitoba Planetarium. This was completed in 1973 as part of the celebrations commemorating the 500th Anniversary of Copernicus' birth.

Under Jack Newton's presidency (1970–72), astrophotography became the thing to do, and plans developed for a Centre observatory. Several members were making Maksutov telescopes and the hope was that some of these skilled people might build the Centre a 51-cm reflector. The telescope makers never took the bait, but undaunted, and under the leadership of Roy Belfield, the Centre set out to raise funds through donations and raffles. By 1975 the members felt confident that the observatory project would become a reality and so they voted to purchase a large commercially-made telescope if the University of Manitoba would agree to erect and maintain an observatory for it. That was not quite what happened but the University of Manitoba did sign a twenty-five-year agreement providing a completely serviced site while the Centre was responsible for constructing and maintaining the building. In 1977 the Glenlea Observatory officially opened with an 20-cm Cassegrain telescope built and lent to the Centre by Bill Peters until such time as the dream

ROY AND PHYLLIS BELFIELD received the Service Award jointly in 1976. Throughout the '70s their dedication was of vital importance to the Winnipeg Centre and to astronomy in Manitoba. The basement of their home was used as a Centre office, library, and informal meeting place, and their backyard observatory was the focus of many enjoyable observing sessions lasting until the wee hours of the morning. They were both excellent astrophotographers and helped many others to develop this interest - Roy usually at the telescope and Phyllis in the darkroom. They prepared numerous popular slide shows for use in schools and thought nothing of travelling far afield to promote astronomy. With their active encouragement, an association straddling the international border flourished in Gretna, Manitoba and Nichi, North Dakota, during the mid-'70s.

In addition to all these activities, Phyllis helped out with mall displays, star nights, refreshments, served the Centre as Librarian, Editor and Vice-President and made major contributions to the History of the Winnipeg Centre published in 1977.

Roy, as Vice-President and President, was the driving force behind the establishment of the Centre's Glenlea Observatory which, in fact, was modelled after the one in his own backyard. He was much sought-after as a speaker by various schools, camps and other groups and as a result was invited to become a part-time lecturer by the University of Winnipeg where he conducted in-service astronomy classes for teachers.

After moving to Victoria, Mr and Mrs Belfield served the Centre there as Vice-President and Treasurer in 1985. Roy was not in good health and died in 1993.

for a larger instrument could be fulfilled. Soon the Observatory was in constant use for meetings of Council and other groups, public and school tours and members' observing nights. Though astrophotography was still the main attraction, some observers took an interest in occultations, variable stars, planetary and solar work.

The solar eclipse of 1979 kept members very busy with public seminars, school lectures both in and outside Winnipeg, and an information booth at the Manitoba Planetarium. Twelve colour slides of the eclipse, taken by Centre members, were sold as sets and the proceeds went towards a 35-cm Schmidt-Cassegrain telescope. Then disaster struck. With the dream of having a well-equipped observatory just about a reality, the Red River flooded. Hours of backbreaking sandbagging were all in vain as the waters rose, and hours more labour went into the clean-up operation after the level finally subsided. Eventually all was well. The C-14 telescope arrived and the official re-opening of the observatory took place on March 29, 1980, with Maestro Piero Gamba, honorary president of the Winnipeg Centre, cutting the ribbon.

Everything was looking good. The Government of Manitoba and municipalities along the Red River solved the flooding problem through a multimillion dollar works project. The Centre produced in 1984 a fifty-page beginner's guide called *Astronomy — A Hobby of Many Colors* to encourage new members, and Halley's Comet came along the following year, giving a boost to public awareness. The only cloud in the sky came in the form of rain damage at the observatory. Once again, members rolled up there sleeves and finally in 1986 after more fundraising and with the help of a government grant, they were able to install a new 4.2-m dome to keep their observatory dry from top to bottom. Glenlea Observatory continues to serve a variety of purposes from visits by school groups to CCD photometry. Computer reduction of the photometric observations was supported by the National Council through a $1,500 grant in 1989.

REGINA

In the summer of 1910, Dr Chant travelled from Toronto to Mount Wilson, California, for a meeting of the International Union for Co-operation in Solar Research. Besides the rare opportunity for discussion of astronomical problems with colleagues from around the world, the conference afforded delegates the chance to inspect the solar tower telescope which was just being constructed. (As it turned out, Chant was hoisted to the top in a large steel bucket and found the view and the rapid descent "somewhat startling.") The RASC, as one of the member societies of the IUCSR, sent Chant as their representative, paying $150 of his expenses, and he, always desirous to promote Astronomy and the RASC, decided to visit some Canadian cities enroute.

One of his stopovers was Regina where a group called the Saskatchewan Astronomical Society had recently been formed as an offshoot of the Regina Philosophical Society, originally established in 1893. Contact had already been made when the group had borrowed slides from Chant to help with their meetings and so they were now delighted to have him address them in person. The chairman of the meeting at which Chant spoke was Regina's police magistrate, William Trant. He had been a charter member of the Leeds Astronomical Society in England and was now the honorary president of the SAS. Speaking in the Collegiate Institute, Chant gave his customary address on "The Universe of Stars" and encouraged Regina to become a Centre. Within a few months that is exactly what happened. On October 11, 1910, twenty-seven members from Saskatchewan were elected to membership in the RASC and the Regina Centre was born, the first outside Ontario. Among the charter members were H.S. McClung, an optometrist who later proposed the corneal reflex test for use in mirror grinding, and

J.A. Covington, who was yet to become the father of Canada's pioneer radio astronomer, Arthur Covington. Another charter member was the Honorable Mr Justice J.T. Brown, who made a marvellous gift to the Centre of an equatorially mounted 11-cm Brashear telescope.

At some point the objective lens was stolen from their telescope and so, in 1913, the Centre courageously spent $460 for a new telescope from J.A. Brashear. An observatory was built for it on the roof of Central Collegiate Institute where the Centre held their regular meetings. Many individuals and the City of Regina contributed financially, but the Centre still had to bear some of the cost. On the heels of these problems, the Great War took its toll and the struggling Regina Centre was not able to survive. The sixteen members in 1915 hoped to keep the Centre alive until more settled conditions returned but they were not successful. Even though the Centre was kept on the books with James Duff shown as the president, no reports were received for 1917 or 1918, and the membership at the last report, in 1920, was down to four. A comparison with Winnipeg Centre, which managed to pull through this difficult time, suggests that Regina's smaller population and the lack of a university were important factors in its demise.

There was talk of a Centre again in the 1930s but nothing developed and the telescope was removed from the roof of Central Collegiate in 1936. It was not until after World War II that a few local telescope makers got together under the leadership of John Hodges and formed the Regina Astronomical Society. They built themselves an observatory behind Regina College to house the Brashear telescope. Dr P.M. Millman officially opened the observatory in 1955 and was full of praise for the group's extensive visual and photographic observations of meteors. In 1956, they organized Operation Perseid in which eighty-two observers at seven locations in Saskatchewan recorded 2,800 meteors visually. They even photographed a couple of meteor spectra which Millman called a "definite scientific contribution." Though the Regina Society remained independent of the RASC, there were some contacts. Members of the Hodges family spoke at General Assemblies in 1962 and 1964 and continued to contribute meteor observations until 1966. RASC Vice-President K.O. Wright spoke to the group in 1963. In 1965, the Observatory buildings were moved to the east side of Broad Street, and about the same time an alternate group, called the Wascana Astronomers, formed. Attempts to organize as an RASC Centre, however, were still unsuccessful, though Calgary Centre tried to keep the door open by inviting the Regina group to attend the 1968 GA and by jointly observing the grazing occultation of Delta Sagitarii with them in 1969.

The Regina Astronomical Society continued to flourish. In the years 1985–89 they reported that

The Regina Observatory and 25-cm McClung reflector (l) in 1955.

their observatory was in use on average 173 days for 200 observing hours and 1,100 visitors annually. The RASC was certainly the poorer without this dynamic group of amateur astronomers. Consequently the National Council was pleased to learn in 1989 that some old personal grudges had been forgotten and the Regina Society now, after all its years as an independent body, wanted to join the family of RASC Centres. Subject to submission of satisfactory by-laws, the Regina Centre was welcomed into the fold on July 2, 1989. President Lloyd Higgs addressed the Centre on September 27, and the details were finalized four days later.

During 1989, the Saskatchewan Science Centre decided to accept a generous offer from the potash firm, Kalium Canada Limited, to finance the development of a public observatory. The University of Regina presented its 15-cm refractor which a number of Centre members helped to install, and the Centre donated the dome of their own Broad Street Observatory. The rest of the structure became part of the Centre's new facility at Davin, a much darker site 40-km southeast of Regina. With the financial help of a grant from the Province, the original observatory from Central Collegiate was rescued and restored, and along with the Brashear telescope was installed at the Western Development Museum in Moose Jaw.

Besides their help with the Kalium Observatory, the Regina Centre supports public education with star nights for Guides, Scouts and Beavers, public viewing nights, and an annual star party at Cypress Hills Provincial Park.

EDMONTON

In the summer of 1910, Chant continued on from Regina to Edmonton, seeing first hand what interest there was in astronomy in western Canada. He found Edmontonians still enthusiastic about the bright Comet 1910a seen to good advantage in January but disappointed with Comet Halley which was obscured by the long twilight during May. Chant paid another visit in 1922, this time en route to Australia for the solar eclipse. He spoke to the Canadian Club in Edmonton about Relativity and how the apparent shift of stars observed during the eclipse could provide experimental verification of Einstein's Theory. By this time Dr J.W. Campbell was established at the University of Alberta, but it took a few more years for general interest in astronomy to grow to the point where he could organize a viable group.

Campbell began discussions with the RASC about forming a Centre in 1931 and called an organizational meeting for January 14, 1932. Interest was high, and at a subsequent meeting on February 3, nearly sixty people signed a petition to the General Council of the Society requesting permission to form a new Centre. Reaction from Toronto was quick and supportive, and by March 15 the new Centre was launched with by-laws adopted almost unchanged from the London Centre, and the first executive and council elected. Professor Campbell was the first president and Professor E.S. Keeping the vice-president.

Cordial relations with the university were thus established right from the start. Centre meetings were held on campus, a very special one in June, 1932, in the Athabasca Lounge of the university. On this occasion, the Centre hosted a banquet to honour a party of British scientists who were on their way to a post at Fort Rae on Great Slave Lake where they were to make meteorological, magnetic and auroral observations during the International Polar Year.

During the early years, it was customary at regular meetings to have short introductory talks featuring a constellation of the month or coming events, based on *The Observer's Handbook*. The main lecture would then follow, and on evenings when it was clear, a telescope was usually set up for observation afterwards. Refreshments at the end of the evening provided an occasion for informal exchange of opinion and information. During the war, cocoa and *Postum* were served and members brought their own sugar because of rationing.

The group began to take an interest in observing as early as 1936. A few members would gather at Cyril Wates' place for a picnic and after supper would enjoy viewing the Moon, planets, clusters and double stars through his 16-cm or 23-cm reflectors. Also about this time, the Centre began to form a small library with the purchase of a star atlas and a few good books on popular astronomy. A subscription to *The Sky* was a popular addition in 1940.

The Centre recognized the valuable support of the university in 1936. Out of gratitude for the use of their facilities, the Centre presented the university with ten glass transparencies — 28-cm x 36-cm photographs from Yerkes and Mount Wilson, which were then illuminated and mounted in the Arts Building.

Again stressing the Centre's strong links with the University of Alberta, Cyril Wates presented them with a new 32-cm telescope in 1942. The authorities accepted his generous gift and agreed to build a small observatory to house it as well as the 10-cm refractor which belonged to the Mathematics Department. The building turned out to be 6.1-m by 9.8-m with a 4.9-m dome at the north end and three piers for smaller instruments at the south end. Observation nights at the observatory got a good response from members and from the public, and on some occasions the crowds were too large to be easily managed.

In the early 1950s, definite observing programs developed in the Centre spurred by the enthusiasm of a new arrival from Montreal, Earl Milton. In 1954, the observing committee consisted of four sections: atmospheric phenomena under Milton, solar and lunar observations directed by Franklin Loehde, planetary and comet observations with Arthur Dalton, and photographic work under Ian McLennan. Four members of the Observers' Group were named honorary demonstrators in Astronomy

CYRIL G. WATES (? -1946) was a maintenance engineer with the Edmonton Municipal Automatic Telephone System. His exceptional mechanical ability led him into telescope-making in 1931 and over the next twelve years he built six telescopes and wrote many articles on the subject for *Scientific American* and the RASC *Journal*. His own observatory overlooking the Saskatchewan River, across the road from his home at 7718 Jasper Avenue, was the site of many Centre gatherings.

His largest undertaking was the construction of a 32-cm telescope, a project which took him five years to complete. He began the rough grinding of the Pyrex mirror by hand, but constructed a Hindle-type machine to do the figuring and polishing. This machine, which he described in the *Journal*, was made from all manner of spare parts from dental equipment to washing machines. When the telescope was finished, he presented it to the University of Alberta along with a 10-cm rich-field finder which he had also made himself. The University agreed to build an observatory and the opening ceremony, attended by the Lieutenant-Governor of Alberta and astronomers McKellar and Pearce from DAO, took place on May 20, 1943. The observatory was named for Cyril Wates, and served the University well for years. Wates received the Chant medal for 1943.

Cyril Wates was frequently the main speaker at Centre meetings and in addition, from 1941-45, regularly gave short talks based on *The Observer's Handbook*. He sometimes amused his audience with humorous verses but he also wrote serious poetry and composed music. He served as President of the Centre in 1938 and in the same year was also President of the Alpine Club of Canada. Strongly attracted by the silent grandeur of the scenery, whether terrestrial or celestial, he delighted in showing others inspiring views of either.

The photograph was taken following the celebration on May 20, 1943, when Cyril Wates presented his 32 cm reflector to the University of Alberta. In the front row (l to r) are A. McKellar, J.W. Campbell, J.A. Pearce, Lt-Gov J.C. Bowen, C.G. Wates.

John W. Campbell (1889-1955) was born in Scotch Block, Ontario, attended Queen's University in Kingston and then obtained his PhD from the University of Chicago in 1915 with a thesis on the three-body problem. Following this, he taught at Wesley College, Winnipeg, and joined the RASC Centre there. He then served with the Artillery in the First World War and was an instructor in Khaki College. After a short term at the University of Iowa, he accepted a position in the Mathematics Department at the University of Alberta in 1920, where he made his career, until retiring as Head of the Department in 1954. His text-book *An Introduction to Mechanics* was quite widely used and his contributions to his field were recognized by his election to the Royal Society of Canada. Though Dr Campbell's work was largely mathematical, he did develop his professional interest in astronomy, working during the summers of 1922 and '23 on spectroscopic binary orbits at the DAO. He also taught a general astronomy course at the University of Alberta for over 30 years.

Campbell was the founder and first President of the Edmonton Centre, and their Honorary President from 1934 until his death. His successor in many respects, E.S. Keeping, wrote of Campbell's "great delight in showing visitors the beauties of the night sky" and his weekly notes on the night sky in *The Edmonton Journal*. Scarcely a year passed when Campbell did not speak to the Centre or give his popular "Handbook Talks" at meetings. From 1945-48 he served the national Society as Vice-President and President.

by the university Physics Department and were added to the university Observatory Committee.

At the meetings in the 1950s, reports of activities at other Centres was a regular feature, especially items which came from Montreal Centre's newsletter, *Skyward*. Edmonton Centre started their own newsletter, *Stardust*, in 1954 as a medium where observing reports could be printed. In 1955, a successful course of lectures on elementary astronomy was organized and held in the "Observatory Annex" with about twenty people registered.

For many years, some members had realized the potential of a planetarium and dreamed of the day when Edmonton might boast of such a facility. Interest surfaced as early as 1936 when the past-president of the Centre, Professor Keeping, returned from a visit to the Adler and Hayden Planetaria and addressed the Centre on this subject. In 1943 a simple travelling planetarium was built in the Education Building by Mr S.A. Linstedt and Professor L.E. Gads. In 1958, the City of Edmonton offered a challenge – What could be built to commemorate the visit of Queen Elizabeth II to the city in 1959? The Centre, with the idea of a planetarium from member S. Frank Page, rose to the challenge and spoke to many civic groups and finally persuaded City Council to build the Queen Elizabeth II Planetarium, the forerunner of many new Canadian facilities which would open in the coming years. With its opening in September, 1960, the city and the Society really began to appreciate the educational and cultural advantages of a good-sized public planetarium. Regular Centre meetings, previously held at the university, moved to the planetarium. Special demonstrations for members were initiated by the director, Ian McLennan, and right up to 1979, the last year in which the Centre held its meetings in the now too small facility, it was still being used to familiarize new members with the sky.

Inspired by the general interest in astronomy which the Planetarium awakened, the Centre also began to reach out to the general public. With Western enterprise, they charged admission to their first star night in 1961 and made a profit of $584. At the next year's event, the Centre put up a large tent for shelter and films. By 1963 the displays had grown to occupy the lower floor of the Jubilee Auditorium where approximately 3,000 people paid to attend. Profits from these events were to go toward the acquisition of an observatory and space museum in Edmonton.

Meanwhile, members maintained their involvement in observing activities. Some participated in planetary observations as outlined by the national Co-ordination Committee, some made regular reports on the aurora and others were engaged in timing lunar occultations for the U.S. Naval Observatory. Light pollution at the university observatory was a severe problem by this time and so the Observers' Group embarked on a multitude of plans over the years to provide a reasonably dark site where members could use their telescopes. Two very successful grazing occultation expeditions in 1971, one a joint effort with the Calgary Centre, supplemented the regular programs of observing and astrophotography. Trophies and awards helped to promote these activities as well as telescope design and construction. After a hiatus of a few years, annual star nights resumed in 1974 with a two day public program of displays, lectures, films and telescope observing jointly presented by the Centre and the Queen Elizabeth Planetarium. In late 1978 a group of Centre members began planning an extensive astronomical exhibit for display in shopping centres and community clubs. Completed the following year, and comprising eighteen panels, *Discover the Universe* drew large crowds wherever it was set up. 1979 seems to have been an especially good year. In April, about 1,000 people turned out to view celestial wonders through an array of telescopes supplied by RASC members while on September 6, at 4 o'clock in the morning, over 250 people witnessed the total eclipse of the Moon at a special Eclipse Starnight. A public Lunar Eclipse Seminar held two nights earlier attracted thirty-five paying customers. In May, over fifty people registered for a general astrophotography seminar. Together, these seminars raised over $325 for the Centre. As might be expected, membership in the Edmonton Centre increased dramatically from 35 in 1973 to 103 five years later.

Average attendance at regular meetings also picked up from thirty-four in the early 1960s to the point where the room at the Planetarium was no longer adequate, so the location was switched to the Public Library in 1979. The music room there provided the needed space, and meetings could now be safely advertised to the public through the local media. One factor in this growth was an annual program of exchange speakers. Starting as a Junior Forum with members from Calgary Centre in the 1960s, the program grew to involve many other western Centres, and eventually developed into a program financed by the national Society.

Commencing in 1978, a series of popular "Observer's Corner" meetings were held every other Sunday afternoon in the comfortable Physics lounge at the university. In the years ahead, these sessions provided a monthly forum for members' interests in observing, astrophotography and telescope making.

A direct outcome of these meetings was the decision to begin working towards an Edmonton Centre observing site. Many weekends were spent searching the countryside for a possible location, various funding schemes were proposed and a five page questionnaire was sent to all Centre members to assist in the organization of the project. By the end of 1979, the Centre made the decision to lease a three-hectare dark site on the south slope of Buck Mountain, southwest of Edmonton, on provincially owned land. They erected a metal prefabricated

Douglas P. Hube, photographed by his daughter, Susanne.

DOUGLAS P. HUBE (1941-) grew up in St Catharines, Ontario, joined the RASC in 1960, and went on to study Astronomy at the University of Toronto. In the course of his graduate work on the radial velocities of B8-B9 stars, he met Joan, a research assistant at the DDO. They married and went off to Africa for seven months where Doug obtained spectra of southern B stars at the Radcliffe Observatory. On returning to Toronto, they both continued their research, Joan co-authoring three papers and Doug completing his PhD in 1968. Following a year on an NRC Post-Doctoral Fellowship at Kitt Peak National Observatory in Arizona, Dr Hube joined the faculty at the University of Alberta.

Hube soon became known at the University for his rare ability to combine a strong research program with outstanding teaching. He greatly strengthened the astronomy component of the Physics Department and oversaw the fabrication and installation of a new 50-cm telescope at the University's Devon Observatory in 1977.

Edmonton benefitted greatly from his talents. He served as Vice-Chairman of the Edmonton Space Sciences Foundation and played an important part in the success of the ESS facility. Hube became known as a vital resource person by the media. He wrote dozens of articles for *Stardust* on current developments in Astronomy; he spoke to the Centre on a number of occasions and was their President in 1972 and 1982.

The broader membership of the Society got to know the Hubes at General Assemblies. A number of his articles appeared in the *National Newsletter* and *Journal*. Douglas Hube was awarded the Service medal in 1982 and accepted the position of Second Vice-President in 1990.

building to serve for warm-up, storage and sleeping quarters in 1981. Buck Mountain served as the site of an annual "celebration" of all-night twilight called "The Summer Solstice Sacrifice and Debauchery" which replaced the former and more genteel annual picnic at the university's Devon Observatory. As an alternative to the ninety-minute drive to Buck, the Centre also had the use of the observatory of member Tony Whyte at Ellerslie, just southwest of the city. His 20-cm refractor was used on numerous occasions for lunar, solar, planetary and deep-sky observing.

The Centre telescope had a rather checkered career. Work on a 44-cm Dobsonian reflector began in 1980 though its completion was delayed by late shipping of the optics. After a few short years of successful operation the objective mirror was stolen and the telescope was out of commission for a year. In the meanwhile, the telescope itself was modified under Bob Drew's direction to improve its portability, and the redesigned club telescope earned awards at both Riverside and Mount Kobau in 1989.

In the 1980s, the Observers' Group held successful sessions at locations inside and outside the city. A few memorable events are all that can be described here. At Elk Island National Park, observers described the spooky experience of being watched by buffalo whose beady eyes glistened like stars in the approaching car headlights. Several members became actively involved in determining the path of a spectacular fireball which streaked across the sky north of Edmonton from east to west on the evening of February 22, 1984. Though no recovery was made, the conclusion was that a meteorite may have fallen in dense bush southwest of Grande Prairie. Comets were a highlight in 1985, especially Halley and Giacobini-Zinner. In 1989 several members attended Riverside, Mount Kobau and the Alberta Star Party, where they earned a number of prizes for astrophotography and telescopes.

A development of great significance for Edmonton, and for the Centre, was the establishment of the Edmonton Space Sciences Centre in 1984. Under the leadership of Dr Douglas Hube, Rod McConnell and Franklin Loehde, members of the Edmonton Centre actively took part in fundraising for the Space Sciences Centre by promoting a star donation campaign. They set up a booth presenting RASC/ESSC exhibits including a model of the ESSC in the Northlands Coliseum during Edmonton's Klondike Days and in various shopping centres. Under Franklin Loehde's leadership, Edmonton Centre members volunteered to prepare packages for the Donate a Star Program which were distributed to 16,000 classrooms. The Centre itself purchased a second magnitude star Alinetak. Many Centre members enthusiastically gave hours of voluntary help in preparation for the official opening in July 1984. In 1987, many Centre members assisted at a two-night casino which raised approximately $45,000 for the ESSC.

The RASC Centre benefitted directly from the new Space Sciences Centre. Starting in September 1983, the regular meetings and library were moved to the brand new classroom there. A series of nine astronomy lectures was held in the Centre's Devonian Imax Theatre between February and April, 1985, and many members benefitted from these talks. Another example of co-operation was a very successful Astronomy workshop which took place on October 27–29, 1989, to help members get the most out of their telescopes.

The Edmonton Centre has always taken pride in its history. In 1972, to commemorate the fortieth anniversary of its founding, Honorary President Professor E.S. Keeping wrote a short book, *The Earlier Years of the Edmonton Centre, R.A.S.C.* Also that year, the Centre's annual banquet, though it had been held for many years at the university, returned to its original locale, the Corona Hotel. In 1979, *Stardust* celebrated its twenty-fifth anniversary with a series on

ERNEST S. KEEPING (1895-1984) came to the University of Alberta from England in 1929. Though his background was in theoretical physics, he joined the Mathematics Department where he taught until 1970, nine years after his retirement as Head of the Department. "Frank" Keeping took an interest in Astronomy, spending the summer of 1945 doing research on Wolf-Rayet stars at the DAO. His wife, Dr Silver Keeping, and their three-year old son, John, accompanied him to Victoria.

He was the first Vice-President of the Edmonton Centre when it formed in 1932 and went on to be elected President the following year, Secretary in 1936 and Librarian in 1940. He was the Centre's Honorary President from 1955 on.

Professor Keeping was often the main speaker at meetings of the Edmonton Centre. His topics covered a very wide range, but his interest in the historical side of science was evident in his speeches on the occasion of Eddington's death in 1933 the tercentenary of Newton's birth (1942), and the centenary of the discovery of Neptune (1947). He wrote short histories of the early years of the Edmonton Centre and of the Mathematics Department at the University of Alberta.

Two trips were especially memorable for him - one to view the Solar Eclipse of 1963 with other RASC members in Fort Providence, NWT, and the other a voyage to the Galapagos Islands in 1971. Much earlier in his life, during World War I, he had served in Iran and Iraq. Over 100 images of his wartime duties are in the U of A Archives.

The Society honoured Keeping with the Service Award in 1965. The University of Alberta conferred an Honorary LID degree on him in 1972.

The photograph shows Sam with his second wife, the former Marie Fidler, also a Service Award recipient (see p. 30).

In this photograph, taken by John Howell at Sulphur Mountain at the time of the 1968 General Assembly, Ken Meiklejohn is at the left with (l to r) Jack Grant (Director of Meanook Meteor Observatory), Bob Nelson (President of Calgary Centre in 1965), Jim Wright (see p. 223) and Ed Kennedy (p. 79).

SAMUEL LITCHINSKY (1907-73) was a charter member of the Calgary Centre and was one of their original Councillors. After serving as President in 1961, he was Editor of *Star-Seeker* from '67 to '69, and was on both the Centre and National Council for several terms. When he was granted the Society's Service Award in 1966, it was noted that "Mr Litchinsky has generously provided the meeting place for the Council and has been the key individual in supplying astronomical information to the Calgary schools, radio and TV stations."

Following his death in 1973, his role in initiating and planning Calgary's Centennial Planetarium was marked by a plaque in the Pleiades Theatre, and the Centre named its 30 cm telescope for him.

KENNETH B. MEIKLEJOHN, a school Principal, was also a charter member of the Calgary Centre and was their first Treasurer. He succeeded Sam Litchinsky as President in 1962 at which time Mrs Meiklejohn (Bessy) was Secretary. He served on the Centre Council for a number of years and was Chairman of the organizing Committee for the 1968 General Assembly. The citation which was read in 1971 when he was presented with the Service Award recalled that "for years he enlivened each meeting of the Centre with an educational feature, known as 'Ken's Constellations', on some interesting astronomical topic, and he also served the school community as a judge of astronomical exhibits at Science Fairs."

the first thirty years of the Centre's history and a special issue that featured the minutes of the General Meeting of October 2004. Won't that make interesting reading a few years from now!

CALGARY

As early as 1918, H.C.B. Forsyth (in later years a Vancouver member) expressed an interest in forming an RASC Centre in Calgary. The war was still on, and Council advised against it. Two years later, however, an independent group under the name of the Astronomical Society of Alberta, was organized by W.P. Taylor of Calgary and J.S. Plaskett lectured to them in the spring of 1920. Forsyth, a member of this group as well as the RASC, placed an ad in the *Journal* for 1920 offering to purchase a 14-cm or 15-cm refractor and by 1924 he was reported to be finishing a 30-cm mirror for a reflecting telescope, the seventh mirror he had ground. By 1934, when J.R. Collins, a past-president of the RASC, made a three month trip west, the Astronomical Society of Alberta was no longer in existence. Possibly Collins' visit stimulated a revival; "The Astronomers of Calgary" formed in 1935, and before the year was out they became known as the Calgary Centre of the RASC. Their president at the time was Harold King who had a paper published in the *Journal* describing his technique for making a 24-cm mirror from a piece of glass only 1.3 cm thick. Meetings were well attended, but with the economy still depressed, the Centre never attracted more than a dozen paying members and the group vanished after about three years.

The Centre which still flourishes today began in 1958. It was a prosperous boom time in Calgary with the oil and gas industry rapidly expanding, and of course interest in space was high in the months following the launching of the first Sputnik. The University of Calgary was not yet established but the University of Alberta had a temporary campus there. One of the professors, Walter Stilwell, placed an advertisement in the *Calgary Herald* inviting anyone interested in forming an astronomical society to an organizational meeting. Over a hundred showed up and about sixty became members of the group which soon became the Calgary Centre of the RASC. For the first three years the Centre meetings were held in the lunchroom of Maclin Motors through the courtesy of a charter member, Ernie McCullough. Generally there were about eight monthly meetings each year, mostly involving speakers, but occasionally featuring a "Students' Symposium," film presentation, field trip for observation or an exchange speaker from Edmonton. In addition to regular meetings there was a series of ten lectures on general astronomical topics given by Stilwell in 1958 and a further series of five talks in the spring of 1959 to acquaint members with *The Observer's Handbook* and its use. In 1960 there were special lectures by national presidents McKellar and Millman, and one by Dr Gus Bakos on the need for co-ordinated observations of artificial satellites as set up by the Smithsonian Astrophysical Observatory in Project Moonwatch.

Moonwatch was an especially successful activity in Calgary with Mr E.M. Rogers organizing photographic and visual observations of thirty or more artificial satellites from two separate stations. The co-ordinator from the Smithsonian Observatory, Mr Alexis Bakeef, visited the Centre in 1961. Dr T.A. Link, Calgary Centre's first Honorary President, donated money and a Unitron telescope.

Maclin Motors was scheduled to be sold in 1960, so the Centre had to look for a new location for its regular meetings. By good fortune, the new University of Calgary was just opening, and for the next seven years the Centre enjoyed the fine facilities in the Science and Engineering Building there. The regular monthly lectures, advertised in *The Starseeker*, were well attended by forty to fifty people and continued to span a wide range of topics offered by a healthy mix of professional scientists, academics and amateurs. The "Students' Symposium" and "Junior Forum," usually held each January, gave the younger members a special opportunity to be heard. The last gathering of the year was generally the Annual Meeting which included a dinner at the Highlander Hotel and a guest speaker.

During this period, the Centre actively reached out to the general public in several ways. In 1964, John Howell headed a committee which arranged an exhibit, called Astro-night, at the Southern Alberta Jubilee Auditorium and this was repeated the following year. Displays from major observatories, and from NASA as well as exhibits by Centre members were supplemented by films, lectures and opportunities for telescope use. In cooperation with the Adult Education Division of the Calgary School Board, a fifteen session course for beginners was offered. Members were also called on to serve as judges of school science fairs, and to prepare Scouts and Guides for their Astronomy badges.

There were some special highlights for the Centre in the sixties. The first General Assembly to be held in the west was in May, 1962. Though the

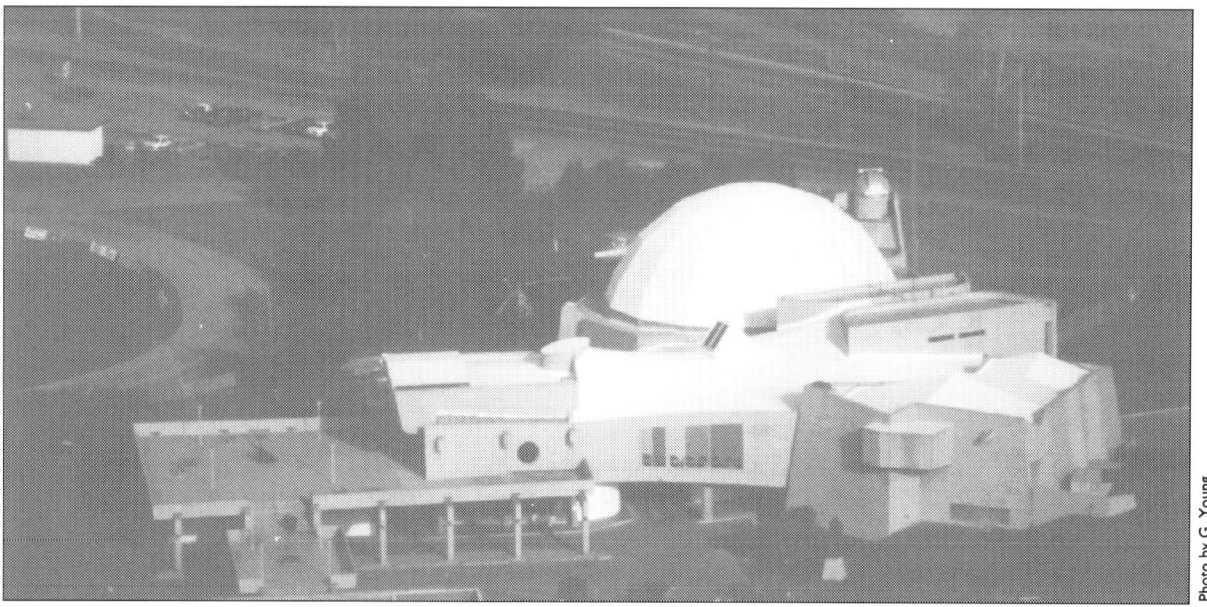

Calgary's Centennial Planetarium (now the Alberta Science Centre).

Don Hladiuk at Calgary Centre's Wilson Coulee Observatory, 1987. Don won the Society's Service Award in 1992.

meetings all took place in Edmonton, on the final day Calgary was host to the delegates on a trip to the Cosmic Ray Observatory on Sulphur Mountain at Banff and to the Calgary Power plant at Seebe. This was but one example of the co-operation between the two Alberta Centres. Speaker exchanges and joint star parties were an annual feature. In 1963, several members travelled 1,400 km by car to Fort Providence, NWT, to view the total solar eclipse. They were lucky with the weather and brought back slides, films and stories of their adventure. But it was plans for Canada's Centennial which launched the young Centre into a its most ambitious and significant undertaking to date.

When the *Calgary Herald* announced a contest in 1963 to determine the best idea for a city project to mark Canada's Centennial, the Centre wasted no time in approving the formation of a committee under the leadership of President Walter Stilwell to promote the building of a planetarium. A detailed brief was prepared, circulated widel, and presented to the city's Centennial Committee. The project won their approval over hundreds of others and in 1967 the Calgary Centennial Planetarium became a reality thanks to the initiative and hard work of the RASC Calgary Centre and the generosity and foresight of three levels of government. The Planetarium has been the official home of the Centre ever since and was the locale for nearly all of its meetings except during the 1980s when the university facilities were used again.

Though the Centre had no observatory in the 1960s and 1970s, for a time there was an observing group which met every Friday during the summer, weather permitting, and there was generally one meeting each fall at the university's Rothney Observatory which opened in 1972 near Priddis. A few hardy souls were undeterred by winter conditions – even temperatures of -34°C. Members' telescopes and the Centre's 30-cm reflector, acquired in 1973, were used for occultations and of course at star nights. A memorable solar eclipse expedition arranged by the Observing Group took eleven RASC members and eighty Americans by charter aircraft to Tuktoyaktuk, NWT, IN 1972. The highly successful trip was the subject of a colour booklet which the Centre sold for $1.25.

There were dramatic developments in 1980, and again these were prompted by an important celebration. In commemoration of the seventy-fifth anniversary of the Province of Alberta, and as a result of a successful grant application, the City of Calgary donated $47,000 to the Calgary Centre for building an observatory. A site was chosen about fourteen kilometres south of the city, and planning and construction proceeded over the next couple of years. The official opening of the Wilson Coulee Observatory took place on 29 January, 1983, the twenty-fifth anniversary of the inaugural meeting of the Centre. Mayor Ralph Klein officiated and several founding members of the Centre were present. The main focus of the observatory is public education, and each year hundreds of school children, Scouts, and others benefit from tours through the facility and viewing through the telescope. The original 32-cm telescope has been replaced by a 36-cm Schmidt-Cassegrain on a Byers drive. This was the result of a successful fund-raising campaign in 1990 which garnered over $10,000 from the membership and corporate sector.

The Centre's public education activities are not solely carried out at its observatory. Star nights are held each year, sometimes at the Centennial Planetarium, sometimes in City parks or at more remote and darker Provincial Parks. Thousands of people came to public star parties at the Planetarium and in many southern Alberta communities in connection with the Mars opposition in 1988. These events were part of a "Mars Observing Program" organized by the Alberta Science Centre with the support of a Federal Government grant and the co-operation of many RASC members who supplied telescopes, time and talent. The Centre's Astronomy program hosted by Don Hladiuk on CBC radio attracts a lot of listeners. Hladiuk, as head of the Centre's Education Committee, has also organized some very popular introductory Astronomy courses with the co-operation of the Department of Continuing Education. Efforts are being made to get Astronomy into the schools as a part of the regular curriculum and the Centre is playing a very important role in providing courses, training and resources for teachers. All this work has brought about strong growth in the Calgary Centre with membership doubling since the observatory opened in 1983.

Members with a yen to see the sky at its best have enjoyed observing nights at Plateau Mountain, a very dark site about eighty kilometres southwest of Calgary, at Mount Kobau in southern British Columbia, and at various provincial parks in Alberta where Calgary, Edmonton and Lethbridge have each taken turns in hosting the annual Alberta Star Party since 1986. The Centre, as an institutional member of the International Darksky Association, is trying to make light pollution a matter of serious concern in southern Alberta. Handouts are routinely given out at public starnights. They have had a committee to deal with this problem since 1988, and got the national Society to follow their example in 1991.

All these activities, supplemented with social events, like the annual summer barbecue, a reception for new members, and the annual banquet at which awards are presented, ensure that there is something for everyone to enjoy.

SASKATOON

While the visit of former president J.R. Collins to Calgary in 1934 may have been a factor in the formation of a Centre there, his stay in Saskatoon was not so successful. He was "surprised to see the beautiful College and University buildings ... and the splendid little domed observatory with its six-inch

J. CAMPBELL FAHRNER (1917-) joined the Calgary Centre in 1970 while employed by TransAlta Utilities as Hydro Plant Superintendent.

He was Registrar for the 1976 General Assembly and was elected to the Centre Council a year later, becoming Treasurer in 1980, a position he held for ten years. During this period he was responsible for the wise management and investment of the $47,000 grant obtained for the Wilson Coulee Observatory. Members outside Calgary got to know and appreciate Cam's warm personality as he was on the National Council in the '80s and attended a number of General Assemblies.

He took an interest in education and presented a paper at the '84 GA on "The Construction of the Wilson Coulee Observatory" and at the '89 GA on "The Mars Observing Program." He has been the RASC representative on SERG (Society of Educational Resource Groups), a large group of organizations that work with School Boards to provide tours to complement in-school programs. For several years, he has been a judge at the Junior Science Fair.

Cam Fahrner credits three events with his attraction to Astronomy and to the Society:

> Early interest was sparked by seeing Saturn with its rings when a geologist set up his telescope at a Scout meeting. [I] listened to a series of lectures by Peter Millman ... on CBC ... [and] Sam Litchinsky invited [me] to an RASC meeting.

Luckily for the Society, these events culminated in a long and effective association, marked by the presentation of the Service Award to Cam Fahrner in 1985.

The above photograph, dated 1979, shows a corner of the classroom in the University of Saskatchewan used for meetings by the Saskatoon Centre.

In this photo, supplied by Jim Young, Gordon Patterson is seen in a familiar role, teaching a class at a Centre meeting.

GORDON N. PATTERSON (1916-) was employed as a departmental assistant in the Physics Department at the University of Saskatchewan. He was the first Secretary of the revitalized Saskatoon Centre in 1969-70 and then became President in 1971-72. For the next decade he was a dynamic force in the Centre, being largely responsible for the construction of the Rystrom Observatory and as an avid astrophotographer, he generously sharing his know-how with other members. He gave lectures on the subject locally and as an exchange speaker to other Centres, and the Pattersons' home was the scene of many less formal classes. He oversaw the Centre's production of a star atlas, he wrote many articles for the Centre and the *National Newsletter*, and was the author of a *Handbook of Astrophotography for Amateur Astronomers* (1st ed -1974, 2nd ed -1988). In 1981-82, Gordon Patterson served a second term as Centre President, during which time he led the organization of the Saskatoon GA. He received the Service Award in 1982 for his vital role in promoting astronomy in Saskatoon.

[later described as seven-inch] Cooke refractor and equipment." But even though some astronomy was taught by Professor A.J. Pyke (who had been a leading light in the early Regina Centre) and James Duff (previously a Regina and Victoria member), no RASC Centre developed in Saskatoon at that time.

In 1947, Professor William Petrie, formerly of the Vancouver Centre and a brother of astronomer Dr R.M. Petrie, took steps to organize a group of thirty-five members into the Saskatoon Centre. Among its councillors was Dr H.E. Johns, son of the immediate past-president of the national RASC. The Centre remained viable for about four years, with the members of the Physics Department giving most of the talks, but when Petrie left Saskatoon for the Defence Research Board in Ottawa in 1952 the group seemed to lose its guiding spirit. For the next ten years, nothing more was heard of the Centre. In the early 1960s some Saskatoon members did report meteor observations and national President Ruth Northcott and Secretary J.E. Kennedy did try to encourage a revival of the Centre in 1962. As they were both going west for the General Assembly in Edmonton, they called an open meeting at the university observatory on Thursday evening, May 17, at 8 pm. Though the proposal appeared as a news item in the local paper, the turnout was disappointing and nothing developed until a year later when an independent group, the Saskatoon Astronomical Society was formed. The founding president of the SAS was David Roger, later the director of Vancouver's MacMillan Planetarium.

The RASC Centre was reactivated finally in 1969. Having received the unanimous support of the National Council in September, the organizers held the first meeting of the new Saskatoon Centre on November 17 in the university observatory. Among the thirty-three founding members was Professor J.E. Kennedy, national president of the RASC at the time. Dr F.A. Holden and Gordon N. Patterson were elected as president and secretary respectively.

Between them, they spoke at four of the monthly meetings during that first year. Films, some observing, and a tour of NRC's Meteorite Observation and Recovery Project, rounded out the season. The monthly meetings of the Centre have continued along similar lines with nearly all of the speakers being local members and with films and videos providing some variety.

The campus observatory was the focus of Centre activities throughout the 1970s. Some members put a great deal of effort into renovating the old 7-inch [18-cm] refractor so that it would be more convenient to operate, and various groups within the Centre used it on a regular basis. The public came in large numbers to open houses every Sunday; over 7,000 signed the visitor's book in 1972. Excellent publicity by the media encouraged even more to come so that open house had to be held on Wednesdays as well, and group tours scheduled for Friday attracted so many from across the province that Thursdays had to be used for this purpose too. While the observatory belonged to the university, it was Centre members who made significant contributions to the programs for the enjoyment and education of the public. In the early 1970s they also assisted the university with the summer school and by speaking in a series sponsored by the Extension Department. On the other side of the coin, the Centre benefitted by having the university facilities available for their own purposes. Spaces for the Centre library, equipment, darkroom and meeting room were all provided gratis by the university, an arrangement which continues to the present time.

Members took an increasing interest in observing. At first they were content to use the refractor following the monthly meetings supplemented by an annual outing to Auckland's farm, 30 km west of Saskatoon. But in 1972 some special groups formed, including telescope makers and astrophotographers, with the result that observing at a dark site became both easier and more desirable. An agreement was reached in 1974 allowing Centre members the privilege of using land belonging to Ed Rystrom, about 8 km southeast of the city. The astrophotographers used this site very successfully to produce a prize-winning star atlas in 1976–77. Patterson had prepared them well through a series of instructional classes at his home. The joys of working together on the photography, developing, printing and layout added to the accomplishment of this excellent group project. Mainly through the efforts of Patterson and Merlyn Melby, a substantial observatory was built on the Rystrom farm in 1978. Its three metre dome houses a 20-cm catadioptric telescope but is big

Saskatoon's Rystrom Observatory in 1979.

enough to accommodate a larger instrument which the Centre plans to acquire in the near future.

The 1972 solar eclipse, though only partial in Saskatoon, aroused a lot of public attention. The Centre performed a useful service, but at the same time capitalized on this interest by selling solar filters for safe viewing of the sun. Proceeds went to the Telescope Fund. Informative bulletins were also prepared and distributed to teachers.

What began as a picnic followed by observing for members at Diefenbaker Park in 1973, developed in later years into a star night for the general public. By 1981, two consecutive nights were set aside each July, and usually at least one turned out to be clear. This annual event generally attracted over a hundred people, some of them families on holidays from various parts of the country. The Centre set up displays at the Hobby Show in 1977 and 1978 and began Astronomy Day displays at shopping malls in 1981. These have included slide shows, demonstrations of mirror grinding and computer programs, and of course a variety of telescopes with members on hand to talk to passers-by. Over the years, members noticed a greater level of sophistication in the questions asked by the general public. In 1990, the Centre held three public star nights, arranged a display at the Western Development Museum, visited nine schools and promoted an interest in astronomy on radio and television programs. Though membership has not grown from its established forty-to-sixty person base, the Centre can be proud of its varied programs to attract and retain members.

VICTORIA

We now conclude this chapter with sections on the two west-coast Centres – Victoria and Vancouver. The earliest contacts between the Society and professional science in Victoria were with the Meteorological Service just as they had been in Toronto thirty years earlier. The federal government opened a weather station in Victoria in 1898. E. Baynes Reed was in charge and handled the meteorological work while his assistant, F. Napier Denison, looked after the seismic equipment and data. Denison had helped to set up a seismograph the previous year in Toronto and so had come to Victoria with experience. He also came as a member of the A&P Society of Toronto, the first in the west. After a few years, and perhaps at Denison's suggestion, Reed became a member of what was by then the RASC, and soon tried to establish a Centre in Victoria. He arranged for some meetings in 1907 but he could not keep the Centre going. Reed did, however, arouse interest in the Society, and in fact the RASC membership list for 1909 showed about twenty members living in the Victoria area.

One of the members from those early days was Arthur W. McCurdy, MPP, who also belonged to a group known as the Natural History Society. (McCurdy's son, incidentally, flew the *Silver Dart* in 1909, the first aircraft flight in Canada.) It was at one of the meetings of the Natural History Society that Napier Denison appealed for better facilities to carry out his scientific work, and substantial results were soon forthcoming as RASC President A.D. Watson recalled a few years later. "As a result of this meeting and much assiduous work done by Mr McCurdy and his associates, an appropriation of $20,000 was made for a seismological station at Victoria." The station referred to was a meteorological and seismological observatory built on Gonzales Hill in 1913. Continuing to quote Watson, "When this enterprise was well under way, the same indefatigable workers secured Dr J.S. Plaskett to give a lecture … on March 6th, 1914. Mr McCurdy was the Chairman that evening and a Centre [of the RASC] was organized with 58 members." The National Council gave their blessing on April 28 and the new Victoria Centre was launched with Plaskett as honorary president, Denison as president and McCurdy as vice-president.

Undoubtedly the source of much of the newfound enthusiasm for astronomy in Victoria was the announcement by the federal government that it would build a major observatory just north of the city. The chosen site on Little Saanich Mountain was the location for the Centre's first social event, a picnic in July, 1914, on the occasion of a visit from Ottawa by W.F. King, Dominion Astronomer, and RASC president, J.S. Plaskett, soon to be director of the new Dominion Astrophysical Observatory.

World War I was hard on Victoria, but it did not deal a fatal blow as it had in other Centres. During 1914, attendance at meetings dropped from about 60 to 30, and paid membership declined from 101 in 1914 to 40 in 1916, though there were 12 others on active war service who were carried as honorary members. Even as the war drew to a close, a worldwide influenza epidemic seemed to be especially bad in Victoria. All public meetings were cancelled after May 13, 1918 and the regular schedule did not resume until some time in 1919. Otherwise during the war years, the Centre managed to hold about six lecture meetings annually, many of them at the King's Daughters' Rooms on Courtney Street.

Meanwhile work progressed on the DAO. The 1.8-m crown glass disk was shipped from Belgium just days before war was declared, the mirror was ground and figured by Brashear and Company in the United States, and the structure was erected by local contractors. The official opening of what was the world's second largest telescope, took place on June 11, 1918. Many distinguished visitors were on hand, some who had been in Goldendale, Washington, for the solar eclipse three days earlier.

In the postwar years, the four astronomers at the DAO generously supported the Centre, hosting annual open houses and frequently speaking at meetings. But quite apart from RASC activities, the public evenings at the Observatory gave hundreds of thousands a taste of astronomical science. In the peak year 1929, the Observatory welcomed 39,000 visitors, including even a team of Japanese baseball champions. Dr William Harper, who would later become director of the DAO, did much to stimulate public interest in astronomy through his biweekly radio talks and newspaper columns.

The Victoria Centre, of course, shared in this enthusiasm and experienced strong growth in the 1920s. One especially large crowd of about 300, including students from Victoria College and Brentwood College, came to a meeting at the Victoria High School in 1926 to hear UBC Professor Gordon Shrum speak on the topic "Some Fundamental Concepts of the Atom." Another example was the Annual Meeting of December, 1927, in the Girls Central School when an audience of about 350 enjoyed ten interesting and instructive reels of astronomical motion pictures interspersed with discussion. Successful innovations at regular meetings included short talks on timely topics prior to the main address and a question box with the queries being answered by various members at a subsequent meeting. Even debates were sometimes held on provocative topics such as the habitability of Mars. Meetings from 1935 were held in the YWCA and sometimes they were a bit offbeat as this account of the Annual Meeting of December 13, 1939 shows.

> As the dinner concluded, Mr James Petrie, father of astronomer R.M. Petrie, sang two bass solos and an instrumental trio composed of Mr and Mrs Darimont and Mr Gordon Shaw performed two enjoyable numbers. The business part of the meeting was dealt with in short order, and then Mrs Beals played two excellent piano solos. After a brief recess, considerable banging and commotion was heard as a man identifying himself as Professor Whosis, the astrologer, entered the room.

Outreach to various community groups was also quite impressive. The years 1927 and 1928 must have set some sort of record as Past-President Elliott showed astronomical motion pictures at most of the public schools in the city and various mem-

ROBERT PETERS (1872-1965) joined the Victoria Centre shortly after his discharge from active service in World War I. During nearly 50 years of membership, he held the offices of Secretary, Vice-President, President (1939-40), and was on the Centre Council for over 30 years. He was one of the original organizers of "Summer Evenings with the Stars", and frequently contributed lectures to this popular annual series.

When the Victoria Centre acquired its 10 cm-refractor, the observatory was moved to the Peters' home in Gordon Head and he was appointed Director of Telescopes. In this capacity he gave generously and enthusiastically of his time, instructing the junior members in the art of observation, and holding "observation evenings" for members of the Society and their friends. For many years he gave regular reports on astronomical phenomena at meetings and accounts of the observatory's activities at the annual dinner meeting.

Esteemed for his genial personality and for his enthusiastic support of all the Society's activities, he was twice the Honorary President of the Centre (1943-44 and 1965) and received the Service Award in 1961. (Adapted from the citation written in 1961 by J.A. Pearce)

Gladys and Bob Peters at the DAO, probably in the early 1950s.

bers spoke to the Farmers' Institute at Sooke, the Victoria Teachers' Association, the United Church of White Rock, the Victoria Gyro Club, St. Columba's Brotherhood, the Aloha Club of Tacoma, Washington, the Royal Yacht Club, St. Mary's Young People's Association and the Masons of Victoria. Both local newspapers covered meetings of the Centre on a regular basis, public star nights began and weekly radio broadcasts were revived by H.B. Brydon and a dozen other members in 1939. During World War II, lectures by Centre members to men in the armed forces proved very popular.

Victoria Centre's long-standing policy of offering short courses to the public has already been alluded to in chapter eight. The first instance seems to have been a series of six talks attended by forty-to-fifty people in March and April, 1925. Three were given by astronomers from the DAO, and three by other RASC members on various aspects of the Solar System. The summer course which still continues to flourish under the name "Summer Evenings with the Stars" began in 1931. At first it was intended for members and was held weekly during July, August and September at Brydon's home and observatory at Oak Bay. But once the public was included, the locale was changed to Victoria College and by 1934 nearly sixty people attended the nine-session course.

For many years, Victoria was the only RASC Centre west of Winnipeg, and many of its members lived far afield, some even in Saskatchewan. So the establishment of Centres in Vancouver (1931), Edmonton (1932) and Calgary (1935) had a negative impact on the membership roll in Victoria as old members transferred their allegiance and new members naturally joined Centres closer to home. Still Victoria Centre held its ground and continued to run a successful program.

Though there had traditionally been open houses at the DAO, and Denison occasionally welcomed the Centre to the Gonzales Observatory, very few members had taken an interest in making regular observations and few had telescopes of their own. The situation began to change when observing sessions were introduced at the conclusion of meetings in 1929. Brydon's fine old 10-cm refractor, which he acquired from A.F. Miller in 1931, got wide usage. Later, with the formation of a telescope-making section under W.R. Hobday in 1934, the first such group in any RASC Centre, many members began to make and use their own instruments.

Revenue from the Summer Course fees, though modest, accumulated to the point where the Centre, with help from contributors, was able to purchase the 10-cm refractor from Brydon in 1944. For the next three years members used the telescope at Brydon's Oak Bay Observatory for observing variables, double stars and occultations but when he sold his home in 1947, the observatory had to be relocated, and Robert Peters agreed to have it on his property in Gordon Head. The official dedication of "The Brydon Telescope" as it was now known and the opening of the observatory at its new site took place on April 26, 1950.

During the early 1950s most regular meetings of the Centre were held in the auditorium of the provincial Normal School, and then from 1957 on, at Victoria College (which later became the University of Victoria). Proximity to the DAO ensured a steady supply of speakers including not only astronomers from the Observatory but some outstanding visitors such as Herman Bondi, Otto Struve, W.W. Morgan, Bart Bok and P.W. Merrill.

Efforts made in the 1950s to increase member participation by organizing sections in telescope making, astrophotography, constellation lore, observing and elementary computation were largely short-lived as were the groups of meteor and auroral observers headed by R.S. Evans as part of the International Geophysical Year program. George Ball was more successful with the weekly observing sessions which he started first at Gordon Head and

GEORGE BALL (1910-) is well-known in the Society for the superb instruments he designs and fabricates in his basement workshop. Usually these incorporate some unique features. His observatory, outside his Victoria home is a case in point. Instead of having the usual rotating dome, his entire observatory rotates. Instead of having wheels running on a circular track, here the track is above the wheels to prevent dirt from falling onto the rails. To form the circular track, Ball built a tool specially for the purpose. Inside the observatory is a 30-cm Schmidt-Cassegrain telescope built entirely by Ball including right ascension and declination scales with verniers, engraved and nickel-plated by himself. The telescope can be smoothly moved along either axis by variable-speed motors easily controlled from Ball's comfortable observing chair complete with a crank by which he can raise or lower himself at the eyepiece. Other equipment he has designed and built include a mirror grinding machine, an aluminizing apparatus, a lensless Schmidt camera, a cold camera and special silicon rubber molds to make pitch laps used in mirror polishing.

Obviously, all this has required a great deal of talent and time, but George Ball has never shirked from helping others. A year after joining the Victoria Centre in 1955, he was on the Council and in the years since has been National Council Representative, Vice-President of the Centre, Director of Telescopes and Observations, and National Co-ordinator for Instrumentation. He has used his own equipment to aluminize many mirrors for others and supervised and instructed dozens of members in the construction of their own telescopes. For years he has hosted public observing nights, arranged for displays at hobby shows and assisted at the DAO on public evenings. He received the Service Award in 1968.

G. Christopher L. Aikman (1943–) is one of those surprisingly rare professional astronomers whose interest in astronomy can be traced back to his days as a youthful member of the RASC. He joined the Quebec Centre in 1958, graduated with a BSc from Bishop's University in Lennoxville in 1965 and proceeded to earn his MSc at the University of Toronto with a thesis on Microwave Observations of HII Regions. He joined the staff at the DAO in 1968 where he still works as a Research Officer though his interests necessarily changed from radio astronomy to spectroscopy of peculiar stars and binaries and more recently to comets.

Chris Aikman received the Service Award in 1983 for his important role in organizing the 1981 GA and for his significant contributions to Victoria Centre as a speaker, member of Council and Secretary. In this latter capacity he took on a great deal of responsibility, helping the Centre with incorporation, grant applications and acquiring charitable status. He has written a number of articles for the *National Newsletter* and several abstracts of papers presented to CASCA have appeared in the *Journal*.

At Hornby Island, September, 1992.

then, in 1968, using the University of Victoria's 30-cm telescope. Also in the 1970s, members played a greater role in the regular meetings as speakers and as participants in activity nights, on panel discussions and on trips to joint meetings with Vancouver Centre. Recent developments have included a new observing group, meeting monthly for stargazing and homemade cookies, and an Education Committee which promises to be an effective means of handling the many enquiries from schools and community organizations for speakers.

Several initiatives were responsible for steady growth in membership from the 60-70 person range in the 1960s to present levels of 120-150 members. The old reliable "Summer Evenings with the Stars" program undoubtedly continued to attract new members as did occasional public star nights in Beacon Hill Park. But at least since 1964 some members encouraged interest in the Centre by looking after a booth on visitors' night at the DAO and in more recent years by also setting up telescopes for public viewing on the Observatory grounds. Also, participation in the annual Victoria Hobby Show began in the sixties, and other venues have included a NASA exhibit on Manned Space Exploration in 1964, mirror-grinding demonstrations to the Lapidary Society in the 1970s, Recreation Discovery Week in 1981 at Mayfair Mall, the Better Living Show in 1983 and Space 1990, a project jointly sponsored by the Royal British Columbia Museum and the DAO.

The Centre got the biggest single boost in its history in 1976 with a bequest of $11,700 from the estate of the late R.S. Evans. After much discussion, plans were made in 1980 to use this money to construct a mobile 50-cm reflector. Leo Vanderbyl was the skilled craftsman who completed the project in 1983, and the Evans-Vanderbyl Telescope has seen much good use at star parties for members and the public in the years since. Discussion is now taking place to see if a permanent site can be found for the telescope and trailer.

VANCOUVER

The earliest indication in Society records of public interest in astronomy in Vancouver dates from 1912 when Napier Denison of the Meteorological Service, gave an astronomy lecture in aid of the widows and orphans of the crew of the *Titanic*. The substantial sum of $93 was raised. Denison was to become the first president of the Victoria Centre in 1914, and in succeeding years, perhaps through his influence, a number of people living in mainland British Columbia took out membership in the Victoria Centre. Dr Gordon Shrum, then of the Physics Department at the University of British Columbia (UBC) and many years later the well-known president of BC Hydro, was one such member who joined in 1926. By 1931 there were twenty-three RASC members living in the Vancouver area, and with prospects of encouraging even more to join, a local Centre was established with Shrum as the first president.

The first open meeting took place at UBC on Tuesday, November 10, 1931, with a good attendance of members, university students and general public. Dr Daniel Buchanan, Dean of Arts and Science, and the Centre's first honorary president, gave the address on "The Making of Worlds." With this auspicious beginning, and the recruiting efforts of Mr H.C.B. Forsyth, the Vancouver Centre signed up seventy-one members in its first year, nineteen of whom transferred from Victoria Centre. Forsyth had just completed a 30-cm stainless-steel mirror which performed well without silvering, and as a result of his expertise, he became the chairman of the Centre's first telescope-makers group.

With so many neophyte members, there was a need for a course of lectures on general knowledge of the sky. This was instituted in 1933. Other special events that year included an excursion to the DAO in Victoria and a debate on the resolution that life may exist extensively in the Universe. "Garden meetings" held on Saturday evenings in July attracted twenty to thirty people. These gave members an opportunity to inspect other members' telescopes and to try them out on double stars and other celestial favourites. The refreshments and socializing which rounded out the evening built friendships and cohesiveness in the young Centre.

As happens not infrequently, the initial enthusiasm was hard to maintain, and the number of members dwindled down to thirty-two by 1939. This decline occurred in spite of some excellent lectures provided by astronomers from the DAO and from professors at UBC. Shrum continued to be one of the most popular. At one meeting he demonstrated gamma-ray emission from radon and also exposed a silver coin to slow neutrons, transforming it into cadmium. This was not really astronomy, but joint meetings with the university Physics Club and the Chemistry Society helped boost attendance at the lectures if not in the Society itself. During the 1930s,

Members of the Victoria Centre gather around the Evans-Vanderbyl Telescope (below). Leo Vanderbyl, the builder of the telescope, is seen in the photo above.

ARTHUR M. CROOKER (1909-1990) was born at Cayuga, Ontario, received his BA from McMaster University and then proceeded to his MA and PhD at Toronto under J.C. McLennan and to further post-graduate work in England. From 1937 until his retirement, he was on the faculty of the Physics Department at UBC, except for the War years, 1941-45, when he was in Toronto in charge of Optical Design at Research Enterprises. In the 1970's he was an NRC Research Associate at NASA's Goddard Space Flight Centre.

Crooker joined the RASC in Toronto in 1934, and transferred to Vancouver Centre three years later. He held many offices in the Centre including the Presidency in 1951, and often spoke at meetings and gave instruction with telescopes. He was helpful in getting the University to waive charges for the Centre's use of campus facilities for meetings, and in 1981, donated a 41 cm f/6.3 mirror to the Centre. He received the RASC Service Award in 1965.

ROBERT J. CLARK (1893-1972) was born in Vancouver, received his BA at McGill and his PhD from Cambridge. He taught at Edinburgh University from 1926-31 and then at the Egyptian University, Cairo. Later Dr Clark taught Physics for several years at the University of Saskatchewan before finally returning home to Vancouver in 1947. His primary responsibility at UBC was in managing the physics laboratory for fourth year honours students.

When Clark received the Service Award in 1962, it was noted that he had occupied all the offices of the Vancouver Centre, including ten years as Secretary. He took a special interest in the young members and managed both the telescope-making and observing sections of the Centre. For several years he generally opened each meeting with a 15 minute talk about some point of astronomy, followed by a brief description of the sky for the coming month.

the Centre held annual dinners and members' nights in May. The meal was usually in Union College Dining Room and was followed by a meeting in the Science Building where demonstrations were sometimes set up illustrating such topics as molecular motion of gases, chemical luminescence, or spectroscopic and optical experiments. These occasions also gave an opportunity for the ever-active telescope makers to display their latest products.

Gradually over the long period from 1940 to 1968, membership in the Centre lurched back to its initial level of about seventy. For many years, members brought their telescopes to the university campus every clear Tuesday evening from May to September, and in 1954 began to set up equipment at the Pacific National Exhibition. The telescope-makers group had ceased to function during the war, but began again in earnest in 1955 with a few members completing mirrors each year. The Annual Meetings, a members' night, often a film showing and about five regular lecture meetings continued to be held each year at the university.

Vancouver Centre got a wonderful new lease on life in 1968 when the H.R. MacMillan Planetarium opened. A close affiliation was ensured by making each member of the Centre an associate member of the Museums Association for a fee of one dollar each. In this way, members were invited to each new planetarium show, and the auditorium of the planetarium was made available for regular monthly meetings of the Centre, a smaller room for use by the telescope makers, and the Members' Lounge as the venue for Council meetings and the social hours following regular meetings. A number of staff also belonged to the RASC. The cordial relationship with the planetarium resulted in a greatly increased interest in the Centre, and membership doubled to 143 between 1968 and 1971 with nearly half being students.

Another great encouragement came with the opening in 1979 of a public observatory next to the planetarium. It was financed by the Gordon Southam family and the provincial government, and members of the Vancouver Centre volunteered to operate the telescope for students and the general public. In return the Centre got to use the Observatory following regular meetings and had the benefit of the adjoining lecture room for Council meetings and to house the Centre's library. A strike by civic workers in 1981 prevented access for a time and this may have been one factor in advancing plans for an independent Centre observatory at a location 50 km southeast of the city at Langley, on the property of member Art Holmes. For a number of years the site was used for the Centre picnic and for star parties, but light pollution spread. In 1987, a new and darker site was found at Aldergrove Lake Park where the Centre now has a roll-off observatory housing a 35-cm Newtonian. This facility is named for Dale McNabb, a dynamic member of the Centre who died quite suddenly shortly after the observatory was completed. In the meanwhile, in 1986, the Centre had become incorporated and had received a 25-cm f/15 Cassegrain telescope on long-term loan from the Southam Observatory. The following year a 36-cm mirror was finished and installed in a temporary mounting and members worked on a trailer for both these scopes.

Not only did facilities improve but the range of members' activities increased dramatically over the years. In 1965, a monthly newsletter was inaugurated mainly as a means of communication for the newly established Observers' Section and for the Telescope-Making Section, both of which held monthly meetings. The newsletter became the now-familiar *Nova* in 1968 and was enlarged and improved in the early 1970s. The Seattle Astronomical Society also started to use *Nova* for their members in 1977. When the cost of producing and mailing the newsletter began to soar, the Centre came up with a unique way to solve the problem. Viewers for the 1979 solar eclipse were

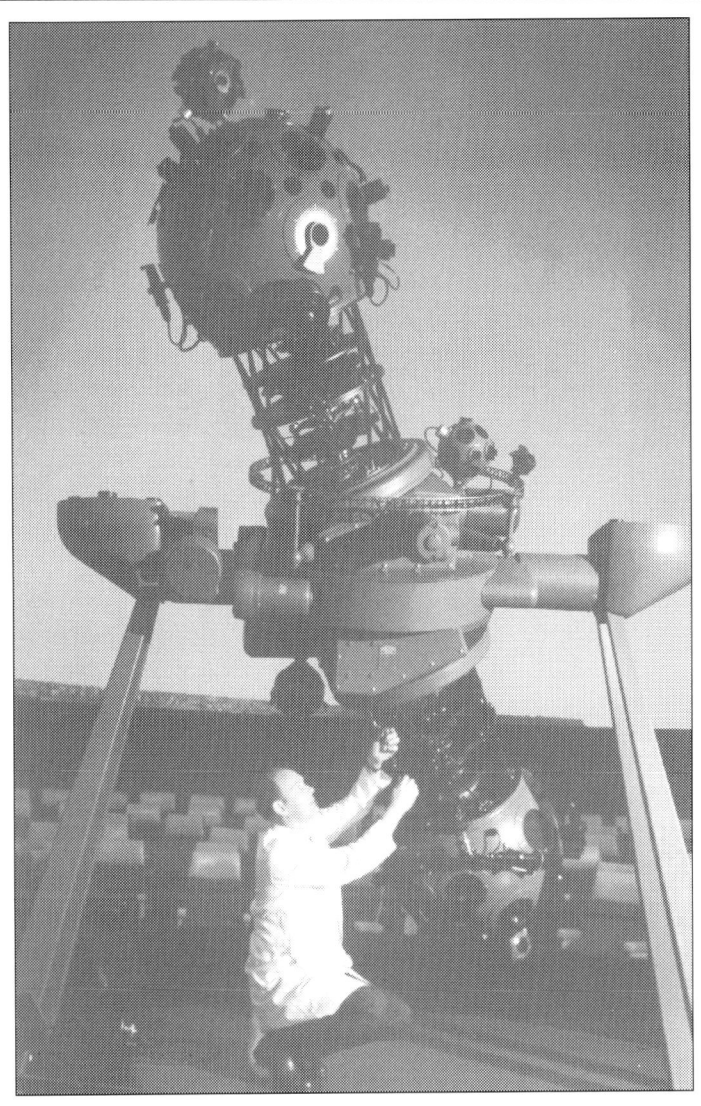

James Wright is shown here working with the projector of the MacMillan Planetarium, to whom credit is given for this photograph.

JAMES F. WRIGHT (1931-) joined the RASC in 1946. As a student at the Southern Alberta Institute of Technology in 1958, he was a founding member and the first Vice-President of Calgary Centre. He was an active auroral observer and contributed to the IGC Visual Meteor Program in 1960. Having built his own observatory, he joined the AAVSO and became the Centre's co-ordinator of planetary observations in 1962.

His expertise in designing electrical and optical systems led him to take an important part in planning the Calgary Centennial Planetarium, and he became its Technical Supervisor when it opened in 1967. For his enthusiastic role in promoting the planetarium and astronomy within the community, Jim Wright received the Service award. Also during Centennial year he served as President of the Calgary Centre and his wife, Norma, was Secretary. They then moved to Vancouver where he accepted a position as Assistant Director of the MacMillan Planetarium. He naturally became involved in the Vancouver Centre, holding a number of Offices including the Presidency in 1974. Norma unfortunately passed away after the Wrights returned to the Calgary area in 1988.

produced and marketed and the venture proved so successful that a fund was created which enabled *Nova* to continue without cutbacks. Another source of revenue for the Centre came from raffles for doorprizes at the monthly meetings in 1988. Now the Centre is into big time money raising at casino nights, with proceeds of several thousand dollars in one night.

The Observers' Group was abetted in 1973 with the arrival of a number of enthusiastic members transferred from Ottawa. They started up a variable star and meteor program and also encouraged weekend trips to dark sites for some really outstanding observing. Six members on excursion to the Kootenays in mid-August, 1975, enjoyed four clear nights in a row, with excellent conditions for the Perseid meteor shower and the maximum of Comet Kobayashi-Berger-Milon. More recently there have been weekend star parties 200 km east of Vancouver at Manning Park and further afield at Mount Kobau.

Visits to and from other astronomical groups have been popular with Vancouver members. The year 1984 was an especially active one with travels to Victoria and to Seattle for joint meetings. Later that year the Centre hosted a combined assembly of several American and Canadian astronomical organizations.

Activities planned for the public have also increased in scope and variety. A successful series of summer lectures was organized each year from 1969 to 1971. Nonmembers were charged a fee of 50 cents per lecture. Of course, public observing nights have continued to be popular, especially in connection with some well-publicized celestial event. In May, 1975, for instance, a lunar eclipse party was staged on the Planetarium grounds. Over two dozen telescopes were trained on the Moon, as well as Venus, Saturn and other objects, and over two thousand people attended. Another successful party late in July that year brought hundreds out to see Comet Kobayashi-Berger-Milon. Cypress Bowl Provincial Park became the site of monthly star parties in the summer of 1982. Over 900 gathered there for a well-publicized Perseid Meteor Watch in 1983, but when the government leased the park to a private concern in 1984 for development as a ski area, the Centre had to look for a new locale. John Dobson and the San Francisco Sidewalk Astronomers visited Vancouver in August, 1982, and gave two special lectures to the public. For the lunar eclipse of December 30 that year, over 800 came to the Gordon Southam Observatory, even though totality was at 3 am. Now the Centre runs its own Sidewalk Program.

In 1984 a video on basic astronomy, called "Stargazer" was prepared in conjunction with Rogers' Cable TV. It was well received with the result that a longer program was produced the following year. In 1986, the Centre co-operated in two big events. One in May was a three day "Space Update '86" sponsored by the BC Science Teachers' Association, where a number of RASC members served as lecturers and demonstrators; the other event took place at Campbell River Park in October and was organized by the Greater Vancouver Regional District. It was called a "Celebration of Nature" and was attended by 10,000 people over a two-day period. The Centre's continuing participation in this event and in "Fraser River Days" each June has earned it high marks from the Parks Board and was probably a factor in the approval given by the Board for the construction of the McNabb Observatory on public land.

An initiative which turned out to have national significance was the Vancouver Centre's production of a calendar for 1992 containing useful information on meeting dates, observing nights, daily times of moonrise and moonset and weekly times of sunrise and sunset. Rajiv Gupta carried out the calculations and supplied the beautiful celestial photographs for each month. The Centre, under President June Kirkaldy, then took the unprecedented step of volunteering to produce and distribute a national version with profits shared equally between the national Society and Vancouver. With the financial backing of two Hamilton Centre members, Anne and Bill Tekatch, the National Council eagerly accepted this generous offer. The result is a very handsome calendar which brings credit and recognition to the RASC and promises to be a moneymaker for the Centre and the Society as a whole.

The H.R. MacMillan Planetarium (l) and the Gordon Southam Observatory (r).

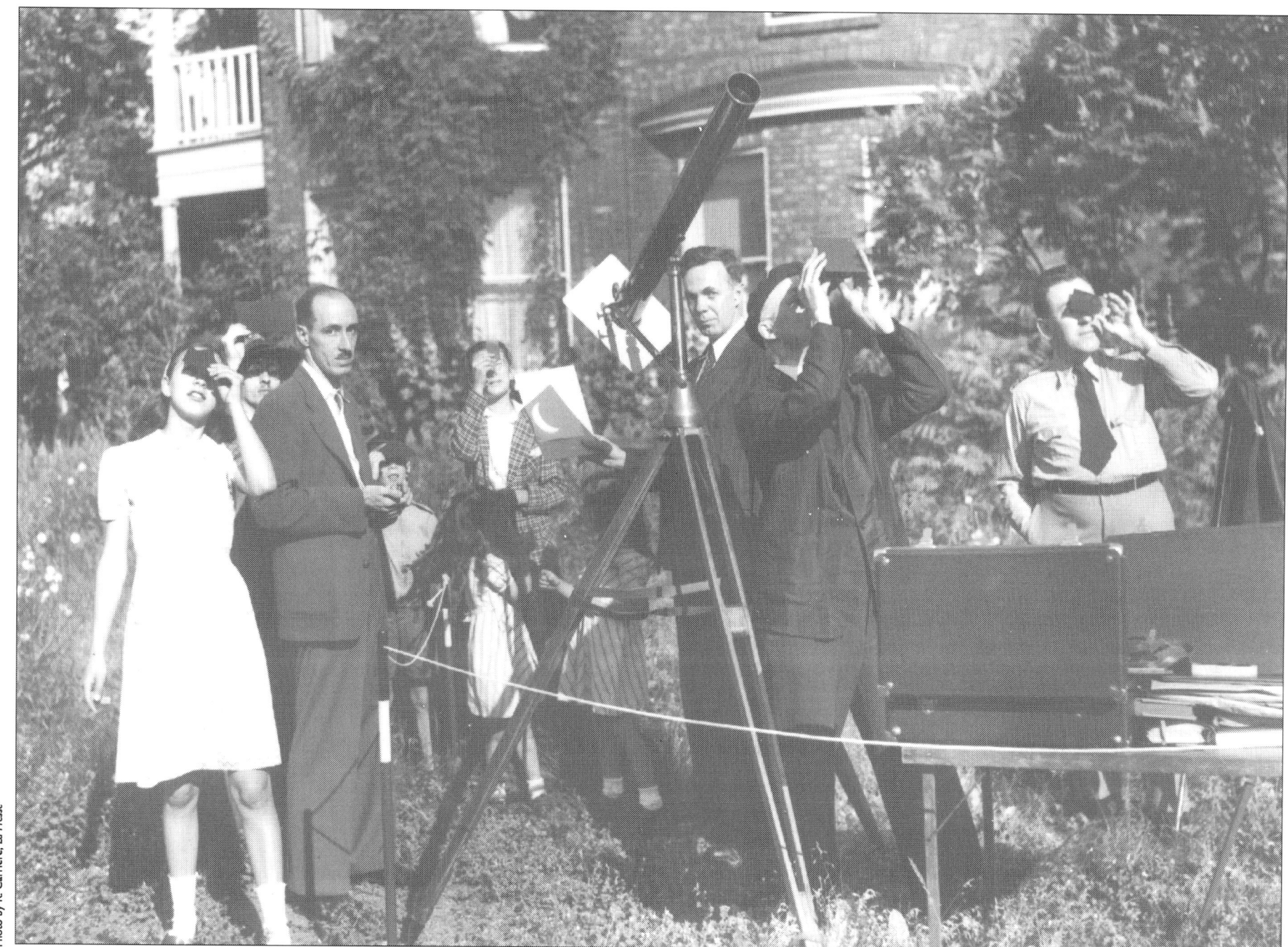

As the Moon slowly moves from west to east across the face of the Sun, a group at DeLisle Garneau's home in Montreal watches the partial phases of the solar eclipse of 9 July, 1945. Let's hope they were using #14 welder's glass.

Photo by R. Carriere, *La Presse*

CHAPTER 13

From West to East

Having now completed our tour of Centres west of Ontario, we turn to the east.

Originally, professional interest in astronomy in Montreal, as in Toronto and Victoria many years later, was an offshoot of meteorological studies. Charles Smallwood began making weather observations in the 1840s at his observatory on Île Jésus and incidentally recorded positions of Donati's comet in 1858 and timed the partial phases of the eclipse of the Sun in 1860. Named an honorary professor of meteorology at McGill in 1856, he transferred his equipment to the campus when a small observatory was constructed there in 1862. Smallwood also operated what is presumed to be the first seismograph to be set up in North America in 1870. Following his death in 1873, one of his students, C.H. McLeod, assumed responsibility for the observatory and developed it into the centre for timekeeping in Canada.

Montreal had a long tradition of scientific Societies aimed at public education. The Natural History Society was founded in 1827, the Mechanics' Institute the following year, and l'Institut canadien was established in 1844. An Astro-Meteorological Association, whose primary aim was to study the connection of astronomy with terrestrial phenomena, first met in 1884 but did not survive past 1889.

As early as 1892, the editor of *The Montreal Star* wrote to Toronto for information about the A&P Society. Nothing concrete developed as a result of his enquiries, but through a sort of diffusion, by 1917 there were several RASC members in Montreal. One of them, H.E.S. Asbury, founder of an astronomical group in Seaforth, Ontario, and a leading light in the early years of the Guelph Centre, now suggested forming a Centre in Montreal. Another member, the Reverend W.T.B. Crombie of Kingsbury, Quebec, was thinking along similar lines and wrote to each of the fourteen members then residing in the Province of Quebec to find out their interest in a Montreal Centre. As a result, a meeting was called for May 28, 1918, in the Engineering Building at McGill. Those assembled decided to form a Centre under the presidency of Monseigneur C.P. Choquette of Ste-Hyacinthe, with Asbury as vice-president and Crombie as secretary-treasurer. They also agreed to hold an inaugural meeting in September with Otto Klotz, director of the Dominion Observatory, as the invited speaker. The General Council of the RASC welcomed these initiatives and by the end of 1918, the Montreal Centre was firmly established with thirty-five members.

Within ten years there were over 100 members. Such strong growth was a result of the excellent programs which were arranged. The co-operation of McGill University was very helpful in allowing the Centre the use of the Macdonald Physics Building for meetings and in facilitating joint meetings with campus organizations like Sigma Xi and the Physical Society. Several McGill people were active in the Centre and a number of them, including professors A.H.S. Gillson, A.S. Eve and A.V. Douglas served as speakers as well as officers. In addition, many outstanding visitors addressed the Society, among them

DeLisle Garneau (1906 - 67) was a son of the well-known historian, Hector Garneau. He remembered his father pointing out Halley's Comet to him in 1910, but found his interest was really awakened by a book given to him as a Christmas present in 1919. Garneau told the fascinating story of his growing interest in astronomy in the *Journal* in 1939. His fine observations of sunspots, lunar occultations, Mars and Comet Cunningham, made with his 10-cm refractor, were also reported in the pages of the *Journal*. Behind his home in Montreal's Notre Dame de Grace district, he built an observatory in 1941 which later housed the Centre's 15-cm Aitchison refractor, and members came there regularly to use it. At the same time he was appointed Chairman of the Centre's Telescope Committee, a position which he held for the next eight years. He very carefully planned a successful campaign for the observation of Perseid meteors in 1942 and again in '43, and was an active auroral and variable star observer. Though employed as an accountant with the Department of National Revenue, Garneau found time to write popular articles on astronomy in the Montreal magazine, *La Revue Moderne*, and a weekly column in the newspaper, *Le Devoir*.

He became the first President of the Centre Français de Montréal in 1947, a position which he held again in 1951, and he spoke frequently at meetings. He also served as national Second Vice-President in 1951-52. The Society awarded him the Chant Medal in 1951.

The photograph shows DeLisle Garneau observing sunspots on a screen with the aid of the 15-cm refractor in his Wilson Avenue Observatory. The photo originally appeared in *L'Oeil*, 15 mai, 1944.

Canadian geologist A.P. Coleman and physicist J.C. McLennan, American astronomers Harlow Shapley and Annie Cannon and world-famous philosopher, A.N. Whitehead.

Amateur members took their part in Centre duties too. Some, like Miss M. Ellicott, the first woman to join, and E.E. Bridgen, gave short talks preceding the main address. Mr Asbury organized a series of informal meetings for members who wanted to learn more about the basics, the constellations and how to use the *Handbook*. Films were shown and in 1921 D.R.P. Coats of the Marconi Company, "kindly mounted a complete wireless receiving set of seven tubes, enabling [the audience] to hear a special radiophone concert." On another occasion he spoke about his famous great-uncle, Richard Proctor, who as a world-renowned author and popularizer of astronomy, had given a series of lectures to Montreal audiences in 1879.

The second decade, through the Depression and leading up to World War II, was marked by little growth. In spite of the large cost of paying expenses for out-of-town speakers, the Centre maintained their ambitious program, and continued to attract some very prominent people. Canadian H.H. Plaskett, American E.W. Brown the lunar expert from Yale, and Georges Lemaître the renowned cosmologist from Belgium, all addressed meetings of the Centre during this period. An audience of over a thousand heard Lemaître's speech on the "Expanding Universe" in 1933. Another highlight that year was a broadcast organized for the Pacific Science Congress in Vancouver in which Ernest Rutherford in Cambridge had a conversation with his former colleague Professor Eve of McGill.

A very significant development in the 1930s was a growing interest in observation. The Centre spent $205 in 1933 to purchase a 15-cm refractor from the family of H.E.S. Asbury and they used the telescope on the roof of the Sun Life Head Office Building where they gathered once a week from May to October, weather permitting. Among the members joining about this time were Frank DeKinder and E.R. Paterson, both keen observers. Paterson organized groups of meteor observers, usually drawing on Boy Scouts at summer camp to count Perseids, and in later years to plot and photograph them. DeKinder headed up a telescope-making section starting in 1936.

Observing developed into a major Centre activity in the 1940s. With the outbreak of war, Sun Life's facilities could no longer be used for security reasons, but a member, Delisle Garneau, came to the rescue and offered his home at 4052 Wilson Avenue. His backyard Ville Marie Observatory became home for the Centre's 15-cm refractor. Every Saturday evening from 7:30 to 9:30, about ten to twenty members would get together for observing if skies were clear or for discussion in Garneau's basement. The planets, some double stars and Messier's nebulae were all popular objects, supplemented by occasional comets, occultations and eclipses.

In 1943, two programs were initiated which over the next few years yielded much useful data. One group, under Isabel Williamson's direction, made systematic observations of the aurora with results reported monthly to Dr C.W. Gartlein of Cornell University who directed the National Geographic's program of auroral research. Other members made variable star estimates and carried out nova searches for the AAVSO. Garneau and John Duffie were the principal contributors, each making hundreds of estimates of dozens of different variables, SS Cygni and R Scuti being the most frequently observed. Meteor observing continued, usually at Lower Canada College, and results from the Perseid, Orionid and Leonid showers were communicated to Peter Millman once the war was over and he had returned to Ottawa.

About 1947, the approaching solar maximum attracted a lot of observers. Coverage was remarkably complete with reports of 23,691 sunspots recorded

FRANK J. DeKINDER (1892-1970) joined the Montreal Centre in 1934. All the lectures at the time were formal affairs given by professional scientists, and he feared that many amateurs were intimidated. He himself was an estimator for a tile company, so he was rather startled when Dr A.V. Douglas, the Centre's Secretary-Treasurer, replied to his request for some lectures of a more popular nature by suggesting he do it. Though he had never given a public lecture and had to rent an evening suit (as Centre protocol required), he reluctantly agreed. DeKinder handled the publicity himself, seeing to it that announcements were made in every high school in the city and in various public places. His maiden speech on current knowledge of the solar system was a phenomenal success, and he was thereafter marked as just the sort of dynamic individual the Centre needed.

From 1935 until 1964, he held various offices in the Montreal Centre, including a term as President in 1939-41, and was on the National Council from 1949-64. Over the years he gave many talks in English and French to both Montreal Centres.

Frank DeKinder was an active observer and it was primarily for this that he received the Chant Medal in 1955. Meteor showers attracted him at first but after 1945 when he built a domed observatory at his home in Sault-au-Recollet to house his clock-driven 10-cm refractor, solar observing became his main interest. From 1950-64 he was Observations Director for Montreal Centre. Though he was a diligent participant in the AAVSO nova search program, it was chiefly his solar work that got him involved on the council of that association. At their meeting in California in 1963, he became Second Vice-President. Because of this commitment which led to his AAVSO Presidency in 1967-69, he gave up all RASC duties in 1964.

Frank DeKinder at the eyepiece of the Montreal Centre's 15-cm refractor about 1957.

by over twenty observers on 259 days in 1948. Sixteen members worked on mirrors that year including Alfred Donnelly who made a mirror-grinding machine to help with the production of a 30-cm mirror. But the group which really started a trend was the Messier Club, originated by Isabel Williamson. The goal was to see and record the appearance of each of the 103 objects originally observed and catalogued by Charles Messier in the late 1700s. Competition was keen among the twenty or so members to see who would be first to complete the list. In the end, Tom Noseworthy was the champion and was invited to spend the summer of 1952 at the David Dunlap Observatory studying periods of RV Tauri variables.

The Ville Marie Observatory got tremendous use. Besides being the site of the weekly observing meetings, it housed the Centre's growing library which logged an annual circulation of several hundred. The Observatory also served a very important role in public relations receiving well over a hundred visitors each year, including Girl Guides, school groups and Air Cadets. Fortunately there is no indication that Garneau's family or neighbours ever complained about the 1,400 members and visitors who came and went in a typical year.

Public star nights began in 1945 on the grounds of Lower Canada College. The following year the site was moved to Westmount Park and plans were more elaborate. Newspapers and radio provided good publicity and the city of Westmount co-operated by screening the street lights and installing a public address system. Libraries and stores arranged displays around astronomical themes and the Centre president, Dr Henry F. Hall, gave a 15-minute radio talk over CFCF. Seventeen telescopes were set up and over 1,500 people attended. The activities of the Centre were continually before the public in a column written for the local weekly newspaper, the *Notre-Dame-de-Grâce Monitor*. All these efforts paid off as membership, which had been reported at 80 in 1941, shot up to an all-time high of 290 in 1946.

The expanding activities of the Centre were promoted in their newly instituted newsletter, *Skyward* (an RASC first), and supported beyond the fees by members' and friends' donations amounting usually to a couple of hundred dollars per year. 2,500 people attended the annual star night in Westmount Park in 1949, and other star nights or talks were sometimes held in outlying areas, at summer camps and at schools. For the members themselves, there were the weekly observing meetings and monthly lectures, where one novel feature was a photo quiz in which twenty-five slides were shown for the audience to identify, followed by a reshowing with answers at the next meeting. In addition, the library continued to grow in size and circulation. An annual social meeting contributed to a fine esprit de corps. Nonetheless, a slow decline in enrollment began in 1947 with the formation of the Centre français de Montréal.

In 1953, Delisle Garneau decided to move and the Montreal Centre had to look for a new location for its observatory. Fortunately the honorary president at the time, Dr Norman Shaw, was also the Head of Physics at McGill University and he knew of an abandoned experimental radar station on campus that was slated for demolition. The Centre was given the use of this building and after a lot of renovation and repair, the new facility opened the following year. Frank DeKinder and Charles Good worked closely together in drawing up plans, obtaining supplies, organizing work parties and supervising the whole job. DeKinder, with his knowledge of construction work, obtained estimates for structural changes to be made in the original building and for the revolving dome. All was made possible by the many members who volunteered to do excavating, carpentry, wiring, plastering and painting, and by generous financial support from annual donations to the Centre, special contributions to the Observatory Fund of nearly $2,700 and a wonderful bequest of $25,000 from the estate of G.H. Townsend, a member of the Centre since 1943.

The bulk of the Townsend bequest, $20,000, was put into a trust fund whose income provided for a continuing series of outstanding speakers over the years. American astronomers, Harlow Shapley, Fred Whipple and Cecilia Payne-Gaposhkin, and Canadians R.M. Petrie and Peter Millman were all among the Townsend Lecturers in the 1950s. The Centre's regular lecture program was supplemented by courses of four or six sessions on a more elementary level which were of interest to new members but which were also intended to attract the general public. And the Observatory at its downtown campus location became even more the focus of Centre activities. Isabel Williamson wrote in 1958:

> At the present time the building is open at least twice a week for observing. The regular observation meetings, to which visitors are welcome, are held every Saturday, year round. Wednesday evenings are reserved for members only – for Messier hunting, lunar and planetary work. ... Meetings of the Council and Observation Committee are held there and in the course of the year many groups of Scouts, Guides and students visit. ...
>
> It is wonderful to have a home of our own. Having all our equipment and records under one roof has done much to co-ordinate the work of the Centre. And it is a good thing for the general membership to have an observatory that can be reached as easily by city bus as by private car.

The success of the Montreal Centre continued unabated through the 1960s. A new Constitution and by-laws led to incorporation under Quebec law in 1963. There were annual trips to the Sundells in Vermont for the Perseids, to Stellafane for the Telescope Makers' Convention, and some joint meetings with Ottawa Centre. In 1962, Montreal Centre hosted the annual convention of the American-based Association of Lunar and Planetary Observers. Expo '67 was of course the site of Canada's centennial celebrations, and it was also the focus for the General Assembly hosted by both Centres in Montreal. The following year, to mark its fiftieth anniversary, the Centre held a banquet at Chateau Champlain and issued an excellent history, entitled *Fifty Times Around the Sun*.

The regular Centre program of eight lectures and/or films was maintained at the Macdonald Physics Building on the McGill campus and from time to time courses at an elementary level were offered to the public and to beginners among the members. The annual star-night in Westmount Park still attracted 700 people in 1969. Annual attendance at the Observatory reached a peak in 1965 of

Two views of the Montreal Centre Observatory on the McGill campus. The photo in the meeting room was taken about 1960.

W.A. Warren with his 15-cm reflector.

WILLIAM A. WARREN (1902 -) became active in the Montreal Centre, first as a member of the telescope-making class in 1950 and somewhat later as meteor observer and participant in the AAVSO Nova search program. He also took part in lunar and planetary work and was chairman of the Centre's fireball and meteor section. Bill Warren was on the Montreal Centre Council from 1955 to '66, being President from '62 to '65. He chaired the Committee which organized the Montreal convention of the Association of Lunar and Planetary Observers (ALPO) in 1962. When he received the Society's Service Award in 1966, he was commended for the many hours he devoted to instructing Boy Scouts and Girl Guides in preparation for their astronomy badges.

2,782 which included fourteen different groups of visitors.

A report from the Director of Observational Activities for 1969 showed a tremendous range of members' interests from meteors to Messier objects. The transient lunar phenomena program got a lot of media attention because observations were scheduled to coincide with the dates of the Apollo missions. Twenty observers made 346 timings of lunar occultations. Thirteen observers made over 100 reports on Mars, Jupiter and Saturn. Several members made sunspot observations, both naked-eye and telescopic, and many participated in the auroral patrol program. Systematic searches for comets and novae were made and reported to the AAVSO. Magnitude estimates of variable stars were also recorded by a number of individuals. Meteors in five different showers were counted. Some work was also done on radio detection of meteors, and solar radio noise was measured. On the downside, interest in artificial satellites began to wane with the re-entry of Echo I in 1968 and Echo II in 1969 and telescope making was also at a low ebb. Though the Messier Club and astrophotography attracted a lot of interest, they were not well suited to the urban location of the observatory, and in fact relocation was discussed as early as 1963.

A changing emphasis in observing interests and the growing problem of light pollution were only two factors which led to serious decline in the Centre in the 1970s. Fees were increased to $15 for regular members and $7.50 for students – twice the national levies. The observatory suffered two attempted break-ins in 1971 and a more serious one the following year when a 15-cm objective lens and a finder scope were stolen. Doubts were raised about the safety of the building itself as reports suggested it was literally sliding downhill. Though sites were inspected for a new observatory and a 51-cm blank was purchased for a new catadioptric telescope, progress was very slow, and a general feeling of malaise hung over the Centre. For a number of years reports were not submitted for inclusion in the Society's Annual Report and those that were sometimes contained bad news. Only two regular lecture meetings were held in 1975, and only one the following year. In fact for four months during the summer of 1976, the Observatory was closed without warning and so there were no meetings, no newsletters, and insufficient contact with members and the general public during that period. Membership dropped to a record low of fifty-eight in 1977.

An important event in the history of "La Belle Province" took place in 1980 when a majority of Quebec voters said "No" in a referendum on sovereignty. Could it be more than coincidence that the Montreal Centre emerged into the new decade with a renewed sense of confidence and purpose? The report for that year referred to substantial improvement in communication and publicity fostered by an Executive that worked extremely well together. Members got busy and made substantial repairs to the Observatory and decided to purchase a new 35-cm telescope. The official opening and dedication of the Townsend Memorial Telescope took place on September 20, 1980, with a crowd of admiring members, professors and students from the Université de Montréal and a television crew from CFCF. Except for the location of the building, the McGill connection, which had been so important to the well-being of the Centre in its early days, was now practically nonexistent.

Some of the lecture meetings now were held at the Dow Planetarium, and almost all the speakers were connected with the Université de Montréal or Observatoire de Mont Mégantic. Once again the meetings were back on a monthly schedule and attracted good attendance by members and the general public as a result of publicity in the media. Many members took part in the less formal gatherings at the Observatory on Saturday evenings and gave talks on a fascinating assortment of topics. Occultations, daytime astronomy and computer enhancement of astrophotographs were among the subjects dealt with in 1989. A really unique initiative was the Galileo Study Group whose twenty-three members hoped through observation to understand the road that led to the scientific discoveries of the past.

Public education continued to be high on the Centre's agenda. School visits involving daytime activities and slide shows were in great demand, and Astronomy Days drew crowds of people of all ages. Even with bad weather the 1990 Astronomy Day attracted 300 people to a tent full of high tech telescopes and demonstrations of computer image processing and astronomical software. The special guest speaker on this occasion, David Levy, delivered two inspiring talks and autographed copies of his book *The Joy of Gazing* which he had originally written for Montreal Centre in 1982. This timeless little book still sold well and the Centre raised $500 from that day.

Another source of revenue for the Centre came from the rental of telescopes. Two of these were donated in 1986 and 1989, one belonging to Charles Good and the other, a 10-cm brass refractor which had belonged to the Centre's founder, the Reverend W.T.B. Crombie. The Centre's observatory was constantly maintained and improved. Though the downtown location was convenient for the Saturday evening meetings, for the library and for some types of observational work, many members felt the need for a dark site. So when they were offered the use of a property called Cedar Crest, near Alexandria, Ontario, (about an hour's drive west of Montreal), they jumped at the opportunity. About fifteen times a year, a number of members would drive out with their telescopes for an enjoyable night under the stars. The Centre finally completed their 50-cm f/3.9 mirror in 1983, eleven years after the blank had been purchased, but it was never mounted. It was sold to the St John's Centre in 1990. As they approach their seventy-fifth anniversary, the Montreal Centre still contemplates a permanent observatory in the country.

CENTRE FRANCOPHONE DE MONTRÉAL

The Montreal Centre, established in 1918, had managed to attract and hold some French-speaking members during its first thirty years, but since its membership was largely anglophone, English was the operative language. For most of that period there was no francophone organization for amateur astronomers in the Montreal area. J. Edgar Guimont, a member of the Montreal Centre, had founded the Institut astronomique et philosophique du Canada in 1926, but it lasted only about seven years. He and some colleagues tried again in 1945 forming a circle known as "Les Amis de la Nature" under whose auspices a highly successful star-night attracted some 3,000 people.

A natural outcome of these efforts was the formation of a Centre français de Montréal. The fact that this did not occur until 1947 is perhaps a reflection of the classical emphasis which characterized Quebec education. The first president, DeLisle Garneau, and some of the original members came from the Montreal Centre. In fact Garneau continued his affiliation with the anglophone Centre as Director of Observations. Among the eleven who gathered for the founding meeting in the library at Saint-Sulpice on the first of May were Frère Robert of the Université de Montréal and Messieurs Garneau and Guimont.

The Centre held regular monthly meetings at the École Polytechnique and in addition arranged courses open to the public. In 1947–48 Frère Robert gave ten weekly lectures based on his recently published book, *Regard sur l'univers*. In the next two years short lecture series were given by Valery Giscard d'Estaing, at the time an engineer at the École Polytechnique de Paris, but later President of France. These latter series were jointly sponsored with the Association Canadienne-Française pour L'Avancement des Sciences in 1948 and with l'Institut Franco-Canadien in 1949.

Garneau made sure that observing was an important aspect of the Centre's activities. In 1951, he hosted thirteen observation meetings at his Ville Marie Observatory on Wilson Avenue where he introduced new members to the principal celestial objects and how to observe them. Undoubtedly there were times when members of both Montreal Centres observed together. Certainly there was a joint star night in La Salle in 1949 and a meeting in 1952 when together they visited Montreal High School to see its planetarium. And of course the Townsend Lectures of the Montreal Centre attracted

J. Edgar Guimont (l) and Jean Naubert (r) on duty at an exposition in the parish hall of the Immaculate Conception, Montreal.

large audiences in the 1950s including members of the French Centre.

On the Centre's tenth anniversary, the mayor of Montreal received members at a dinner at the restaurant Hélène de Champlain at the Île Ste Hélène after which members and guests enjoyed stellar views through several telescopes that had been set up.

The following year the Centre moved to Montreal's Botanical Gardens (near what is now Olympic Stadium), where the Parks Department of the City allowed the Centre the use of a lecture hall, laboratory and the grounds. In the years ahead this became a wonderful headquarters for the Centre's library, and for meetings, telescope making and star nights.

The next ten years (1958–68) were marked by remarkable growth and a very ambitious program. Membership soared from about 50 to 260. Courses were given in some years, the annual meeting and a dinner were held each year, regular meetings were held monthly featuring films or lectures (some by Hubert Reeves) and special meetings were held weekly. These latter sessions tended to deal with more technical and advanced topics, and concluded with observing whenever weather permitted. A bulletin was inaugurated in 1960 which included the celestial events of the month as well as Centre activities. A telescope-makers group formed in 1961 and subsequently produced some very fine instruments.

Some observing highlights during this period included the timing of the partial lunar eclipse of August 25, 1961, and the study of the lunar eclipse of December 18, 1964 at three different observing posts. Meteor groups and the Messier Club were also very active. Each autumn, soirées d'astronomie populaires attracted a couple of thousand interested spectators. There were visits to various schools, convents and religious communities, and there were slide and film presentations to Scouts and other organizations.

JEAN NAUBERT (1888 - 1968) won the Chant Medal in 1953 for his exceptional photography of the Moon and his original and expert telescope making. He was one of the founders of the Centre français de Montréal and served for a number of years in the '50s as their Director of Observations. By 1972 he had figured no less than four 41-cm mirrors one of which he donated to the Centre in 1968.

JEAN ASSELIN (1905 -) held various offices in the Centre français de Montréal between 1949 and 1967. He was President in 1949-50 and again in 1956. By profession he was an engineer. In astronomy, his interest and talks to the Centre were mainly about solar system objects. When he received the Service Award in 1962, he was praised for his effective promotion of astronomy through radio, TV and published articles. The generosity and hospitality of Monsieur and Madame Asselin was important in furthering the interests of the Centre.

An evening meeting of the Centre français Council at the home of the Guimonts, 6634 rue St-Dominique, 10 February, 1949. Standing (from l to r) Jacques Desjardins, J. Edgar Guimont, Gerard Beaudry (Recorder), Roger Bonin (Librarian). Seated (l to r) Jean Naubert, Joseph Leduc (Secretary-Treasurer), Jean Asselin (President) and Amedée Buteau (Vice-President).

FLEURANGE LAFOREST, a legal secretary by profession, was a founding member of the Centre français de Montréal. She was its Secretary-Treasurer from 1951 to '65 and became Director of Public Relations the following year. She spoke to the Centre on a number of topics including a new theory of the surface of Mars, the star of Bethlehem, and Galileo. Her interest in astronomy dates back to her youth. Fleurange spent nine of her formative years in Chile, where she spent many hours reading astronomy books belonging to her father, a civil engineer.

In the citation drawn up for the presentation of the Service Award to her in 1960, it was stated that Miss Laforest "worked tirelessly for the Centre, recruiting members, organizing programmes, expeditions and observing meetings. The members acknowledge her as the spirit and binding force which has kept the Centre alive and flourishing in good years and bad."

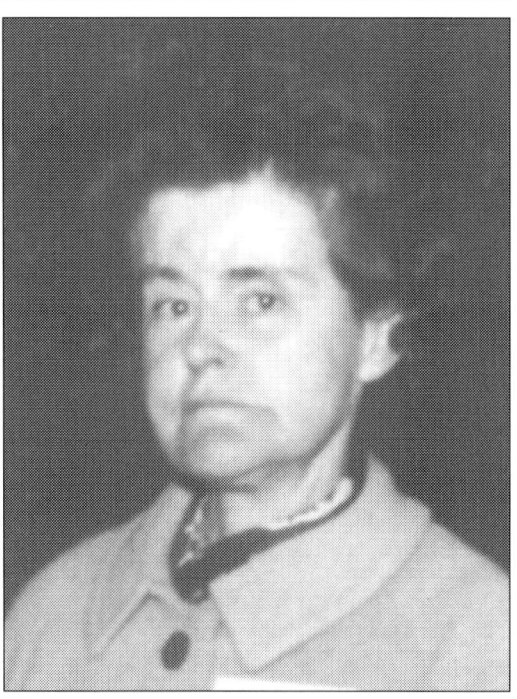

PIERRE LEMIEUX (1904 -) was President of the Centre français from 1958 to '60, Treasurer through most of the '60s, and the Director of Observations from 1960-65. As Treasurer he was instrumental in getting an annual grant from Quebec's Ministry of Cultural Affairs for the Centre's library. He promoted astronomy by visiting religious communities and arranging observing nights; he spoke to the Centre on space travel and eclipses, and designed and explained a new version of the *Almanach graphique*. For his devotion to amateur astronomy in Montreal, he was presented with the Service Award in 1966.

The photographs, by Roland Choquette, were both taken at a soirée populaire in 1965.

In 1968 the group was legally incorporated and changed its name to La Société d'Astronomie de Montréal. It now became possible for members to belong to SAM without joining the RASC, and many chose to do so. In fact membership in the Centre itself plummeted to sixty-six in 1970, comparable to the level ten years earlier. While a loss of close to 200 members (about 7 percent of the total RASC roll) was a real blow to the Society, it was an understandable development. For most of these people, the RASC *Handbook* and *Journal*, being almost entirely in English, were of doubtful value. Instead, they could now receive the *Annuaire Astronomique de l'Amateur*, a French version of the *Observer's Handbook* published by the Centre français since 1964, and their own newsletter, *Le Québec Astronomique*. The RASC did try to stem the flow, first by providing translations of a few pages of the *Handbook* into French for distribution as a supplementary pamphlet to the francophone centres, and eventually, in 1969, by agreeing that any French-speaking members could receive, at the Society's expense, the *Annuaire* in lieu of the *Handbook*. Unfortunately these arrangements were a bit too late to stop the exodus of so many from the RASC.

There were other complications with the operation of the Centre français. The by-laws of the RASC naturally required, for instance, that Officers of the Centre be members of the Society. But frequently those who took leadership roles in SAM were not RASC people, and so it was necessary to have an almost fictitious arrangement of nominal officers for the Centre. Nonetheless it was encouraging to see that a substantial number of members of SAM did retain their RASC connection.

After 1968, it is difficult and perhaps meaningless to report on activities of the Centre d'Astronomie de Montréal (as it was now known) as a separate entity from SAM. In fact for the next three years, the Centre apparently regarded itself as a convenient fiction and sent no reports to the RASC.

However, in 1972, the twenty-fifth anniversary of the Centre, a comprehensive account of the year's activities was submitted by the secretary, Roland Noël de Tilly. From this it was clear that SAM was growing by leaps and bounds. There were now 472 members of which 67 belonged to the RASC and 42 to the Société Astronomique de France. An average of ninety members attended weekly meetings at the Botanical Gardens. The program that year included eleven lectures by notable guests, generally astrophysicists from the Université de Montréal, thirty-six talks by members and one annual meeting. A course in twenty-seven installments for beginners preceded many of these meetings. Fourteen presentations were given to school groups and campers, and the annual soirée at the Botanical Gardens attracted some 5,000 people. An optical workshop complete with aluminizing apparatus operated under the direction of experienced members and a remarkable total of ninety mirrors were completed in 1972. Several telescopes won prizes at Stellafane in the next few years.

A 41-cm mirror was ground and polished by the Centre's optical section in 1969 and a piece of land at St-Valérien, 60 km south-east of Montreal, was purchased in 1974 as the site of an observatory. The project was never completed but the site was used from 1979 to 1981 for a concours annuel des fabricants de télescopes d'amateur (CAFTA) which the Société continues to hold at a variety of locations.

The 1970s were very good years for SAM. Membership continued to grow to 806 by 1980 of which 94 were RASC members. (Interestingly, the proportion of RASC members in SAM stayed within the 12–15 percent range while the proportion belonging to SAF became almost insignificant.) During the 1970s the Société grew to the point where paid help was needed. An executive secretary was hired to work twenty hours per week to answer enquiries, maintain membership records, co-ordinate material for the publications, supervise the library and look

LUCIEN E COALLIER (1911-1986) joined the Centre français de Montréal in 1955 and soon began speaking to the group about his favourite aspect of astronomy – the Moon. In 1969, he opened his own observatory which he named "LUNO" after himself, Lucien, and his wife, Nolita. This handsome stone structure in suburban Duvernay housed two 15-cm and 12.5-cm refractors. Coallier was the Centre's Director of Observations from 1969-71, and President of SAM in '78-'80 and again in '82-'83.

Lucien Coallier was widely known for his articles under the heading "Club des Observateurs" in *Le Québec Astronomique*. He was an inspiration to amateurs of all levels, constantly encouraging them to observe, and especially to make and report observations of use to the astronomical community.

In 1979, he initiated the Concours Annuel de Fabricants de Télescopes Amateur (CAFTA) which is still going strong. He received the RASC Service Award in 1981 and the Méritas trophy of the AGAA in 1983.

On the left, a photo of Lucien Coallier in 1984 by Marc Gelinas (courtesy of Jean-Pierre Urbain). On the right, Coallier is seen at the opening of his observatory on June 14, 1969.

after the store which sold telescope accessories, books, slides, posters and so on. The $4,000 profit realized by the shop in 1978 went a good way towards the salary of the secretary.

Weekly meetings continued to be very well attended; 100 was the average figure reported for 1978. The library, comprising about 300 volumes, saw a circulation of over 200 each year. The annual dinner at which awards were presented was generally attended by over fifty members and guests and a bus trip to the Mont Mégantic Observatory (jointly operated by the Université de Montréal and Laval) became a popular annual excursion. The secret of the Société's success was attributed to good publicity which was largely provided by an umbrella organization funded by the Quebec government called la Fédération Québécoise du Loisir Scientifique; of course the enthusiasm of the members in carrying out the program was vital too.

As several autonomous groups formed around the province, SAM began to see its role as a conduit for co-operation among them. Plans were laid to improve the monthly magazine, Le Québec Astronomique, so that it would be appealing and useful to these other groups. The Annuaire was already widely circulated with 1,000 copies sold or distributed in 1974. These developments led to the formation in 1975 of the Association des Groupements des Astronomes Amateurs (AGAA).

Then came some important changes. In 1980 an agreement was reached with the AGAA whereby their newsletter, Magnitude Zéro, and SAM's monthly bulletin would be merged. The name Le Québec Astronomique was retained, but AGAA was to look after the publication which it was hoped, would contain news and articles by members of the twenty or so Astronomy clubs around the province. The FQLS still provided printing and distribution facilities but things did not always go smoothly. Funding cutbacks almost caused FQLS to close and Le Québec Astronomique had to be suspended for three months. The Annuaire was frequently late; the 1983 issue did not appear until May. The Société even lost the Botanical Gardens as its headquarters. These circumstances and the establishment of several local clubs in the province which attracted former SAM members caused membership to fall dramatically by 1983 to about a third of its 1980 levels.

Spirits did not fail, however. A display was designed and set up at various places and times to inform the public about SAM and its activities as well as important events like Comet Halley and the Voyager exploration of the outer solar system. The monthly lecture program featuring popular talks by professional astronomers now took place at the Dow Planetarium and the weekly meetings, run by the amateur members, at the Centre Loisir St-Mathieu. CAFTA gatherings continued to attract hundreds of participants and electricity was installed at the St-Valérien site for the convenience of members. The 20-cm telescope ("C8") was used increasingly by members until it was stolen in 1989. The Annuaire improved in size and scope and now contained much material which was not found in the Observer's Handbook, and once again, SAM started up a newsletter for its own members, this time under the name Astro-notes. A highlight of 1988 for nine members was a trip to seven astronomical sites in the Soviet Union, but many more enjoyed the slides and presentation after their return.

By 1993, there was yet another reorganization of amateur astronomers in Quebec. The AGAA, an association of groups, ceased to exist. In its place was the Fédération des astronomes amateurs du Québec (FAAQ), to which individual members of local clubs could belong. At the time of writing, members of SAM were also offered the additional option of joining FAAQ and receiving a subscription to Astronomie Québec, as it was now known, or membership in the RASC, or both.

QUEBEC

Why would a group of Francophone amateur astronomers in Quebec City want to associate themselves with the Royal Astronomical Society of Canada? And especially why in 1942, the year in which a national plebiscite on conscription put the country under severe strain with the majority of Quebecers voting no, and the majority in the rest of the country voting yes. Within the RASC itself, prior to 1941 there had been no papers in French in the *Journal* and there was no indication that the Society was interested in becoming bilingual. Yet, for whatever reasons, the Quebec Centre was established in 1942, with fifty-seven members on the roll by the end of the year. Presumably the prestige of association with a society which by now was over fifty years old and which had built a considerable reputation outweighed any political feelings to the contrary. Besides, Paul Nadeau, who was the virtual founder of the Centre, strongly supported the war effort. The new Centre was naturally welcomed into the fold and a few papers in French, written by Quebec members, began to appear in the *Journal*. Victoria Centre made a nice gesture by donating a set of *Bulletins* of the Société Astronomique de France for the young Centre's library.

As usual, an organized group of amateur astronomers preceded the formal establishment of the RASC Centre. Nadeau had spearheaded the formation of the Cercle Astronomique de Québec in 1940. This group of thirty-five wished to popularize practical observing and had optimistic plans of erecting a public observatory. The Cercle, or Société Astronomique de Québec as it soon became known, obtained the use of the Martello Tower in the National Battlefields Park and got a grant of $100 from the provincial government to assist in the construction of a dome for it. The 1941 opposition of Mars was the occasion for a widely publicized star night when about 300 people turned out to hear young Albéric Boivin give an illustrated lecture entitled "Mars, La Planète Étrange," and to view the planet through the 10-cm refractor mounted in the tower.

The Centre was very fortunate to have a building which served as a library and office as well as an observatory and to enjoy annual financial support of the Quebec government. In addition to grants amounting to $300 by 1947 and $400 in the 1950s, the Provincial Ministry of Forests kept Nadeau on the payroll as Astronomer. Later on, in the 1960s, the government paid for a number of students each summer to work as Nadeau's assistants on computations, mirror grinding and in public education programs. Right from the start Laval University also encouraged the Centre by allowing them to hold some lecture meetings on campus and

Standing outside the Québec Centre Observatory on the Plains of Abraham are (l to r): Robert Proulx, Gérard Lafontaine, Jean Paul Boudreau, Joseph Matte, Peter Millman (national President), Guy Delorier, Joseph Bouchard, Paul-H. Nadeau.

by lending them a 40-cm reflecting telescope. Made in 1867 by Secrétan, it was ordered by the Reverend J.B. Bolduc of the Quebec Seminary at a price of $1,400 and was used by him until Laval University became the owner in 1885. Though it was a large and historic instrument, its optics were less than ideal and its wooden tube was not rigid enough. Nonetheless, it was installed under the dome on the upper level of the tower and was used for variable star work, for planetary observing and for public viewing. A clock drive was added in 1960.

Public interest in astronomy was aroused by a weekly newspaper column prepared by Nadeau and other members. Over half of the columns in 1944 were translations of sections of Chant's book, *Our Wonderful Universe*, which Nadeau subsequently translated completely and the Centre arranged to have published as *Notre Univers Merveilleux*. The Centre also distributed star maps consisting of two sheets, each about a metre square, showing 9,700 objects and stars down to sixth magnitude. They were drawn by a member Raymond Fortin, a draughtsman, who estimated that he put 500 hours of work into their production. Still another initiative which began at this time was *Le Graphique du Ciel* (later known as *l'Almanach Graphique*), a French edition of the Graphic Timetable prepared by the Maryland Academy of Sciences and adapted for a north latitude of 47°. Eight hundred copies were distributed without charge in the first year, 1945, and the numbers quickly rose to 5,000 by the 1950s boosted by a growing demand from amateurs and teachers in France.

At the Observatory every clear evening, the public was admitted for one hour, and usually there were over a thousand visitors in the course of a year. In the early years, the guest book showed many armed forces personnel and names from all parts of the world, even Australia and New Zealand. Of course, the historic Martello Tower in its magnificent situation overlooking the St Lawrence River itself attracted many strangers but most enjoyed the added bonus offered by a look through the telescope. Not content to be only receptive hosts, Centre members travelled to Girl Guides camp, St-Vallier and to Shawinigan Falls where they went with telescopes and movies to put on a star night for about 300 citizens.

Starting in 1947, much of the effort of the Centre went into planning and fund raising for a new telescope and observatory as the Martello tower was only considered a temporary site. The government grants barely covered the operating and maintenance costs of the observatory, so additional sources of money had to be found. The newspaper columns which members wrote for *L'Action Catholique* furnished the material for a series of *Feuillets d'Astronomie Populaire*. Sale of these leaflets bolstered the telescope fund by $275 in the first year alone and all profits from the sale of *Notre Univers Merveilleux* also went to the fund. By 1950 it stood at $1,232. Certainly the Centre deserved good facilities. They continued to welcome the public in large numbers, and initiated school visits in 1952 with groups of fifteen students at a time. Members' activities included some telescopic work like variable star estimates, occultations and solar photography, but also involved naked-eye observing of meteor showers and aurorae and telescope making. Nonetheless, in spite of years of planning and saving, the Centre never got its new observatory.

Lectures, except for an address at the Annual Meeting, were almost nonexistent in the 1950s and early 1960s as members put their entire efforts into work connected with the observatory. There were, however, meetings for discussion and observing every Saturday evening at the observatory.

In the late 1950s several groups of amateur astronomers formed under the wing of the Quebec Centre at various localities across Quebec, including Trois Rivières, Lévis and Chicoutimi. Pierreville and Montmagny soon followed with Montmagny's

PAUL-HENRI NADEAU (1910-90) grew up in Quebec City, took classical studies at the Seminary and graduated in Chemistry from the Université de Laval. Though his interest in astronomy was piqued by reading Flammarion's *Astronomie Populaire* when he was 19, studies and work kept his enthusiasm dormant. Employed as a chemist for the Quebec government, some years passed before he was able to take a sabbatical leave to study astronomy at an American university.

Nadeau began writing articles on astronomy (and chess) for the newspapers in 1940, and soon had a regular page every Friday in *L'Action Catholique*. His writings stimulated a great deal of interest which led to the establishment of the observatory on the Plains of Abraham and to the formation of the Quebec Centre of the RASC. Because of his success, the Quebec government took the unusual step of setting up an Astronomical Service and naming Nadeau as the Director. This appointment allowed him to devote his full energies to writing, public education and observing. Being unmarried, he put heart and soul, and evidently a lot of personal money, into his work even to the extent of establishing a time service. He continued his weekly astronomy column until about 1970, edited an *Annuaire Graphique* from 1944 to 1988, and translated C.A. Chant's popular book, *Our Wonderful Universe*, into French. The proceeds from this latter project went to a fund for a future public observatory which was a life-long dream of Nadeau.

As a teacher of telescope making and an active observer of variable stars, sunspots, auroras and meteors, he was always at the observatory every clear night, for public viewing in the early evening and observing later. His only failing was that he took everything on his own shoulders and was reluctant to share responsibility with others.

Within the Society, Nadeau was Secretary of the Quebec Centre from 1942 to 1966, Treasurer from 1947 to 1966. He served on the National Council from 1943 to 1946. Paul-H. Nadeau received the Chant medal in 1945. He was acknowledged for his expertise in being named by the Society to the National Committee for Canada of the IAU in 1952 and by the Quebec government to a Committee in 1965 to inquire into the state of astronomy in the province. He was proud of his nephew, Daniel, who became an astronomer at the Université de Montréal.

The photo shows a delegation from the Quebec Centre presenting a proposal for an observatory at Saint-Romuald to the Honourable Paul Comtois, Minister of Mines and Technical Surveys in Ottawa, in June, 1959. Standing from left to right are architect Roland Mainguy, Hon. Jacques Flynn, MP, Lionel Galichan (a member of Council), Drolet, Nadeau, and Centre President, Alfred Dumont.

MAURICE DROLET (1925-), a technical director at Quebec's Sacre-Coeur Hospital, received the Chant Medal in 1957 for his important work in popularizing astronomy on radio and television, and for organizing a very special eclipse expedition in 1954. Known as L'expédition Fleur de Lys, it was the first French-Canadian astronomical expedition and was hailed in the papers as a decisive step in Quebec science. Two years of preparation by Drolet and other members of the Quebec Centre preceded the event. Drolet planned to photograph the corona with a camera he had designed and built himself. The party of nine flew to the Ungava district in Quebec's far north in a plane provided by Hollinger Mines; the Quebec government and an anonymous donor provided additional funding. The undertaking achieved a great deal of publicity for the RASC, but ended in disappointment under overcast skies. The mayor of Quebec City hosted a civic reception for Drolet on his return from the expedition.

Photograph by Frédéric Marmet

PAUL MARMET (1932-) has had a distinguished career as a physicist, mainly at Université Laval until 1984, and since then at the Herzberg Institute for Astrophysics and the University of Ottawa. He was appointed to the first Associate Committee on Astronomy of the National Research Council in 1971 and played an important role in negotiations for the Canada-France-Hawaii Telescope. Among the several prestigious awards he has received, he was named an Officer of the Order of Canada in 1981.

From 1966 on, he was an active member of the Quebec Centre, as a telescope-maker, observer and frequent speaker at meetings. Dr Marmet led the renaissance of the Quebec Centre in 1966 and served terms as President and Vice-President during the years 1966-71. He was largely responsible for restoring the meetings, the Centre newsletter, and the observing nights to a regular schedule. He was of great assistance to the Centre in making calculations for the *Almanach-Graphique*, for comet ephemerides, eclipses and grazing occultations and edited the French language articles for the *National Newsletter* from 1977-82. He received the RASC Service Award in 1977.

25-cm telescope made by members in the Quebec Centre's workshop. Spurred on by the prospect of thousands of eclipse watchers converging on the province for the 1963 solar eclipse, some of these groups banded together in 1961 to form the Fédération des Groupements Astronomique du Québec and as such shared in provincial government funding previously enjoyed solely by the Quebec Centre. Montreal felt left out, and once the eclipse was over, the Fédération folded. A new umbrella organization, l'Association des Groupes Astronomes Amateurs (AGAA), emerged some years later, in 1975.

The year 1966 marked a new beginning for the Quebec Centre. Though Paul Nadeau had achieved remarkable advances for popular astronomy in Quebec, he had run the show for too long. He had been secretary of the Centre since its beginning, and had been treasurer as well since 1947. The national president, M.M. Thomson visited Quebec Centre in 1966 to try to assist with transitional difficulties, such as obtaining keys and files, and before long the problems were straightened out. As the new secretary, Yvon Dufour, reported, elections were held at which the great majority of the executive was replaced. The new Council met at the beginning of each month, in a democratic manner conforming to the by-laws.

Monthly meetings for both the general public and the members resumed at Laval University, frequently with a related film as part of the program. The old observatory at the Martello Tower was still used by members and the public, though less frequently than before. About 1970 it closed for renovations, the historic telescope was reclaimed by Laval, and the tower was never again used by the Centre. This was a pity from the point of view of ease of access for the general public, but light pollution pretty well impeded any deep-sky observing. Anyway the Centre now had the opportunity to use Yvon Dufour's own observatory and that at the Collège de Lévis.

Membership in the Centre had always been in the fifty to one hundred range, but in 1975 there began a period of rapid growth with numbers reaching 135 in 1979. The monthly programs were well publicized on the radio and in the papers, members were interviewed from time to time about astronomical items on television, exhibits were set up at Expo-Loisir and Expo-Sciences, and the *Almanach Graphique* was still being distributed in the thousands each year. But these efforts were hard to maintain and membership slipped back to fifty-four by 1985. Perhaps another factor was a change in publications.

The Centre had issued a monthly bulletin since 1967, but starting in 1979, this was phased out. The AGAA now issued their own periodical, *Magnitude Zéro*, and it contained most of the material which had previously been in the Centre's bulletin. The only problem was that in order to receive *Magnitude Zéro*, the Quebec Centre members had to pay a subscription fee to the AGAA. This was the start of a confusing system of divided loyalties which led eventually to the Quebec club offering its members the option of being members in the RASC (and receiving the publications of the national Society) or subscribing to the AGAA's *Magnitude Zéro* (changed to *Québec Astronomique* in 1980) or both or neither.

Mainly younger members with fresh ideas and a keen interest in active observing comprised the new Council elected in 1985. They reduced the number of formal lectures by professional astronomers and replaced them with discussions of practical importance to amateur observers such as hypersensitization of films, different methods of studying the planets, instrumentation and astrophotography. A formal agreement was reached with the Collège de Lévis which permitted Centre members to use the College's 35-cm telescope at St Nerée on moonless weekends. The Centre itself had a 15-cm f/15 telescope, and the AGAA put a 20-cm Schmidt camera at the disposal of the Centre. Some members, of course, had their own equipment and some built prize-winning telescopes. An Observation Committee was formed in 1990 which laid plans for disseminating information and ephemerides electronically, got discounts on films, provided forums for discussion of practical problems and solutions, and arranged for observing evenings.

The public continued to benefit from star nights. The most popular were attended by hundreds of people and were held at Jacques-Cartier Park at the time of the Perseid meteor shower in August. In 1988 there were talks and displays earlier in the day, including photographs, drawings, and meteorites with observing starting after dark. The event in 1989 was a three day affair encompassing the lunar eclipse of August 16 as well. Four members gave astronomy courses in co-operation with the towns of Beauport and Charny. And of course, the one constant over all the years, the distribution of the *Almanach Graphique* continued as a public service to Astronomy in Quebec. This was made possible in recent years through the technical co-operation of Dr Paul Marmet and the financial support of the RASC's Ruth Northcott Fund.

In 1990, the Centre moved to a new home at Domaine Maizerets where there is a large hall for meetings, sufficient space for the Centre's large library, and a suitable area for public star nights once a month near the time of the first-quarter Moon. The future of the amateur astronomy in Quebec City looks very promising. The future of the RASC Quebec Centre is less certain as members can choose among various optional affiliations. We can only hope that many will continue to see the wisdom of their forebears of 1942.

HALIFAX

Perhaps someday a historian will find that an astronomical society flourished in Atlantic Canada even prior to Confederation. After all, sea-faring people traditionally have a good knowledge of the sky. General education, certainly in Nova Scotia, was

MICHAEL W. BURKE-GAFFNEY (1896-1979) was born in Dublin and graduated there in Civil Engineering. Following the First World War, he joined his brother in Manitoba and was soon designing bridges for the province. But, he recalled, he began to lose satisfaction with "just making money" and decided to enter the Jesuit order in 1920.

At the time there were no Canadian Jesuit astronomers, though astronomy had long been a tradition in the order. So, as he told it, after three years teaching chemistry, "the Father Superior looked around for someone to turn into an astronomer, and spied me." Years of study culminated in a PhD in Astronomy from Georgetown University in Washington, DC, in 1935. His thesis was the basis for a book, *Kepler and the Jesuits*. Dr Burke-Gaffney then came to Toronto where he joined the RASC and taught Astronomy at Regis College. Following a short period as Professor of Astronomy at St Paul's College in Winnipeg, he and a group of Jesuit priests were sent to Halifax in 1940 to assume responsibility for St Mary's University. There Burke-Gaffney served with distinction, first as Dean of Engineering, then as Dean of Science and after 1955 as Professor of Astronomy.

A member of the IAU and other learned societies, he published a number of papers, some in the *Journal*, and addressed the Halifax Centre many times over the years. For his enthusiastic support and his key role in establishing astronomy in Nova Scotia, he was presented with the RASC Service Award in 1964. The M.W. Burke-Gaffney Observatory at St Mary's opened in 1972, largely the result of a generous benefaction from one of his many admirers.

The photo, taken at Hamilton in 1958, shows Reverend M.W. Burke-Gaffney as a member of the National Council flanked by Reverend Norman Green and Dr Helen Hogg, President.

well-advanced by the 1860s and at least two universities, New Brunswick in Fredericton and King's College at Windsor, Nova Scotia, had included astronomy in their curriculums. The country's earliest observatories were established at Louisbourg, Castle Frederic and Windsor, NS, and at Fredericton, NB. The founding of the Nova Scotian Institute of Science in 1862, and Mechanics Institutes at several locations, show that there was a healthy interest in science. Yet organized amateur astronomy seems to have been slow to develop "down East." Perhaps climate had something to do with it.

The first tentative enquiries with a view to establishing a Centre in the Maritimes came from Halifax in 1936, though there were no developments for many years. A group known as the Nova Scotia Astronomical Society was formed sometime prior to 1951; that year, Dr C.S. Beals, RASC president and himself a Maritimer, had a visit with their officers. Three more years passed until correspondence between Beals' successor, Dr J.F. Heard, and the group's honorary president, Father M.W. Burke-Gaffney, resulted in recognition of the Halifax Centre by the RASC National Council on January 15, 1955. The first president was B.J. Edwards.

Later that same year, the Centre's first secretary, Donald Crowdis, who was the curator of the Nova Scotia Museum of Natural Science, persuaded the Museum to purchase and install a small Spitz planetarium. The dome was an ingenious design, made of flannelette sections sewn together by the renowned sailmaker, Randolph Stevens of Lunenburg. Right from the start the planetarium was the regular meeting place of the Centre and members served as demonstrators.

In the 1950s and 1960s, the Halifax Centre was a small but enthusiastic group. An average of about twenty people came out to the regular monthly meetings, which comprised lectures, planetarium demonstrations and films. They ran a newsletter, called *Star Gazer* (later named *Galaxy*), organized a telescope-making group, held public star-nights when the weather co-operated, and prepared Girl Guides and Boy Scouts for their astronomy badges. Perhaps it was all too much for the few who shouldered most of the work. 1963 was the last year for which a report from the Centre was published; by 1967 the national secretary regretted to state that the Centre was having difficulty retaining its membership and faced an uncertain future. One dedicated member, Mary King, kept the planetarium in operation until the old Museum closed in 1970. For many years, no provision was made in the new Nova Scotia Museum for the planetarium.

Mainly through the efforts of Barry Matthews, newly transferred from Ottawa, the Halifax chapter bounced back. Thirty-seven people attended the opening meeting of the revitalized Centre at the new Museum on September 18, 1970. A close working relationship redeveloped which still continues, with the Museum providing the Centre with a mailing address, a place for meetings and the library, with assistance in newsletter publication, and in answering phone enquiries. On the other hand, members of the Centre were of considerable help to the Museum's education department and eventually resumed their role as lecturers for the planetarium, when it was finally installed at the Nova Scotia Museum after years in storage and a temporary set-up at Dalhousie University. There are now weekly planetarium shows thanks to the efforts of many Centre members who donate their time. But the little old Spitz projector is really quite antiquated, and the Centre has formed a "Nova Scotia Planetarium Advisory Committee" to investigate the possibility of a professional planetarium facility for the province.

Also in the area of public education, the Centre began public observing nights in 1977, with Astronomy Day being a convenient focus for these since 1982. Various formats and locales have been used, generally with considerable success. Films,

RANDALL C. BROOKS (1948-) joined the Society as a high school student in Moncton, NB. He studied for his BSc at Mount Allison University, worked for a summer at the DAO and went on to the University of Waterloo where he received his MSc in 1971. He then returned to the Maritimes, accepting a position as Research Assistant and Technician at St Mary's University in the Astronomy Department. The Halifax Centre had just been revitalized and Randall became one of its most active members. He was responsible for organizing the 1975 GA, was the Editor of the Centre's newsletter, Nova Notes, from 1975-79, and during the next few years served terms as Secretary, President, and then Treasurer. He was the main speaker nearly every year at Centre meetings; he arranged displays for the annual Societies Show and did much to encourage the development of astronomy clubs in outlying regions. His wife, Diane, served the Centre as Librarian, Secretary and Vice-President, and was presented with the Centre's Burke-Gaffney Award in 1981. In 1986 the Brookses left for England where Randall studied for his PhD in the history of astronomy at the University of Leicester.

After receiving his degree in 1989, Dr Brooks continued at St Mary's University for a while and then in 1991 accepted a position at the National Museum of Science and Technology in Ottawa, where he has been able to develop his historical interests to the benefit of the public and the Society.

The citation for the RASC Service Award referred also to his frequent guest appearances on radio and television and his regular astronomy column in a Nova Scotia newspaper.

The 1980 General Assembly was also organized by Randall Brooks. Henry Lee took this photo of him at that time in the Burke-Gaffney Observatory.

displays and planetarium shows have at times complemented astronomical viewing at locations including the Museum and Dalhousie University in town and Bridgewater, about 100 km west of the city. The return of Comet Halley in 1985–86 stirred up a lot of interest. Dr Roy Bishop gave a public lecture on the subject in 1985 and several talks were given to Cubs, Brownies and students. About 100 saw the comet at Uniacke House in November and another 500 came out to the Halifax Exhibition Park on Prospect Road to see it in December. On March 26, 1986, beginning at 4:30 am, forty-five members and guests went to Purcell's Cove to have a last good look at the comet.

Summer evenings at a campground make an ideal opportunity to introduce young and old to the heavens above. About 200 visitors to Kedgemakujik National Park shared in a Centre observing and camping weekend in August, 1985, and since 1987 the Nova East star party at Fundy National Park (New Brunswick) has been an annual August highlight. Co-ordinated jointly by the Halifax Centre and the Saint John Astronomical Club, the public presentations in both English and French, and observing sessions, have been extremely well attended and thoroughly enjoyed by park visitors from the Maritimes, Quebec, northeastern U.S. and Bermuda.

School children benefitted from Saturday morning activities arranged for them in the early 1970s, and a Junior Astronomy Club which flourished in 1973 and 1974 attracted over one hundred to each of its first meetings. The Centre arranged an astronomy camp for children at the Museum in 1987 involving lots of hands-on activities and a planetarium show. At the present time members frequently give talks and slide shows to school classes, Cubs, Scouts, Guides and Brownies. For several years each spring beginning in 1976, the Centre participated in the annual Societies' Show at the Nova Scotia Museum and since 1987, Doug Pitcairn has written a weekly column called "Sky Scan" for the Halifax *Chronicle-Herald*. The Reverend Ted McLeod's Wednesday columns began the following year in *The Daily News* and Fredericton *Gleaner*. Norman Scrimger and Randall Brooks also contributed a column to a monthly paper, *Rural Delivery*, for several years during the 1980s.

For members, special observing meetings were scheduled once a month at a variety of locations including Beaverbank just north of Halifax, Shubie Park, Dartmouth, and further afield at the Castle Frederic site, Mount Uniacke House and Blomidon Provincial Park. Sometimes there were themes, as in 1985 when the emphasis in the spring was on the Virgo Cluster and in the fall on Comet Halley. Monthly observing sessions were replaced in 1988 by monthly "observing windows." Any observer was welcome to meet others at the Beaverbank site at dusk on any clear night in the "window" period. This had a very positive effect on observing activities. In addition, of course, there are individual members who do their viewing from various dark sky locations around the province.

Popular camping-observing weekends have been a regular summer Centre event since 1977, sometimes held at Trout Lake at the cottage of the Centre's honorary president at that time, Dr William Holden. In 1986, for the first time the camping-observing weekend was out of the province, being held at Norma Fraser's property at West Point, PEI. As already mentioned, Nova East was first held the following summer.

In 1985, the Centre purchased a 20-cm Schmidt-Cassegrain telescope for the purpose of lending to members. Rules were drawn up and it was soon in continuous use. Since then, eleven Centre members have demonstrated their observing expertise by receiving Messier certificates. In 1987 a permanent observatory, a dream for many years, again became a topic of conversation and a committee was formed to plan a course of action but the difficulty of where to obtain land and how to ensure adequate security at a remote site have so far held back any concrete developments.

Although many members live a long way from Halifax, the regular monthly meetings usually attract about 50 percent of the total membership. This good representation is partly due to innovative features. On March 20, 1987, there was a festive air with a Spring Equinox Party featuring Sun Punch, Sun Cake and a pie decorated with the Solar System. Another idea which proved to be very convenient was the showing of video tapes and films, and on some nights introductory lectures for beginners, from 7 to 8 pm prior to the regular meetings. This allowed executive meetings to be held at the same time.

The Centre's newsletter, *Nova Notes*, was credited as a prime factor in the increased activity and membership in the Centre during the 1970s. A feature which was found useful in 1985 was a calendar for the complete year printed on the inside back cover of each issue showing proposed events for the Centre.

Preparing a newsletter and carrying on the activities of the Centre not only takes time and talent but requires money. An important fund raiser during the 1980s was the selling of *Observer's Handbooks*. Nearly 200 copies were sold in 1987 and 1988, bringing in almost $600 each year. Miscellaneous sources of revenue resulted from book and "junk" sales held after meetings, and from the sale of pins, crests, hats, bumper stickers and T-shirts.

Throughout the Maritimes, the Halifax Centre is in close contact with several astronomy clubs. For a while in 1978 it looked as though a Moncton Centre might materialize. Though it didn't, the communication was all to the good. Antigonish, Bridgewater, Hebron, Sydney and Yarmouth have all enjoyed visits to the Halifax Centre or by Halifax members. As well, there have been get-togethers with the Summerside Astronomy Club of PEI, a flourishing group of about forty members, and with the active Saint John Astronomical Club in New Brunswick, many members of which also belong to the Halifax Centre. These small clubs received "satellite" centre status. Each received some financial support from Halifax in proportion to the number of members who also belonged to the RASC and their members were sent copies of *Nova Notes*. With travel time between these places and Halifax taking the best part of a day, such outreach is really remarkable. But to go one better, the Barbados Astronomical Society recently expressed an interest in co-operating with the Halifax Centre. Rumour is that there's no shortage of Haligonians willing to volunteer to liaise with the Barbadians in the winter months!

ST JOHN'S

The St John's Centre was conceived in 1965. Captain J.J. Strong of the College of Fisheries, Navigation, Marine Engineering and Electronics wrote to the national secretary expressing the hope that there might be sufficient sustained interest to form an RASC Centre in Newfoundland. The letter was read to an interested Council on October 16 and two days later an organizational meeting was called in St John's. The group adopted a constitution in November and officially became the seventeenth Centre of the Society at a meeting of the National Council on January 8, 1966. By September, there were twenty members on the roll.

Unfortunately, even these numbers could not be maintained, and it is only very recently that the membership has exceeded twenty. Some felt the group would be better to join the Natural History Society and they very nearly did so in 1969, but in the end, they decided to stay with the RASC.

Originally the Centre held its meetings at the College of Fisheries and moved with it to new and larger quarters in 1985 when the College became the Newfoundland Marine Institute. When these long-standing links were cut in 1987, the Centre was fortunate to obtain a meeting place through the co-operation of the Physics Department at Memorial University.

DORA RUSSELL (1913-86) joined the Society as an unattached member in 1960 and became a founder and the first Secretary of the St John's Centre five years later. For most of the next 20 years, she served in some official capacity, including a total of four years as Centre President. She contributed steadily to the Centre's newsletter and over a dozen of her delightful articles appeared in the *National Newsletter*. When Dora Russell received the Service Award in 1977, the citation spoke of "the many interesting, educational and thrilling moments ... [provided by her] lectures, planetarium shows, discussions and observing sessions." On account of this award, she was honoured at a civic reception.

Mrs Russell was well-known outside the RASC too, as a piano teacher and author. She became the first woman's editor of the St John's *Evening Telegram* in 1945 and many of her "Day by Day" columns were collected into book form and published in 1983. Later, she wrote a weekly column in the same newspaper entitled "All About Stars"; she worked with the Provincial Parks Department preparing a program and booklet *Newfoundland Skies in Summer* for park visitors and she was largely responsible for the creation of an astronomy badge for Girl Guides and drafting the criteria for awarding it. For her extensive work with the Girl Guides, Dora Rusell received a Queen's Silver Jubilee Medal in 1977.

Following her death, the St John's Centre renamed their newsletter *All About Stars* in her memory and also named their library to which she had donated over 80 books, the Dora Russell Library.

One of the attractions of the College of Fisheries was a small planetarium there which the Centre found useful a couple of times a year. Usually there were about eight meetings annually with local speakers but occasionally a visiting lecturer, perhaps the national president, would talk to the group. Excursions to the satellite tracking station at Pouch Cove added a bit of variety in the 1960s.

Observing was also difficult, not only because of the small number of members but also because of the notorious Newfoundland fogs. Still there has been a growing interest over the years from three sessions in 1966 at Donovans to eleven in 1990 at a quarry site just 15 minutes from town. A highlight was the arrival in 1970 of Harold Povenmire who travelled all the way from Florida to observe and time the grazing occultation of a couple of the brightest stars in the Pleiades. Six members and seven other interested people assisted him for the event on June 30. In 1971, Ken Chilton came to St John's and, on behalf of the Society, presented the Centre with Ruth Northcott's 8-cm refractor. Three years later, Jack Woods, the Centre's first president, donated a 15-cm Newtonian reflector. These along with various individual's own telescopes, some made by an active group in 1984-85, meant the Centre had no shortage of modest-sized instruments. In 1991, they acquired a 30-cm f/5 Newtonian telescope from the Montreal Centre at a bargain price and reached an agreement for the use of land at Butterpot Provincial Park, about a half-hour's drive south of St John's.

The achievements of this small Centre in public relations and education are remarkable. In 1966, members visited all the high schools in the area and distributed leaflets encouraging interest in astronomy and in the Society. In 1967, as a Centennial Project, they sponsored an essay contest. The ninth grade girl who won became the proud owner of a pair of binoculars and a carrying case. An Open House at the College of Fisheries in 1970 gave the

Centre the opportunity to exhibit astronomical charts, books, pamphlets and information on that year's solar eclipse. Co-operation with the Newfoundland Parks Department led to a very useful pamphlet called "The Newfoundland Sky in Summer" which was widely distributed to campers. 40,000 copies were distributed in sixty provincial parks in 1973. Similarly, co-operation with the Newfoundland Museum and Parks Canada at the time of Comet Halley's appearance produced information sessions, a fact sheet and a viewing night at Cape Spear. Over the years talks have been given to Girl Guides and Boy Scouts throughout the province, occasional courses in elementary astronomy have been offered at Memorial University and displays have been arranged in shopping malls for Astronomy Day. Media coverage has been excellent. Dora Russell's column "All About Stars" ran from 1979–86 and others have appeared since. The Centre produced Cable TV shows in the early 1980s and some members have appeared fairly often on CBC radio and television, especially for events like Comet Kohoutek, Comet Halley, eclipses and meteor showers.

Though always one of the smallest Centres, St John's has shown lots of spirit. Its recent growth to thirty-eight members is an indication of its vitality and a good sign for the future. Members across the country are keenly looking forward to the first RASC General Assembly in Canada's easternmost province in 1994.

Photo, with names, originally published in JRASC **46**, 172.

CHAPTER 14
The Enterprise Returns

While new Centres were being established in the West and the East, others were popping up in the Society's original home province of Ontario.

Of course, there was interest in astronomy long before the RASC came along. A number of examples could be cited as evidence of early astronomical activity in southwestern Ontario. William Barker of London claimed, falsely as it turned out, to have discovered the nova T Coronae in 1866 and actually got reported in the pages of two British journals, the *Observatory* and the *Monthly Notices* of the RAS. In 1867, Edward Flood, also of London, was trying to sell his father's telescope – a 14-cm Fitz refractor with a mahogany tube, equatorially mounted with 10-cm setting circles. The observatory at Woodstock College, equipped with an 21-cm refractor, opened in 1879. During the 1880s, a Mr H. Petit of Belmont had a large telescope and corresponded with A.F. Miller and other members of the A&P Society of Toronto. Teaching of Astronomy began at the Western University in London in 1883 and in the 1890s, Charles Clark, an Associate member of the A&P Society, tried unsuccessfully to start a branch in London.

As happened in a number of other places, however, the eventual organization of the London Centre awaited the spark of a dynamic individual who had already promoted astronomy in some other part of the country. Like Marsh in Peterborough and Hamilton, and Asbury in Guelph and Montreal, Professor H.R. Kingston moved from Winnipeg where he had been secretary-treasurer and president of the RASC Centre to London, Ontario, where he became the motive force behind the establishment of a new Centre. Soon after taking up his duties in 1921 as Head of the Mathematics Department at Western University, Kingston gave some popular public lectures in astronomy which aroused great interest and enthusiasm. As a direct result, in February, 1922, the London Centre of the RASC was formed with thirty-two charter members. By the end of the year, the numbers were up to fifty, and in fact, for almost the entire history of the London Centre, membership has remained in the thirty-two to fifty range.

During the first year, Kingston lectured at two of the six meetings. To get the members off on the right foot, he prepared typewritten lists of all the astronomy books available in the university library and invited members to borrow, read and learn. In the years ahead he would educate, entertain and encourage his audience in a wonderful variety of ways. In 1924 he illustrated his address on "The Recent Eclipse of the Sun" with moving pictures taken on the Island of Catalina and in Mexico. In 1925, his talk entitled "A Month on the Moon" was followed by an enjoyable game of jumbled astronomical names. During 1926, he gave five of the seven lectures. He hosted many council meetings at his home, he gave radio talks over CJGC, and he prepared a booklet of star maps. His energy seems to have been boundless and he was much in demand, not only in London, but also in Toronto and elsewhere as a public speaker who could exemplify how to popularize astronomy. Yet he did not dominate the Centre to the point of excluding others. In fact

HAROLD R. KINGSTON (1886-1963) was born in Picton, Ontario. After high school, he taught for a few years before commencing university studies. Following his MA at Queen's University in Kingston, he again taught, this time at an Indianapolis high school. He embarked on a PhD program at the University of Chicago and completed it one year after accepting a position in the Department of Mathematics, University of Manitoba, where he was Lecturer and Assistant Professor from 1913 to 1921. He then moved to Western University in London where he had a distinguished career as Head of the Mathematics Department, Dean of Arts and Science and Principal of University College. Queen's and Western both honoured him with LlD degrees in 1953. Throughout his career he maintained a strong interest in education at the high school level, served as President of the Ontario Educational Association and wrote (with J.E. Durrant) widely-used Geometry textbooks. His only son, John, was a Professor of Mathematics at the University of Washington in Seattle.

Kingston also taught some Astronomy courses and consequently became actively interested in the Society. During his years in Winnipeg he was Secretary-Treasurer and President of the Centre there, and in London he was the founding President from 1922 to 1930 and Honorary President thereafter. Through his influence, Astronomy was put on a solid footing at Western and the Hume Cronyn Observatory was established. In the national Society he was Vice-President in 1927 and President in 1930-31. His role in the London Centre would be hard to over-estimate as he was consistently involved in meetings and activities for nearly 40 years. As a result of a bequest from his estate to the London Centre, many outstanding speakers have come to London to deliver the H.R. Kingston Memorial Lecture.

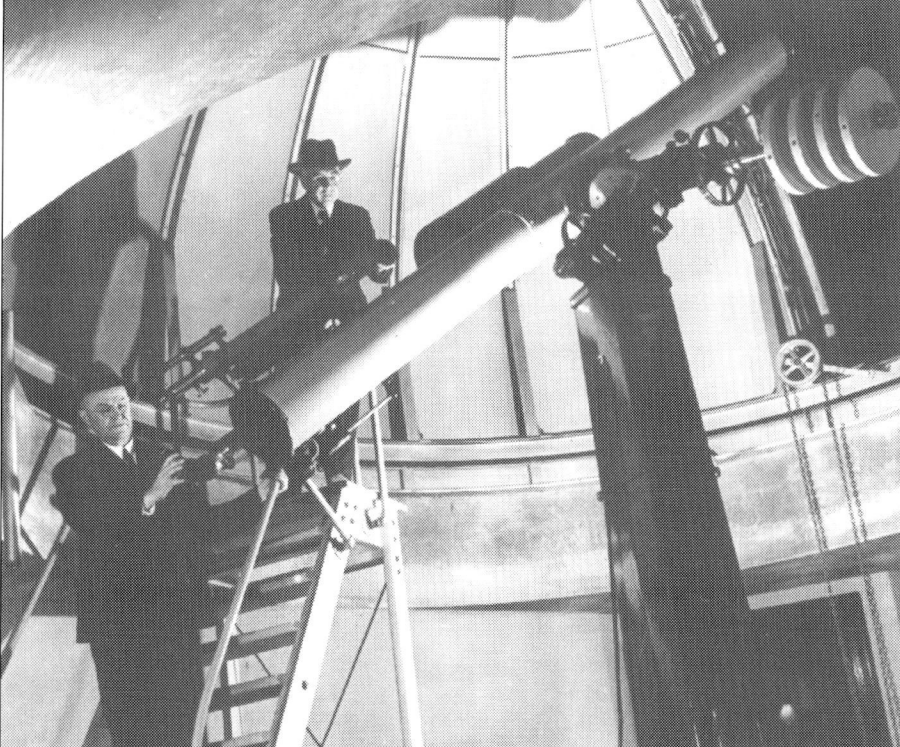

H.R. Kingston is at the eyepiece of the 25-cm refractor of the Hume Cronyn Observatory. W.G. Colgrove is at the top of the ladder.

Photo by A A Gleason, Jr., Regional Collection, D B Weldon Library, University of Western Ontario, London

REVEREND W.G. COLGROVE (1872-1958) grew up on a farm adjacent to what is now the campus of the University of Western Ontario. From childhood he took a keen interest in nature; he had a gift for drawing as well as an innate mechanical talent. His first employment was as a lithographer where his artistic ability was highly regarded, but he abandoned this career for further academic studies. After receiving his BA degree at UWO, he went to McGill and the Boston Theological Seminary for post-graduate work. He served a church in Worcester, Massachusetts before returning to south-western Ontario where he was minister of several United Church congregations.

Two spectacular events started his life-long interest in astronomy - the great comet of 1882 which he saw stretching across the sky day after day, and a meteorite fall in 1885 which he witnessed. A gift of a telescope about this time allowed him to pursue his hobby on a more regular basis. Colgrove joined the London Centre soon after it formed; certainly he was on the Council in 1928. He served the Centre as President in 1939-40 and Vice-President in 1951-53. During this period of 25 years, he contributed much to the popularizing of astronomy through talks, articles and booklets. His unique contribution which led to his receiving the Chant Medal for 1942 was the design and construction of a series of excellent instruments which demonstrated relationships and motions of solar system objects, stars and the Galaxy. H.R. Kingston described them as being "of inestimable value in the teaching of astronomy." Following Reverend Colgrove's retirement from the ministry he acted as unofficial curator of the Hume-Cronyn Observatory, giving demonstrations and lectures to many groups. It was also through his initiative that the Observatory acquired the Dresden meteorite, a treasured possession.

the London Centre has always had a strong involvement by a large number of members and in the early years many novel ideas like songs, games, contests and debates characterized the meetings.

While some of the Centre meetings took place at the university, many were held at the Normal School and the Public Library in the 1920s. Starting in 1927, a member connected with London Life Insurance Company arranged for Centre meetings in the company conference room with observing afterwards from the roof of their newly built head office building. Annual Meetings and social evenings were sometimes held at such local establishments as The Blue Dragon Inn and Wong's Cafe. This latter spot was the venue for entertaining visiting lecturers W.E. Harper and S.A. Mitchell who came to town in 1930 and 1931. An audience of 360 turned out to hear Mitchell's after-dinner lecture on Solar Eclipses in the London Life Auditorium, attracted by his renown for delightful speeches on this subject and by plans for the Canadian eclipse the following year.

Centre picnics followed by an evening of observing were an annual tradition. For the first few years they were held at the farm of Dr and Mrs W.E. Saunders at Pond Mills whose fine 10-cm refractor (later donated to the Centre) was put to good use. Dr Saunders was a well-known druggist and naturalist. Mrs Saunders became Centre vice-president in 1925. In subsequent years other members, including Dr Ainslie at St Mary's and Thomas Wonnacutt at Delaware, were hosts for the picnic. Wonnacutt's farm was the scene of a joint get-together with the Ontario Field Naturalists in June, 1938, when everyone enjoyed a campfire talk by Dr Kingston.

As in most Centres, observing in the 1920s and 1930s usually meant that a few members would set up telescopes following the regular meeting and everyone would enjoy a couple of hours viewing the planets, the Moon or some double stars. The one planned program was meteor observing which became popular in the mid 1930s as a result of Dr Millman's enthusiasm. In London, John Middlebrook organized teams of meteor observers, usually drawing a dozen or so participants for the Perseid shower in August.

Programs for the public were especially successful in 1939 thanks to the spectacular arrival of the 40 kg meteorite at nearby Dresden and extensive publicity in *The London Free Press*. One of six star-nights that year alone attracted 3,000 people who came to view Mars, Jupiter and the Moon through members' telescopes.

A very important development for astronomy in London was the dedication of the Hume Cronyn Memorial Observatory on October 25, 1940. This came about as the result of a $40,000 donation to the University of Western Ontario by Mrs Cronyn in memory of her husband who, as a Member of Parliament during World War I, had pressed the government to establish what was later to become the National Research Council. The Observatory was equipped with a 25-cm refractor and a Schmidt camera, and was intended "to aid in the teaching of Astronomy in the University, to encourage the work of the London Centre of the RASC, and, in general, to stimulate in the public mind an interest in Astronomy."

For the next twenty years or so, the Centre met monthly at the Observatory. A short discussion on some aspect of *The Observer's Handbook* normally opened each meeting and observing with the 25-cm refractor was scheduled after the meeting, weather permitting. The main speaker was usually a London member, but Canadian professional astronomers generally visited once or twice each year, a tradition which continued until very recently. Joint meetings were an innovation in the 1950s and early 1960s, sometimes with Windsor or Hamilton Centre, sometimes with another organization like the Canadian Association of Physicists or the Geophysical Group of the university. Co-operation with these latter groups brought outstanding scientists of the calibre of

Harold Urey who spoke on the origin of the Solar System.

The 1960s saw a number of changes in the London Centre. There was a brief span from 1966–68 when little was done to attract new members or to retain old ones. The number of meetings declined from five, to three, and then only to two per year. The number of members fell to an alarming low of eighteen, but the Centre, under strong leadership, rebounded with programs offering more member-involvement than ever before. Three groups were formed in 1969 – one for observers, another for telescope makers and a third for junior members. The Centre newsletter which began the same year did a lot to develop camaraderie and cohesiveness. Each year, at least one members' night provided an opportunity for a number of speakers to talk for a few minutes about some aspect of their own astronomical activities. The tradition of an annual banquet began in 1973, and there was a picnic and a barbecue in 1974 and 1975. Excursions added some excitement as trips were planned to the Ontario Science Centre and McLaughlin Planetarium in Toronto, to the Abrams Planetarium and the Observatory at Michigan State University, to astronomy meetings in Syracuse, New York, and Stellafane in Vermont, and even to Cape Canaveral for the first launch of the Space Shuttle.

The regular meetings continued to be held on campus, though not at the Hume Cronyn Observatory. Following Kingston's death in 1963, a bequest from his estate provided for a guest lecturer nearly every year. Perhaps the most outstanding of these occasions was the sixtieth anniversary of the Centre when Bart Bok addressed the London members and several guests from other places on the subject "The History of Milky-Way Research 1920–82." In recent years there have been fewer lectures delivered by visiting professionals, but more by exchange speakers coming as amateur members from other Centres.

On the observing front, more members started to take an active role in the late '60s. The university opened another small campus observatory containing a 15-cm refractor in 1966 and members made some use of it, especially for the series of Pleiades occultations in 1969. But with greater interest in telescope making and dark-sky observing, there was a growing desire for independence. With meetings every other week, the Amateur Telescope Makers produced a number of reflectors for personal use and undertook some larger projects for the Centre. The head of the group, Peter Andreae, offered courses at a community college. A 40-cm f/13.5 Cassegrain instrument with 25-cm guide scope was completed in 1974. A 20-cm Dobsonian was completed recently, and it, along with a 10-cm refractor made by Joe O'Neil and a 40-cm commercially made telescope which was purchased second-hand, are all available for rent by members at a rate of ten dollars per month including eyepieces and sky atlas. Interest in observing picked up in proportion to the greater accessibility of telescopes. Deep-sky and variable star work were the most popular, but meteor showers, eclipses and occultations attracted some attention. A number of dark sites were used not far from the city, but on occasion, members would go further afield on camping weekends at Mount Forest.

Public education was also enhanced by the great increase in telescopes and observing experience. Starting in the 1970s there were public star-nights in parks and conservation areas, educational sessions at the library, information booths at shopping malls, talks to Guides, Scouts and recreational groups, and at schools, camps and at the London Regional Children's Museum. Project Zubenelgenubi ran during at least three summers. This government-funded endeavour was highly successful in bringing astronomy to the general public not only through presentations, displays and star nights, but by means of a 100-page programmed learning guide prepared in 1973 and a series of television programs produced in 1980 by five students and based on the Astronomy and Aerospace industry in Canada. No one did more for public astronomy education in London than Peter Jedicke. Besides being committed to the regular programs of the Centre, he initiated a popular course at Fanshawe College, spoke often on CKO radio and hosted a cable television show called "Telescope" from 1976 to 1984.

All these activities were signs of a healthy Centre and indeed membership reached a record sixty-eight in 1980. Though numbers have not matched that level in the decade since, the Centre looks forward with confidence to a permanent observatory and sustained growth in the years ahead.

WINDSOR

In the fall of 1943, Cyril Hallam, a science teacher at Patterson Collegiate organized the Astronomy Club of Windsor. After a year of successful monthly meetings, the group was encouraged by the enthusiastic Dr H.R. Kingston of London and by Dr J.A. Pearce of the DAO to become an RASC Centre. Their inaugural meeting was held on Wednesday evening, December 13, 1944, with twenty-five of the twenty-nine charter members present.

Meetings were well-publicized. *The Windsor Star* carried announcements on the Saturday preceding each meeting, a poster was put up in the library inviting the general public to attend, and Hallam saw to it that all the Windsor secondary schools were notified, with the result that there was good attendance by students. Membership reached seventy by 1948, a high-point in the history of the Centre.

Most of the meetings consisted of a lecture and a short review of astronomical events of the current month with observing through 10-cm refractors preceding or following the meetings. A friendly interaction developed between the Detroit Astronomical Society and the Windsor Centre, which led to frequent visits by members of each to the other's meetings. Detroit members on the Windsor Council provided invaluable assistance in arranging for guest speakers from Michigan including E.R. Phelps and Willard Parsons of Wayne University, Orren Mohler, Lawrence Aller, Leo Goldberg, and Dean McLaughlin of the University of Michigan. Of course, Windsor Centre members took their turns too, as did Canadian visitors like Ruth Northcott, H.R. Kingston, William Wehlau and Helen Hogg.

In the 1960s, the Centre met regularly about seven times a year. "Current events" still formed part of most meetings and films were usually shown a couple of times during the season. Occasional joint meetings were held with London Centre and a star night at Dan Bawtenheimer's Poplar Lodge was a tradition until 1967. Annual excursions by bus were a

For most of its first thirty years, Windsor Centre held its meetings in the Art Gallery of the Willistead Public Library. It was a very pleasant room where a fresh exhibition of paintings hung on the walls each month. The beautiful old building itself was the former home of the Walker family after whom Walkerville is named. Books and magazines on astronomical subjects were always on display and could be borrowed from the Public Library until the next meeting.

Photo of Willistead Manor courtesy of Municipal Archives, Windsor Public Library

DANIEL C. BAWTENHEIMER (1900-1981) received the Service Award in 1965, the first member of Windsor Centre to be recognized in this way. As a Boy Scout he had his first look through a small telescope at some of the planets and the Milky Way. His interest in astronomy stayed with him throughout his life, though he earned his living as a furniture salesman. Every June for many years, he and Mrs Bawtenheimer were hosts at their "Poplar Lodge" home for an Observation Evening when members could enjoy viewing the heavens through his 25-cm reflector. He was the first Secretary of the Windsor Centre, and subsequently served as Vice-President, President (in 1948 and again in 1955) and Librarian. He conducted Students' Nights and frequently spoke at Centre meetings about current events.

Members of the Windsor Centre gather in June, 1953, for their ninth annual Observation Night. From left to right are Cameron Montrose (the Centre's first President), Robert Bradley (Secretary and later Vice-President of the Centre), Mrs Helen Huneault, W. Almon Hare (an original Councillor and President of the Centre in 1953-4), Lambert Huneault (later President), Danny Bawtenheimer and Gus Nyberg (Councillor starting in 1953). They are standing in front of Bawtenheimer's observatory with its 4-metre aluminum-clad dome.

CYRIL HALLAM (1906-1988) was the main founder of the Windsor Centre, and its first Treasurer. He served for a total of 12 years as Treasurer, was on the Council for 23 years, three of these years as President. His career as a teacher fitted him well for his many educational activities in the Centre. From an account of the 1945 solar eclipse which he and Mrs Hallam witnessed in Wolseley, Saskatchewan, to a description of his trip to Kitt Peak in 1983, his talks over the years spanned many topics. He was presented with the Service Award in 1984.

Three successive Windsor Centre Presidents, (from left to right) Cyril Hallam, Dan Bawtenheimer and Charlie Bell discuss plans for a meeting in 1950.

special feature of the 1960s, with the destination usually some point in Michigan such as the planetariums in Bloomfield Hills, Adrian, or East Lansing, the Fermi reactor in Monroe, or the various astronomy facilities of the University of Michigan.

For a while during the midsixties, little was done to publicize the Centre or to attract the public except for an occasional star night or display at the local Fun Fair. The results were predictable and exactly parallel to what happened in London at the same time. Attendance at meetings fell to an average of twenty-two in 1965, and there were, in fact, only four meetings held in 1967. Membership dropped to its lowest point ever, with just twenty-five on the roll in 1967 and 1968.

Fortunately, the Centre recovered quickly and showed healthy if somewhat erratic growth in the years since. The program resumed its normal frequency and format though a noticeable difference was the greater role of amateur members beginning about 1971, with only rare visits from professional astronomers. The annual excursions continued until 1978 after which they were replaced by a summer observing program in Point Pelee National Park and Wheatley Provincial Park.

In 1973 the Centre got the opportunity to take an active role in St Clair College. Long-time Windsor Centre member, Bert Huneault, taught there and became involved in the building of an observatory which housed a 20-cm telescope in a small dome just south of the college's main building. Through this connection, the Centre came to use the observatory regularly and to hold its monthly meetings at the college. They formed a special committee to deal with observational matters and purchased, with the help of a grant from the national Special Projects Fund, three eyepieces and a solar filter for the college telescope. In 1976, a number of members produced, on location at the College, a 16-mm film about observing. Starting in 1978, plans were laid to invite the public every clear Thursday evening, when a

HENRY LEE (1920 -) joined the Windsor Centre in 1947 and in his first year participated in a panel discussion "Is there life on Mars". He soon became Secretary of the Centre, an office which he held for three years, and was subsequently Treasurer for three years, Vice-President for three years and President for four years. He is well known outside the Centre having been Windsor Centre's National Council Representative for over 20 years and the National Recorder from 1988-91. He and his wife, Mamie, are regular delegates at General Assemblies. At the 1984 GA he accepted the Service Award. Since graduating from Wayne University in Detroit, where he studied Electrical Engineering, Physics and Astronomy, Henry Lee has worked in industry, becoming Vice-President, Manager and Plant Engineer for his company. As a result of these connections he secured many well-qualified speakers for the Centre. In addition, he himself promoted astronomy through lectures, star nights and displays. His other interests include swimming, Tai Chi and photography.

Henry Lee taught astronomy at night school in Windsor. In this 1969 *Windsor Star* photo, he is shown with three of his adult students examining a celestial globe under the tailpiece of the Windsor Centre's 15-cm refractor. From left to right are Mrs Patrick Hughes, Robert Bouteiller, Henry Lee and Miss Margaret Janisse.

member of the Centre would be on duty. The Centre virtually operated the St Clair College Observatory until 1984 when local lighting became excessive. Since that time, Randy Groundwater, another key member of the Centre, has been using the observatory in connection with the Astronomy course he teaches for the College. Several people have joined the Society as a result of his efforts.

The Centre advanced through the 1980s with real vitality. Contacts were made with a number of other organizations including the Chatham-Kent Astronomical Club, the Warren Astronomical Society and another group in Michigan called the University Lowbrow Astronomers of Ann Arbor who held an annual "Freeze-Out" conference each February. Astronomy Day, first celebrated by the Centre in 1982, developed into a great success with the help of good media attention. Displays were arranged at the main Public Library, and at the University of Windsor, telescopes were set up for solar and nighttime viewing and by 1988–89 hundreds of people enjoyed these events. Public education also took off. In 1983, presentations were made to hundreds of students and parents at schools in the Windsor-Essex area. In 1985, a public relations program resulted in a string of interviews from newspapers, magazines, radio and television. The same year the Centre made nearly a hundred presentations to Boy and Girl Scouts, primary and secondary schools and to other organizations. In 1987, 400 Cubs and their leaders heard an educational talk and got to look through telescopes at the Essex District Cub Camp. Special public meetings, planned in conjunction with other groups, brought astronomy writer Terence Dickinson and the discoverer of Pluto, Clyde Tombaugh, to town in 1987 and 1988.

Attendance at regular meetings increased sharply due to the quality of the presentations and the tremendous efforts of presidents who arranged not only for an interesting variety of local speakers but also planned for an exchange of speakers with some other RASC Centre each year. The location of the meetings moved from St Clair College to St Mary's Anglican Church in 1984, then to the University of Windsor in 1989, and just recently to St Stephen's Church a short distance outside the city.

More members took an interest in observing too as more telescopes became available. Over the years the Centre had fallen heir to two old refractors, a reflector (Nyberg's) and to the mirror of Bawtenheimer's 25-cm reflector; these were carefully refurbished and rebuilt with the help of a national Special Projects grant. A number of members built quite impressive telescopes themselves. No less than four trailer-mounted instruments of Windsor members travelled to Stellafane and Starfest in 1987 – the 40-cm and 36-cm telescopes were mounted under domes similar to conventional permanent sites while the 32-cm and 25-cm reflectors were on open platforms. In 1990, Vice-President Dan Taylor completed a giant 50-cm f/5 Dobsonian 'scope.

These instruments were put to good use at starnights at a variety of locations. In 1986, for example, members observed at Patrick Langan's residence in Rochester Township, at Point Pelee National Park, and at the Ernie Warwick Conservation Area (known simply as "Eagle") in Elgin County. The latter site was used on eight weekends closest to new moon. In 1988 at Point Pelee, Canada's southernmost tip, Dan Taylor and some other members spotted the famous globular cluster, Omega Centauri, at declination 47° south. Good luck to any other Centre trying to match that!

NIAGARA

Long before there was a Niagara Centre, there were members of the Society in the region. Two who deserve special mention were J. Miller Barr of St Catharines and David W. Rosebrugh of Niagara Falls, an employee of Ontario Hydro and the son of a well-known University of Toronto Engineering professor.

Barr's observations of variable stars have already been referred to in chapter nine. More significant was his finding that the apparent orientation of orbits of spectroscopic binaries was not random. Though he was not a professional astronomer, and presumably never photographed a stellar spectrum, he must have read the professional papers on the subject with some interest to have arrived at his fascinating conclusion. It is an anomaly that still is not fully explained, though it is understood to be due to gas streaming from one component of a system to the other.

Rosebrugh made his first contribution to the Journal of the Society in 1924 in connection with the transit of Mercury which he observed. Though he moved to the United States three years later, he continued to write papers for the Journal until 1947. He became very active in the AAVSO and served as their president in 1948.

The Greater Niagara Astronomical Society began in 1958 when eight people met at the home of Bob Nelson. The president of Hamilton Centre, Norman Green, came to explain RASC policies, but as twenty-five members were required for recognition as a Centre, the Niagara group had to strive and grow for a couple of years before joining the RASC family. Following the approval of the National Council in the fall of 1960, Vice-President Ruth Northcott and National Secretary Ed Kennedy visited the Niagara Falls Centre to discuss the operation of the Society and its Centres.

For the first seven years the monthly meetings were held in Niagara Falls Collegiate and Vocational Institute and the public was always welcome. The speakers were generally local members, though films were the main item a couple of times each year. An annual turkey dinner, or banquet as it later came to be known, often featured a speech by the national president. Field trips to Walter Semerau's Observatory in Kenmore, NY, to the Board of Education Planetarium in Niagara Falls, NY, and to

HUGH N.A. MACLEAN (1915-) was a founding member of the Niagara Falls Centre in 1960 and ever since has been involved at one time or another in every phase of the Centre's activities. He gave many talks and slide shows at Centre meetings; he was active in school visits and star nights, telescope making, and in the production of the Centre newsletter, *Whirlpool,* and he served in a number of offices including the Presidency from 1966-68 and 1978-79. In 1984, Niagara Centre accorded Hugh Maclean the unique honour of being their only member ever nominated for the national Service Award.

Hugh Maclean and Al Kindy (wearing glasses). This photo, by Ron Roels, is part of a larger group picture which was originally published in the *Niagara Falls Evening Review,* 14 November, 1963. The photo below shows members and friends at the dedication of the Al Kindy Memorial Observatory, 1 July, 1992.

the Buffalo Museum of Science added variety to the programs of the 1960s and speakers from the Buffalo Astronomical Association sometimes addressed the Niagara Falls Centre.

Though members numbered only in the twenty to thirty-five range during this period, they owned at least fifteen telescopes, some of them built within the telescope-makers group. These scopes were used at summer star nights for the public and for Girl Guides, Cubs and Scouts of the district. Other events for the public included a display arranged by the Centre in 1962 which attracted 417 people and a popular program presented to the St Catharines YMCA in 1967.

In the fall of 1967, the location of meetings was changed to the Drummond Hill Public Library. For the next nine years the meetings continued much as before. Ties with Hamilton Centre were strengthened as W.J. McCallion, Ken Chilton and Robert Lang were among speakers who came to Niagara and both Centres enjoyed a joint field trip to the McLaughlin Planetarium in Toronto in 1972 and a picnic together the following year. General Assemblies were hosted jointly by both Centres in 1971 and again in 1984.

Public star nights continued annually, with the Good Friday lunar eclipse of April 12, 1968, attracting over 200 spectators. A few members gave talks to various local groups, one of the most rewarding being an astronomy program for the Niagara Falls Boy's Club begun in 1968.

In 1976 the Centre moved its meetings to a third location – the new Centennial Library on Victoria Avenue. There followed a period of healthy growth to fifty-three members by 1980. This was attributed to several shopping mall displays held throughout the summer of 1979 and to a weekly astronomy column in the *St Catharines Standard* written by member Bob Winder. The launching of the Centre's Newsletter, *Whirlpool*, kept members well informed of events and there was an excellent turnout at the monthly meetings, at two annual banquets and on several excursions. These included trips to the David Dunlap Observatory, the NFCAAA meetings, the Syracuse Summer Seminar, and Stellafane. A new Observers' Group was organized in 1978 and its monthly sessions were well attended. Co-operation with Brock University also began at this time. Speakers from the Physics Department addressed the Centre and members of the Centre reciprocated with star nights at the university, an annual tradition which is still maintained.

Niagara, more than any other Centre, draws on visitors to speak at its meetings. Every year at least three and as many as six guest lecturers come from other Centres in Hamilton, Toronto, London, or from Buffalo or Rochester in the States. The November dinner meeting has become a joint Centre-NFCAAA affair with an invited speaker, sometimes the national president of the RASC.

A fairly active Observer's Group usually manages to find a few clear evenings each year for their meetings and generally there are three or four public star nights as well. These are most successful when some special event catches the attention of the public. Halley's Comet, of course, was a big drawing card. The Centre held five star-nights in 1985–86 with as many as 400 people showing up. The lunar eclipse of August 17, 1989, also attracted a good crowd to a special Star Night at Lester B. Pearson Park in St Catharines. Displays at shopping centres, hobby and craft shows, and at the St Catharines Centennial Library also stimulate some public interest in astronomy and in the Society and bring in a few new faces to the Centre. In 1983 and 1984, some members made a point of liaising with schools in the district. The students of Maplewood school, on a week of outdoor education, enjoyed a special star night set up for them in 1983 and the following year, several St Catharines schools benefitted from presentations and star nights provided for them by Charles Fassel and Hugh Maclean. Astronomy Day 1984 was actually held at Stamford Collegiate, a school with an active Physics Club and a small reconditioned planetarium.

Plans for a Centre Observatory were years in the making. A fine start was made when a member offered to donate land for this purpose in 1973. But before the Centre could contemplate owning property, they realized the need to incorporate. This was a much more complicated process than anyone had imagined and was not accomplished until 1978. Members approved a surcharge to help pay for a building in 1986, a 46-cm telescope was purchased the following year and an Al Kindy Memorial Building Fund was established. Various fund-raising activities were used including raffles and sales of RASC jackets, bumper stickers and other paraphernalia. By 1989 the site originally proposed was threatened by lights and development. A new location was secured and the Centre officially opened its own observatory in 1991.

KINGSTON

William Findlay retired as RASC president in 1939. In his address to the Society that year, he noted:

> [Kingston] is one important university city where I feel sure there is a possibility of a good branch being organized. ... We already have a small group of capable members in that vicinity.

He felt even more secure in his prediction because Dr A. Vibert Douglas had recently moved to Queen's University from McGill. She had been a vital force in the Montreal Centre and had just been elected second vice-president of the Society. But many years would pass before Findlay's hopes would be realized.

There had, of course, been Society connections in Kingston much earlier. John F. Baker had written to the Toronto group in 1892 regarding the desirability of teaching astronomy in schools and making it more important in universities. At about the same time, Lieutenant-Colonel McGill of the Royal Military College wrote about examination papers in Astronomy. Professors Matheson and Dupuis, both of Queen's, were elected to membership in the RASC in 1907, and as Findlay said, a few others had also joined the ranks by 1939. Kingston had had an observatory since 1855 and courses in Astronomy were offered at Queen's as early as 1863. Though the old university observatory was demolished in 1946, by 1958 plans were in place for a new facility with a 38-cm Cassegrain telescope.

This new equipment, housed in a roof-top observatory on Ellis Hall, the Queen's Physics Building, gave a boost to the teaching of Astronomy at Queen's and in 1961 a group of enthusiastic students met with Dr Douglas and Dr G.A. Harrower to discuss the possibility of forming an astronomy club which would give them access to the instruments. In the words of the secretary's report for that year:

> Before the end of the spring term, an Executive had been elected and a Constitution drawn up. The application to form a Kingston Centre of the R.A.S.C. was approved by the National Council at its June meeting.
>
> In September, meetings were organized on a bi-weekly basis. Programmes ... included slides ... films ... and a discussion by Dr Harrower on "Sidereal Time and the Use of the Telescope." Several meetings have included observing sessions.

The Kingston Centre's promising beginning did not last. Membership in the first year was twenty-three, but dropped to seven the following year. National presidents Millman and Northcott came to speak, and Vibert Douglas did her best to keep interest alive but with all the pressures of university life, the students who virtually made up the Centre found additional commitments hard to sustain. In fact the Kingston Centre became almost a fiction since the

DAVID LEVY (1948-) began studying Physics at McGill and Geology at Acadia before switching to English Literature in which field he got his MA degree from Queen's University in 1979. His love of Astronomy sprang from summer evenings at his grandfather's cottage on Jarnac Pond, deep in the Laurentians northwest of Montreal, and his observing skill was honed by membership in Montreal Centre's Observing Group. Starting in 1965, he was a keen but unsuccessful searcher of novas and comets, and a diligent observer of meteors, Messier objects and variable stars. By the time he received the Chant Medal in 1980, David Levy had become the most prolific observer in the AAVSO, with over 10000 estimates made in one year.

Levy has always enthusiastically shared his love of the sky with others from children at camp to inmates in prisons. He edited Montreal's newsletter, *Skyward*, spoke at meetings of Halifax Centre while he was a member there, and designed and gave a 12-week course for the benefit of Kingston Centre. He has written regular columns for several magazines, including *Sky and Telescope* and the *RASC Newsletter* and is the author of a number of books, *The Joy of Gazing, The Universe for Children, The Sky: A User's Guide*, some topical observing guides and recent biographies of Clyde Tombaugh and Bart Bok.

In search of darker skies and a better climate, Levy moved to Arizona in 1980. He made his first visual comet discovery in 1984 after 19 years of fruitless searching. He now has seven comets to his credit and eleven more shared with the Shoemakers as an Observer for the Palomar Asteroid and Comet Survey. He has received awards from the Astronomical League, the Western Amateur Astronomers, the Association of Lunar and Planetary Observers, the ASP and in 1988 the IAU named an asteroid in his honour.

David Levy, with part of his former telescope collection.

members all belonged to the Queen's University Astronomy Club which was constitutionally and fiscally accountable to the student government on campus. With membership rarely above twenty, requests for outside speakers could hardly be justified, and so the meetings usually comprised informal discussions or films with an occasional talk by a graduate student or professor from the University's Astronomy Group. In some years the club was successful in maintaining a biweekly schedule of meetings alternating with observing in the dome. At other times there were only about eight meetings per year including a couple of evenings with the 38-cm telescope. The Club also utilized the university's little planetarium for one meeting each year from 1970–73.

The students who kept the Centre alive for its first fifteen years deserve a lot of credit especially since their's was probably the only campus Astronomy Club in the country. There aren't many students who would be willing, as this group was in 1976, to get up before 5 a.m. for several mornings in a row to admire Comet West, as magnificent as it was.

The Centre changed direction in 1976 when a membership drive in the fall attracted new members from outside the university. Since then the Queen's students have played a smaller role in the Centre though their continued participation was encouraged at the university's Clubs' Nights and Open Houses 1979–82.

In spite of the change, meetings of the Centre have stayed at the university. Visiting speakers have again appeared on the program and recently there have been as many as six each year – occasionally a professional astronomer and the rest amateurs from other Centres. Kingston members also speak, of course, and a couple of members' nights in the season give everyone a chance to show a few slides, describe some observations or demonstrate some new software.

Summertime brought a variety of activities. For several years, trips to the nearby Holleford meteorite crater were popular for Kingston members and for outside groups who appreciated having local members as guides. In 1977 and 1978, Centre members watched the Perseid meteor shower. The following two summers there were jaunts to David Levy's summer cottage in the Gatineau Hills. A summer picnic has recently become an annual event.

A few members continued to use the Queen's 38-cm telescope until 1979 when the Centre built its own 25-cm reflector with purchased optics. The telescope, named for Dr Douglas, had the advantage of mobility. It, along with a number of privately owned instruments, travelled on excursions to dark sites where members could enjoy observing deep-sky objects. Sunspots, planetary phenomena, occultations, auroras and fireballs also attracted interest and all figured in reports published in the Centre newsletter, Regulus, which had begun as a two-page leaflet in 1976.

A number of public education activities were initiated in 1979. Members set up a shopping mall display with photos and telescopes and had a public star party that evening in a nearby park. They participated in an Astronomy Day sponsored by the Kingston District Science Council, an event which drew several hundred children. David Levy spoke on teaching astronomy to children at the IAU meetings in Montreal and put on a slide show for inmates of Millhaven Penitentiary. The continuation of Astronomy Day displays for the public soon bore concrete results as membership jumped from eighteen in 1980 to thirty-eight in 1981. And with more members, the Centre was able to expand its efforts even further. Star-nights were provided for Cubs, Scouts and campers, judges and prizes were offered for a local science fair, and the Astronomy Day displays became more ambitious with posters, solar viewing, games and door prizes.

At the present time, membership has never been higher and there is a strong interest in observing. Thanks to the generosity of some members, the

---

LEO ENRIGHT (1943-) lives on the shore of Sharbot Lake, some 60 km north of Kingston, Ontario. By day he teaches English in the district high school and on clear nights he enjoys viewing the sky from his backyard observatory. Some of the most memorable events he has seen and photographed required no telescope, however, and included a series of observations of the zodiacal light 1984-86, an unusual auroral outburst in August, 1985 and a very bright fireball (magnitude -10) in 1983.

Combining his astronomical and writing talent, Enright has reviewed many popular astronomy books for the *Journal*, has contributed many interesting pieces to the *Newsletter/ Bulletin*, and from 1977 to 1988, edited *Regulus*, the newsletter of the Kingston Centre. *The Beginner's Observing Guide*, which he wrote for the Society in 1991, is a natural outgrowth of his enthusiasm in helping Scouts and other young people become familiar with the sky.

Enright was primarily responsible for expanding the Kingston Centre from a campus club to a true regional Centre of the RASC, especially in 1978 when he was Centre President. He advocated, arranged and manned mall displays, and promoted Astronomy Day as a means of getting the public informed and interested. As a result, the National Council appointed him Astronomy Day Co-ordinator, a position in which he worked very hard and effectively to get all centres involved in the annual program. Council also benefitted from his meticulously prepared minutes during his term as Recorder from 1982-88.

Leo Enright received the Service Award in 1986 for his many contributions to the Society. He and Denise Sabatini, whom he married in 1988, have continued to enrich the Society in many ways since.

Leo Enright patiently answers questions at a display in Cataraqui Mall, 1984.

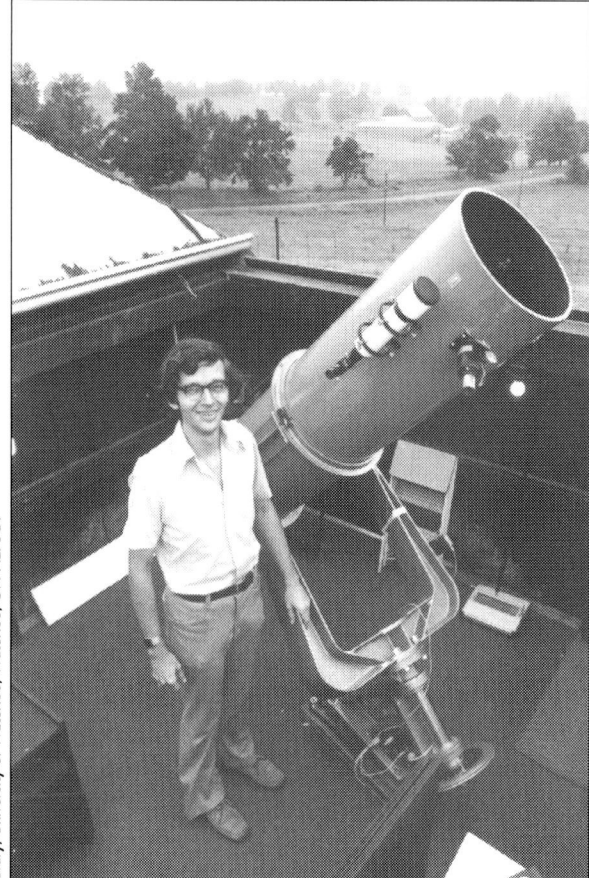

Douglas telescope has a fine new Dobsonian mount and a good choice of eyepieces, and the Centre's library is growing. The Centre is justly proud of the high proportion of its members who continue to play a significant role in the Society and in the advancement of Astronomy.

KITCHENER-WATERLOO

Twenty-two members of The Grand Valley Astronomers, a group which had originated in 1952, decided in 1980 to form the Kitchener-Waterloo Centre. There was already a 2-storey domed observatory, built in 1973 near Ayr by Hans Schilter, with a 32-cm f/5 Newtonian telescope. Adjacent to this was another structure with roll-off roof known as the Dance Hill Observatory which was owned and operated by three members. There they used a 20-cm Schmidt camera for astrophotography and were planning to do photoelectric photometry using a 36-cm telescope.

The enthusiasm of the new Centre is best captured in the words of the first annual report, written by Secretary Murray Kaitting:

> We have seen a 40 percent increase in membership since our enrollment into the Royal Astronomical Society of Canada and have just begun to realize the benefits of being part of this Canadian-wide network for amateur astronomers.

The new Centre, with their considerable background and experience, planned a gala grand opening to mark the beginning of their new status. Continuing to quote from Kaitting's report:

> Dr Ian Halliday, President of the RASC, formally welcomed us as the 19th centre and spoke on the national character of the Society and its membership of more than 3,000 astronomers. Ceremonies commenced under the direction of [Centre President] Professor Ray Koenig of WLU [Wilfrid Laurier University]. Our Honorary President and founder of the astronomy club, Carl Arndt, gave a slide show on the 28 year history of the Grand Valley Astronomers. Following an introduction by Clifford Cunningham, Dr Helen Sawyer Hogg of the University of Toronto and Honorary President of the RASC delivered the Centre's inaugural address on the topic of Variable Stars in Globular Clusters. In appreciation of her visit, Gerald Kennedy presented her with a plaque on behalf of the Centre.
>
> Ending the ceremonies at WLU was a brief slide show of the two observatories near Ayr, south of Kitchener. About 50 people then proceeded to the observatory site where Dr. Hogg cut a red ribbon and declared the facilities open.
>
> The event proved to be a great success, with representatives from five Centres present, as well as two visitors from New York. In all, more than 100 people attended.

In these photographs, taken in September, 1980, Clifford Cunningham is seen in the Dance Hill Observatory and at the entrance of the original 1973 observatory.

By the following year the 36-cm telescope was in use for prime-focus photography, Cunningham and Kaitting were doing photometry on RS CVn variables and on asteroids at the Dance Hill Observatory, and Kennedy had added his own building at the site where he did astrophotography with a 20-cm Schmidt camera and a 35-cm telescope. Members of the Centre had purchased no fewer than five telescopes of aperture 20 cm or greater and their interests ranged from planetary to deep sky observing.

In the years since, the observatory has continued to be an important focal point for many Centre activities. Besides the members who use it for their own projects, Scouts, school groups, and the general public have all enjoyed coming to open houses and starnights. Most years the Centre has a picnic at the site preceded by a general clean-up and followed by an evening of observing. The 32-cm telescope has been frequently upgraded. Currently there are plans for a computer-assisted drive and other improvements.

Recently, a 25-cm Dobsonian telescope was built from a scope donated by the Schilter family. It is very convenient for use away from the observatory, for instance at Starfest which a number of members attend annually near Mount Forest, and at public starnights in Kitchener-Waterloo, Guelph, Brantford and Woodstock. Each year, members have hosted star nights for the local field naturalists, given talks at conservation areas, participated in hobby shows and set up mall displays for Astronomy Days.

On top of all this public relations work, the Centre has maintained an interesting program of monthly meetings provided almost entirely by its own members. These are held at Wilfrid Laurier University. The year usually culminates with a Christmas Dinner at a local restaurant followed by videos and entertainment at the home of Professor Ray Koenig, five-time president of the Centre. Members are kept well-informed through the Centre's newsletter, *Pulsar*, and a members' handbook (which includes a brief history of the Centre) a membership list and directions to meetings and the observatory.

In the past few years there have been several new telescopes in action and some members have begun to work with CCDs. There have been frequent discussions with respect to building a radio telescope but so far nothing concrete has occurred. There is also a possibility that an observatory may be placed on a member's piece of land near River Place where Starfest is held.

SARNIA

In 1981, the year after Kitchener-Waterloo Centre was established, Sarnia joined the ranks as the twentieth RASC Centre, an outgrowth of an informal group of astronomy enthusiasts who had been meeting for a couple of years under the name Sarnia Astronomy Club. Some of the group were unattached RASC members and some were members of the London or Windsor Centres. Their first president, Zdenko Saroch, had been a member of the London Centre, and with his knowledge of the Society and encouragement from other London members as well as from Randy Groundwater of Windsor, the decision was made to apply for recognition as a Centre. The National Council gave their formal approval on September 26, 1981. Like Kitchener, Sarnia experienced an increase in membership (to twenty-five) during their first year as a Centre.

Until 1985 the Centre held monthly meetings consisting of speakers, discussions and films at Lambton College. Ties were maintained with neighbouring Centres through visits and exchanges. A Centre newsletter, *Urania*, was published at least once a year. Observing meetings were held at irregular intervals depending on the weather and some members enjoyed solar eclipse trips and excursions to various gatherings like Stellafane, Starfest and Astrofest.

In its first few years, as Zdenko Saroch has written, the Centre had the right people who were able to keep the group functioning well, but unfortunately some of these people got relocated or went to school in other cities. A difficult period ensued. The former meeting place was no longer available, and only four meetings were held in 1986. Copies of *Urania* were no longer mailed out and the Centre's postal box was cancelled, though their financial statement showed a reasonable surplus for the year. The secretary reported "sometimes hostile public opinion" on a radio Hot-line show and "much unfair criticism" regarding their display in the shopping mall. Yet the Centre provided some very worthwhile assistance to youth groups including an observation night for foster children at the Children's Aid Society camp.

Membership declined to only eight in 1988. The Centre might have been headed for extinction but they fortunately found a fine new meeting place in the Town of Clearwater municipal buildings. It was free of charge and came equipped with audio-visual equipment and a kitchen. More importantly, new members joined who had the right chemistry to work together for the good of the group. Observing was done informally at members' homes and in conjunction with other clubs at Starfest, at Windsor Centre's Eagle site, and with the Essex Wayne Observers.

There are now, within the group, telescope makers, computer buffs and active observers. One member has a permanently mounted 41-cm Newtonian telescope. In the area of public education, members provide instruction for Brownies and Cubs, and recently made themselves available as resource persons for elementary teachers in Lambton County who were tackling a new unit called "Out of this World" in fifth grade. With positive steps such as these, Sarnia seems on track for growth and renewal.

THUNDER BAY

As early as 1892, it was reported by the A&P Society of Toronto that the headmaster of the High School at Port Arthur, H.W. Law, agreed to make monthly auroral reports. No other astronomical activity from the Lakehead area was reported in the Society's publications until 1954 when it was noted that a couple of members from Port Arthur, F.T. Matthews and Warren Creighton, went to Mattice, Ontario, for the June 20 solar eclipse. Four years later, fifteen people attended a meeting at Vickers Heights Social Centre and formed the Lakehead Astronomical Society. Max Shapiro and Ralph Ferch were president and vice-president and Creighton was secretary-treasurer. All three and some others were members of the RASC but no Centre of the Society formed at that time. The Lakehead Astronomical Society sent delegates to the 1962 General Assembly, and judging from correspondence with M.M. Thomson, was still doing well on a small scale at least until 1966. The club had regular meetings and members had access to an 20-cm reflector belonging to Bill Gardner housed in an observatory built by Dr Quackenbush. But as many members reached retirement age and moved away, the club ceased to exist. The observatory was donated to Lakehead University but was never used. The telescope is still in Bill Gardner's possession.

On September 24, 1988, an unattached member from Thunder Bay spoke to the National Council about the possibility of forming a Centre. The gentleman was Bob Bishop, an instructor in Astronomy at the Community College there. How this came about is interesting. Bob Bishop and his wife Beverly

Astronomy Day Display, 1990. l to r: Bev Bishop, Harold and Shirley Stackhouse.

Photo by Bob Bishop

were attempting to select an evening course at Confederation College, but found none to interest them. Bev suggested that if Bob couldn't find a course to take, he should teach one and so, in the spring of 1987, with the agreement of the head of Adult Education at the College, he offered his first Astronomy course – seven sessions of three hours each. By the time he came to the Council meeting in Toronto a year and a half later, his course, called Backyard Astronomy, had attracted 120 people and his star nights were enjoyed by the Thunder Bay Field Naturalists and by campers at Sibley Provincial Park. These indications of interest by northern Ontarians looked promising, and Bob Bishop indicated to the Council that he had found seventeen members ready and willing to form a Centre. There was some discussion because the pending new by-laws required a minimum of twenty members for a new Centre, but in the end, the Thunder Bay Centre was welcomed. Hindsight shows it was a good decision as membership went to forty the following year.

Within a month of Council's decision, National Secretary David Tindall went to Thunder Bay where he joined sixteen members for Sunday brunch in a discussion of the formation and direction of the new Centre. The inaugural meeting was held on October 23, 1988, in the home of President Bob Bishop. A newsletter, *Northern Sky*, was started and a couple of television appearances brought the new Centre to the attention of many northerners. There are now active members in communities hundreds of kilometers away from Thunder Bay – in Dryden, Atikokan, Nipigon, Geraldton and Marathon.

The emphasis of the Centre's activities is to enhance members' enjoyment of the night sky and to encourage them to share their pleasure and knowledge with others outside the Society. To this end, the executive held a series, "Finding Your Way in the Night Sky" designed primarily to help beginners. Other indoor meetings centred on photos taken at previous star nights. Special events, such as the lunar eclipse of August 16, 1989, brought out the public and the media. Each summer, the Centre holds a couple of star parties at Provincial or National Parks in the area when campers enjoy sharing an evening of observing and a slide show put on by members. The annual Astronomy Day sparks a lot of interest in the Centre's display of photos, equipment, books and charts in Thunder Bay's biggest shopping mall. During Astronomy Week 1990, three different talks at the public library were given in addition to the mall display. A highlight of 1991 was a two-day visit and invited talk by Terry Dickinson.

TORONTO

By concluding with a history of the Toronto Centre the developments of the national Society will be, in a sense, summarized since Toronto members have always made up a significant fraction of the whole. In fact, even with the ever-increasing number of Centres, Toronto gradually expanded its share of the total membership from about 20 percent sixty years ago to approximately 30 percent recently.

Originally, of course, nearly all members of the Society lived in Toronto and for a long time, even after many other Centres had been established, the city had a unique status in the RASC. All meetings there were considered meetings of the Society generally. The Executive and Council elected by all the members not only ran the national Society but managed local Toronto affairs as well.

The evolution towards separate Centre status was gradual. The first indication came in the financial statement for 1920 when "Local Expenses in Toronto" were shown separately from the general Society expenses. The same year the Society's Constitution was amended and the secretary and treasurer now became known as the general secretary and general treasurer to distinguish them from local officers. At some point, perhaps also in 1920, a committee of the General Council began to be responsible for the ordinary meetings in Toronto, but as the chairman of the Committee was generally the president or vice-president of the Society the distinction was still rather blurred.

A decisive step was taken in 1927. Council resolved to form a separate body to carry on the local ordinary meetings. At the opening session in the fall of 1927 RASC President A.F. Hunter announced Council's decision but there was some opposition especially from A.F. Miller. The matter was deferred to the next meeting which took place on November 1, at which time Hunter, J.R. Collins and A.R. Hassard were appointed as a Board to conduct the ordinary meetings until the end of the year. In 1928, Dr W.E. Harper became RASC president and as he lived in Victoria, he certainly could take no direct responsibility for the ordinary meetings in Toronto. During his term, several local members with Collins as Chairman comprised the Board for Toronto. At a meeting on December 17, 1929, the members present not only elected a Board but named them as the following officers:

Honorary President – Sir Frederic Stupart
Chairman – J.R. Collins
Vice-Chairman – R.A. Gray
Secretary – E.J.A. Kennedy
Treasurer – Miss Evelyn Watt
Recorder – J.R. Gibbs
and Standing Committees for Finance, Programme, Membership and Building.

Four weeks later, at the Society's Annual Meeting, Toronto was referred to as a Centre and the secretary for Toronto reported that "A more definite organization for the Society's activities in Toronto was thought advisable." It is clear in the treasurer's report at that same Annual Meeting, i.e. for the year 1929, that an entry showed for the first time that "Toronto Centre" got a grant for its operations just as the other Centres did, but no formal motion recognizing the "Toronto Centre" as a separate entity seems to have been proposed or passed.

There were some lingering problems. The Toronto secretary's report made mention of the Building Fund and how it would give a great impetus to the science of astronomy in Toronto . The Fund, of course, belonged to the Society as a whole and members from outside Toronto must have had some doubts about the propriety of Kennedy's report. Then there was the matter of who was entitled to the annual grant paid by the City of Toronto to the Society. In the end the Society kept the Building Fund and the Centre got the City grant (for its three remaining years). This helped with expenses, which in 1930 included speakers' travel costs, $83; printing of programs, $17; rental of the lecture room, $65; and advertising in the daily papers, $56.

Up to this time, then, all RASC members who happened to live in Toronto belonged to the Society but not to any Centre. This anomaly was resolved later in 1930 when Council passed a by-law amendment recognizing "Members-at-large," a new classification of those who did not wish to be attached to any Centre, in most cases because they did not live near one. As a result of this reorganization, 104 names were transferred to the new category from the "Toronto" list, leaving Toronto Centre with 164 members for 1930. Nonetheless, the Toronto Section, as it called itself, continued to elect members-at-large until 1936.

Even when Chant wrote his history of the Society in 1940, he still seemed to think along the old lines and wrote nothing about Toronto Centre specifically even though he gave every other Centre at least a couple of paragraphs. This perhaps could be explained by the fact that it was not until 1945 that the Toronto Centre drew up its own Constitution and by-laws and the old "Board" with its "Chairman" was formally replaced by a Council headed by a president as was the case in the other Centres.

The establishment of the Toronto Centre, whenever it formally occurred, was not marked by any dramatic change in the activities of the members or the format of the meetings. At least since World War I, about twenty meetings were held each year in the university's Physics Building. The speakers included professional astronomers, other scientists and amateur members in about equal proportions. In addition, some member would often give a short talk on a selected constellation. Usually a distinguished visitor delivered one lecture in the season (Barnard, Shapley, Eddington, de Sitter among them), sometimes at a joint meeting with the Royal Canadian Institute or, on occasion, with some visiting group like the American or the British Association for the Advancement of Science. The season generally ended in May or June with an open-air meeting on campus when the public joined members at their telescopes or at the 15-cm Cooke refractor for an hour or two of enjoyable viewing.

From time to time innovations were introduced, some more successful than others. To try to increase member participation, alternate meetings, starting in 1931, were arranged by and for amateur members. A question box was also initiated that fall and tried occasionally in later years. Members were invited to submit astronomical questions in writing at one meeting and the answers would form the basis for discussion on a subsequent evening, but the idea was never too popular. On the other hand, motion pictures were shown for the first time in 1926 and this innovation sometimes (as in 1934) attracted an audience of a couple of hundred. The influence of radio was apparent in a "newscast" of recent events in the world of astronomy delivered to a 1934 meeting by "commentator" Fred Troyer and in a 1939 "quiz show" where two teams of six competitors tried to answer astronomical questions previously submitted by the audience.

Following the opening of the David Dunlap Observatory in 1935, its staff took great interest in the activities of the Society and of the Toronto Centre in particular. Each year, except during the war, they hosted a behind-the-scenes tour of the Observatory and its facilities, a tradition which is still a popular part of the Centre's annual program. Professors Heard, Millman and Frank Hogg offered short courses for new members in 1937, 1938 and 1939 as did Ruth Northcott in 1944, and Heard gave a series of five short talks in 1945 on the spectroscope and its uses in astronomy. Virtually everyone at the Observatory was a member of the Centre and all contributed substantially to it as officers, councillors or speakers.

On the observing front also, the DDO astronomers tried valiantly to interest members in useful pursuits. In 1937 an ambitious Observation Committee was formed comprising four sections – variable stars and novae; meteors, fireballs and meteorites; surface features of Sun, Moon and planets; and occultations, comets and asteroids. Each section had at least one professional astronomer among its members. The variable star observers were under the direction of Shirley Patterson, Gold-Medallist and graduate student, and included Ruth Northcott (employed at that time as a computational assistant at the Observatory), amateurs Bert Topham and Neil McNabb with Helen Hogg as an advisor. The meteor observers were inspired by the enthusiastic leadership of Peter Millman, also at the DDO at the time. These two groups were the only ones to file reports though all four were still in existence five years later.

The Centre had made earlier attempts to encourage members to observe. The constellation talks which were a standard feature of most meetings were intended to stimulate interest. Observing Committees were formed in 1930, 1932 and 1935, but no reports from them ever appeared in the *Journal*. The first meeting each fall was traditionally set aside for members to report on observations made over the summer, but these usually amounted to casual descriptions of meteor showers, auroral displays or some atmospheric phenomena. In fact the Secretary noted in 1935, "a falling-off in the number of useful observations reported at meetings during recent years and suggested the formation of an amateur telescope-making group as a way to stimulate interest."

It was not a completely new idea. Even in 1929 four members had given a talk on the topic "Telescopes We Have Made" but an organized group was slow to form. However, by 1946 it was reported that Toronto had a group of about fifty telescope makers who met under the leadership of Raymond Pearce.

In the postwar decades, the Telescope-Makers Group was the largest and most active in the Centre. Under Jesse Ketchum's direction in the 1960s, meetings were held consistently every other week in the basement workshop at the Society's headquarters on College Street. Generally about two dozen members were working on mirrors at any one time. 1967 was a banner year with forty-two telescopes completed, including four 32-cm instruments.

Though the telescope makers became very active right after the war, interest in observing only began to revive in 1954 when Raymond Broadfoot organized an observing and study group. They met at least once a month for Messier observing, lunar studies and some computational work in celestial mechanics, and they issued a bulletin as part of the Centre's newsletter describing what they had observed and what should be watched for. In 1956 there were reports of Lunar and Planetary Sections with frequent meetings. The following year the Secretary reported that:

> several meteor nights were held in connection with the IGY programme and many members made auroral reports. The two bright comets of the year and the artificial satellites were frequently observed by Group members, the comets photographed and the Sputnik's radio signal tape-recorded. The Lunar Section under R.V. Ramsay's direction held monthly meetings and carried on its own observing programme.

This latter Section exchanged information with the Irish Astronomical Society.

As many as sixty members belonged to the Observers Group and in the early 1960s they usually held at least four observing nights and six indoor meetings each year. In 1960 they observed a lunar eclipse, followed Comet Burnham closely, plotting its path over several nights, and made a catalogue of sketches of all the Messier objects. More advanced members were given the privilege of using the 48-cm telescope at the DDO. In addition to holding their own meetings, the group usually contributed to a couple of members nights as part of the regular program, arranged displays for the Annual Meeting of the Society and prepared a newsletter called *Scope* on an occasional basis.

There was some excitement in 1963 when a member offered an unfinished 32-cm telescope to the Centre provided it could be completed and suitably housed. Unfortunately, negotiations with the Metropolitan Toronto Conservation Authority for a site in the Boyd Area, though promising at first, foundered in 1964, largely because the Centre was not incorporated at that time. The observatory project could not proceed and members seemed to lose heart, with the result that the Observers Group was not heard from again for a few years. Nonetheless, individuals and informal groups continued to observe variable stars, meteor showers, grazing occultations and Mars.

Though interest in observing waxed and waned, attendance at the regular meetings of the Centre was always good. The Society's assistant secretary-treasurer, Miss Budd, had instituted a new procedure in 1944. She agreed to supervise the opening of the national library for a short period of time after each lecture meeting. This custom became a very pleasant social hour with tea, coffee and cookies and stimulated increased use of the library, especially after the Society moved to its permanent quarters in 1956. The meetings in this postwar period continued as before with mostly local talent, some films and an occasional diversion such as quiz nights when members had to identify phenomena from the photos posted around the room, a field trip to the Forest Hill planetarium or the one in Hamilton, and a memorable evening when "The DDO Players" enacted Galileo's Dialogues. Among the notable visiting lecturers during this period were George Gamow, H.C. van de Hulst, Nancy Roman, Adriaan Blaauw, G.P. Kuiper and Sir Harold Spencer Jones.

About 1960, the regular meetings moved to another campus location, the Wallberg Building, which was even closer to the national office and library making the social hour even more convenient. The innovative spirit of earlier years seemed to evaporate perhaps because people were less inclined to make their own entertainment with television in nearly every home by this time. The annual schedule now comprised about a dozen meetings at which an average of four or five visiting astronomers would lecture (including one or two from elsewhere in Canada), a DDO astronomer would speak, there would be a couple of members nights and an occasional lecture by an amateur member of the Centre. Of course, the Observation and Study Groups held their own meetings, so overall there was not much change from the traditional biweekly frequency. Traditionally, there have not been regular meetings during the summer as members have put their efforts into public relations activities.

The summer sky gazing parties which had been held on the university campus for years, expanded in 1960 to include five evenings in city parks. An estimated 3,000 people came to look through the telescopes and to enjoy a film presentation. Though attendance varied depending on the weather and what particular objects were visible, the Centre continued to hold several public star nights each summer, usually drawing hundreds of people. Members of the Telescope-Makers Group were always happy to bring their latest products along on these occasions.

R. VERNON RAMSAY (1925-91) joined the Toronto Centre in 1948, after service in World War II. He was active in the Telescope Makers and in the Observation and Study Group which he directed in 1958-60. His special interests were the Moon and Mars. From 1961-64, he was Chairman of the national Committee on Observational Activities.

Vern Ramsay was the founding Editor of the Centre newsletter '*Scope* and served on the Toronto Centre Council for many years, including three years as President, 1966-67 and 1974. He received the Society's Service Award in 1972.

DIARMUID J. FITZGERALD (1930-) shared many interests with his long-time friend Vern Ramsay. They were both "founding fathers" of the Toronto Centre's newsletter '*Scope*. Both were active in trying to establish a Centre Observatory, and both were part-time lecturers at the McLaughlin Planetarium when it first opened. "Fitz" followed Vern as Director of the Observers Group in 1962-65, joined the AAVSO in 1964, and sent in observations of variable stars for a few years. He succeeded Vern as Vice-President and President of the Centre. During FitzGerald's Presidency, he and his family gave many hours of time to the organizing and co-ordination of activities at the 1969 GA.

D.J. FitzGerald was National Secretary 1974-77 and received the Service Award in 1975. Since retiring from his position as a senior analyst working in computer systems design in 1989, he and his wife have been enjoying the fall and winter months in Florida.

From l to r: Ruby and Vern Ramsay, a 30-cm Newtonian telescope on the roof of the Strasenburg Planetarium, Rochester, NY, Diarmuid and Anne FitzGerald.

FREDERIC L. TROYER (1912-) joined the RASC as a schoolboy in 1924. His membership lapsed for a time but he was soon back to stay and in 1933 he became a life member. These facts are of some importance, since they make him presently the longest-standing member of the Society. What's more he has held office for nearly fifty years and is still a valued member of the National Council. His first job was as Assistant Curator in 1925 which really meant that he got to operate the lantern slide projector. For 23 consecutive years, starting in 1941, he held office in the Centre as Recorder, Secretary, Vice-President and President. The Recorder had the large responsibility in those days of summarizing every meeting of the Centre for the *Journal* and Troyer's reports were always examples of clarity and style which he had developed in his profession as an editor and science writer for the *Toronto Star*. Nationally, he served as Recorder from 1952-64 and as Librarian from 1977-82.

For many years Troyer organized Toronto Centre's public star nights in city parks and at the Canadian National Exhibition. These events gave thousands of people a chance to look through a telescope and, as a result of pamphlets he had printed by the *Toronto Star*, gave them some information about the sky and the Society.

His extensive contributions to the Society were recognized as early as 1960 when he received the Service Award but he has since continued his amazing record of dedication. Since 1972, he has served his Centre almost continuously as Recorder, Secretary, and National Council Representative. Toronto Centre accorded him a unique honour by naming him Honorary Councillor in 1982.

HARLAN CREIGHTON (1946-) joined the Toronto Centre in 1956. He became active in public education and chaired the Public Star Night Committee in 1967. In 1970, he became a Lecturer, and subsequently Curatorial Assistant at the McLaughlin Planetarium and assumed a prominent role in Society affairs. Within the Centre he was a member of Council, Recorder, Editor of *'Scope*, and President (1972-73). He helped organize expeditions to observe total solar eclipses. Nationally, he served as Librarian and Recorder and was an important force in the evolution of the *National Newsletter*, first on the NNL Committee, then as Editor and subsequently as Assistant Editor. His name was on the masthead of the NNL and its successor, the *Bulletin*, for 20 years. His well-reasoned and strong opinions were felt on many local and national committees. Harlan Creighton received the Service Award in 1981.

Since 1972, when he joined the staff of Seneca College in Toronto, Creighton has been associated with Community Colleges. At Seneca he taught a general education course in Astronomy and established the College's planetarium as an educational facility for both college and community use. In 1977, he moved to The Pas, Manitoba to join the faculty of Keewatin Community College where he taught adult basic education and post-secondary mathematics and science, as well as an evening course in general Astronomy. In 1979 he led an expedition of students, staff and members of the public to observe the total solar eclipse from Arborg, Manitoba. In 1985, Creighton received an award from the College in recognition of teaching excellence and outstanding service. Since then he has been a learning assistance instructor at Wascana Institute of the Sasakatchewan Institute of Applied Science and Technology in Regina. He continues to be interested in astronomy and astronomy/science education.

ALFRED W. SCOTT (1904-90) joined the Toronto Centre in 1953. He was their Treasurer 1972-78 and 1980-81, and was Membership Secretary 1979-80. As a member of National Council from 1973-77, he served on its Executive, Finance and Property Committees.

Alf, as he was known to many Centre members, was an active solar observer. During the 1970s he made sketches of sunspots on every clear day, observing with a small refractor from his balcony. His keen sense of humour found expression in his cartoons of things astronomical. He was a great booster of the RASC, actively recruiting new members and encouraging them to become involved in Centre activities and their own observing projects. He received the Service Award in 1979.

(adapted from Obituary of A.W. Scott by B.R. Chou)

The 1976 Toronto Centre Council. Front row (l to r): Jack Newton, Ann Scott, Christine Clement, Ralph Chou, Alf Scott. Back row: Fred Troyer, Robert Pike, Ian McGregor, Michael Watson, Jim Cobban, Harlan Creighton, Art Pelletier, Ian McLennan, Bill Peters. A portrait of Colonel R.S. McLaughlin, benefactor of the Planetarium hangs on the wall of the boardroom. Jack Newton's biographical profile appears on page 61. Ian McGregor and Michael Watson were Service Award winners in 1991 and 1992 respectively.

Suburban parks in Scarborough and Etobicoke were added to the list of locations in 1964, and star nights were held 100 km north in Orillia in 1966 and 1967. A special effort was made in Centennial year when over fifty members contributed to the success of five star nights which, in total, attracted 3,700 visitors.

Every Saturday evening from April to October, the David Dunlap Observatory has traditionally opened its doors to the public, and at least since 1961, members of the Toronto Centre have set up their own telescopes on the lawn in front of the dome so that the public could see more than they could in a ten-second peek through the eyepiece of the 1.88-m reflector. This continuing practice not only adds to the enjoyment of DDO visitors but provides a wonderful opportunity to publicize the RASC.

The University of Toronto also got the Centre involved in another public awareness project which each year brought thousands of people into contact with the Society. Back in 1921, at the invitation of the Canadian National Exhibition, the Astronomy Department, along with other departments, set up displays to demonstrate some of the work of the university. This practice continued for a few years, but when the university bowed out, the Department of Astronomy suggested to the Toronto Centre that it might consider doing something similar. So began the Centre's participation at the CNE, an annual commitment that would last until 1967, requiring members to set up telescopes every clear night for a two-week period at the end of August. From such a brightly lit location, the Moon and planets were about all that could be seen, but even so several thousand people usually lined up each year for a look through the telescopes. From 1940 on, the Exhibition gave Toronto Centre a grant of $75 ($200 in later years) of which only a small part had to be used for telescope maintenance, insurance and transportation costs. These amounts were put aside in a Special Purpose Fund which amounted to over $4,500 by 1967. For twenty years, The Toronto Star co-operated by printing an illustrated four-page leaflet called Sky Facts which told a bit about the current night sky and the Society and provided a convenient means of advertising the newspaper. This popular handout, revised repeatedly, was used not only at the CNE, but at other public gatherings as well.

It was not unusual for the Centre to welcome a hundred new members in a year, many of them attracted by public star nights in parks, at the DDO and at the CNE. In addition, there was contact through hobby shows from time to time and regular involvement through instruction and testing Boy Scouts and Girl Guides for their Astronomy Badges. The Toronto Centre's reputation in the field of public education was thus well established by the 1960s and was a major factor in the steady membership growth from about 200 in the early 1940s to nearly 600 in the 1960s.

In 1968, the Toronto Centre moved the location of its meetings to the newly opened McLaughlin Planetarium. In a way, this was but one more step separating the Centre from the national Society. It was no longer convenient for the Centre to use the national library for social gatherings after its meetings or for the telescope makers to use the facilities in the basement of 252 College Street. Toronto Centre now had the use of a lovely new lecture theatre and adjoining optical workshop in the Planetarium building as well as the use of the Planetarium library before meetings. The final step which legally put the Toronto Centre on its own feet was its incorporation, achieved in 1974.

Since its move to the Planetarium, the Centre has pretty well maintained its schedule of meetings every two weeks, except during the summer. There are generally about seven lectures by professional astronomers, with somewhat more reliance in recent years on people from Toronto, either from the University of Toronto or York University, or from the Canadian Institute of Theoretical Astrophysics. Carl Sagan and Bart Bok did address the Centre in the 1970s, but there have been fewer visits by pre-eminent astronomers from abroad in recent years. Perhaps the days of the outstanding individuals in the world of science are past; perhaps the speakers are just as eminent as ever but time has not yet accorded them the recognition won by their illustrious predecessors. Or perhaps, because there are more Canadian scientists, there is less need to invite foreign visitors. Regardless, the quality and variety of lectures is every bit as high as it ever was and the joint meetings with the RCI, which still continue once or twice a year, sometimes bring speakers from the United States or overseas.

Most of the rest of the annual program is made up of members nights, organized since 1971 by the Observational Activities Committee. An indication of the breadth of interest can be seen from the 1990 annual report where the list of topics presented included: astronomy and the public, astronomical equipment, light pollution and the quest for dark skies, "An Evening with Charles Messier," aurorae, the Finland solar eclipse, comets and meteors, astrophotography and solar astronomy. Sometimes a special planetarium demonstration is scheduled to precede these meetings. In recent years, a picnic and Awards Banquet have rounded out the yearly program. All in all, it is an ambitious schedule, which attracts good support from the membership. Though attendance figures are not often given, the report for 1970 indicated that a total of over 3,200 members came to the meetings that year.

Aside from the regular meetings, special groups

met too. In the 1970s, a Study Group, chaired by Ian McGregor, met on alternate Saturdays for instruction and discussion in basic and practical astronomy. Usually attended by fifteen to twenty members, these group meetings were especially helpful for new members and young people.

The Telescope-Makers Group also had about twenty members meeting on Saturdays though at times interest was so strong that meetings had to be held twice a week. Participation has varied, but at present a few mirrors are completed each year.

Observing activity has also had its ups and downs in the Toronto Centre. After being dormant since 1964, an Observational Activities Committee was started in 1971 to provide advice, assistance and co-ordination. Once again there were some negotiations with the Conservation Authority this time over a site in Albion Hills, but again there were no positive results. Under Robert Pike's leadership in 1974, three observing programs were stressed — Jupiter's surface features, three variable stars, and a light pollution project. Manuals were prepared in these areas to guide members in what and how to observe and the sky brightness survey led to some particularly interesting results which were published a couple of years later. Interest was not confined to these three topics, and some members observed meteors, solar flares, grazing occultations, Messier and deep-sky objects.

An observatory for the Centre was finally opened in 1976 on a site belonging to Mr and Mrs R. Kelsch, near Schomberg, about 60 km northwest of the city. At first it was equipped with a 25-cm Cassegrain telescope, on permanent loan from the Ontario Science Centre, and the observatory log showed about 300 user-nights per year. Mechanical problems in 1979–82 kept the telescope out of use, though the site with its warm-up hut was used by a few members with their own telescopes. In 1982, after five years of work, the Centre's own 32-cm telescope was installed and again, for about three years, the Observatory was functional. At the present time, the Schomberg site is closed and the Centre is making plans for a new observatory. For now, observers are on their own, or participate in members' star parties in or near the city or at the Long Sault Conservation Area, 300 km east of Toronto. A number of members enjoy gathering with other observers and telescope makers at Stellafane and Starfest each summer and eclipse chasing to all parts of the planet has been a popular project ever since 1979 when 106 members and friends flew to Gimli, Manitoba, for Canada's most recent spectacular solar eclipse.

That observing has never been a consistent focus for Toronto Centre activities is hardly surprising considering the great distances that members must travel to escape the umbrella of light pollution that extends far beyond the borders of Metro. That is not the whole story, of course, as the examples of Montreal and Ottawa have clearly shown that urban astronomers can be inspired to support an extensive observing program by a few dedicated leaders.

The Toronto Centre's great strength is in its public relations work. Members continue to set up their telescopes for the public as they have been doing for a century. Star nights in city parks and at the David Dunlap Observatory on Saturday evenings are as popular as ever. From time to time, there are special events like eclipses or Mars oppositions which bring out crowds in large numbers. The ultimate event, of course, was Halley's Comet in 1985–86. An estimated 15,000 visitors braved the cold at three viewing sites set up by the Centre, crowds packed the Medical Sciences auditorium for a public lecture organized by Michael Watson, and members were interviewed on local and national news several times.

Many new initiatives have developed in recent years. Since 1973, members have been giving talks at public libraries, sometimes in combination with star nights. Participation in Astronomy Day began in 1976 with a set-up outside the Planetarium attracting 1,500

visitors in spite of poor weather. Over the years, plans have become more and more ambitious and in 1990, to mark the Society's Centennial, an award-winning Astronomy Week involved over a hundred volunteers. Displays in shopping malls have been a part of the program since 1978, with about three usually arranged each year. From 1981 to 1989, Randy Attwood, with the assistance of many other members, hosted "Astronomy Toronto," an excellent series of cable television programs which were aired not only in Toronto but in several other areas of Ontario. These efforts have ensured a flow of new members to the Centre which now has a membership roll equal in size to that of the entire Society about 1940.

It is remarkable that the Toronto Centre, with close to a thousand members, operates entirely with volunteer support and no office. The responsibilities of the Centre today certainly exceed those of the national Society fifty years ago which at that time employed an assistant on a half-time basis. True, the national Society did publish ten forty-page issues of the *Journal* and *The Observer's Handbook* each year, but the Centre now publishes, prints and mails out six issues of its newsletter *'Scope*, and distributes the *Handbook* to its members. While the national Society administered the library, which was a larger responsibility in the 1940 than it is today, the Centre's role in organizing at least fifty events of all sorts each year cannot be compared to anything the Society faced fifty years ago.

The acquisition of a computer in 1984 was a tremendous help, not only in preparing and distributing *'Scope*, but in the administration of membership records. Other initiatives which led to improved retention of members involved mailing of renewal notices, follow-up telephone calls and a telephone answering machine with recorded messages about forthcoming events and numbers for further information. This machine now handles about 130 calls per month with about a quarter of these requiring further attention. In years of especially large growth, new members were hosted at a special meeting, were issued with name tags and received a special supplement to *'Scope* outlining the services offered by the Centre and the Society. *Scope* itself has evolved into a high-quality newsletter. It is a strong unifying force in one of the largest locally based astronomy clubs in the world.

Astronomy Day, 1986, at Toronto's Harbourfront.

Final Thoughts

Having completed our journey through time and space we may reflect on the broad features of the Society's past. Constant throughout its history is a commitment to public education and personal development of its members. Other characteristics have changed.

During the Society's first phase, prior to 1905, the amateurs who totally comprised it were full of confidence that they could advance astronomical science. They put the Society on the world map by circulating its publications and appointing internationally renowned honorary and corresponding members. From 1905 to 1970, Canadian professional astronomers strongly influenced the course of the RASC. The amateur members, still in the vast majority, seemed not only willing but proud to let the professionals strengthen the Society through lectures, papers and leadership. Observations were made in the hope that they would be useful to the experts. Since 1970, in the wake of the space age and the establishment of CASCA, most amateurs don't take themselves so seriously. Most take a recreational approach toward astronomy and describe their interest as a hobby and their local Centre as a club. This terminology and outlook may, in large part, be due to the widespread influence of popular astronomy magazines from the United States where amateur astronomers have not had strong national ties with one another and with professional astronomers. An enhanced appreciation of nature and an environmental movement which has become a powerful force in the past twenty years have also strongly influenced the attitudes of amateur astronomers who prefer to look at the universe aesthetically rather than scientifically.

History is not much help in predicting the future. Perhaps we are on the threshold of an electronic revolution in astronomy in which charge-coupled devices and computers will make serious scientific accomplishments a greater reality for amateurs. Perhaps the confidence in the role of science so evident a century ago will return.

If the Society is to continue flourishing, all members must strive to ensure that the tremendous richness in the ethnic and cultural demographics of the nation will be reflected in an equally wide spectrum of interests among Society members. Whatever happens, the RASC must eagerly welcome aboard all who wish to share in a fantastic journey of discovery. To the Society which makes this possible, let us toast her health with the wish of Captain Kirk of the Starship *Enterprise*: LIVE LONG AND PROSPER.

GENERAL BIBLIOGRAPHY

The following is a brief chronological listing of some book-length histories of Canadian astronomy and some histories of Canadian science with relevance to astronomy during the period since Confederation. As most of these books contain many references to papers on specific topics, they form a good basis for anyone wishing to pursue the history of astronomy in Canada in greater depth. Another excellent starting point for further research is *The Canadian Encyclopedia* published by Hurtig, 1988.

Harper, W.E. "The History of Astronomy in Canada," in *A History of Science in Canada*, H.M. Tory, ed. Toronto: Ryerson Press, 1939, p. 87–99.

Thomson, D.W. *Men and Meridians* (3 Vols.) Ottawa: Energy, Mines and Resources, 1966–9.

Northcott, R.J., ed. *Astronomy in Canada, Yesterday, Today, and Tomorrow*, Toronto: University of Toronto Press, 1967. (This is a special edition of *JRASC* **61**, no.5.(October 1967): 211–352.)

Thomson, Malcolm M. *The Beginning of the Long Dash: a history of timekeeping in Canada.* Toronto: University of Toronto Press, 1978. (The Society has a copy of the original manuscript, containing a considerable amount of material which was not published.)

Stretton, W.D. *From Compass to Satellite A Century of Canadian Surveying and Surveyors 1882-1982.* Ottawa: The Canadian Institute of Surveying, 1982. This was a 362 page issue of *The Canadian Surveyor*, **36**, No.4.(December 1982).

Chartrand, L., R. Duchesne and Y. Gingras. *Histoire des Sciences au Québec.* Montreal: Boréal, 1987.

Dotto, Lydia. *Canada in Space.* Toronto: Irwin, 1987.

Richardson, R.A. and B.H. Macdonald, compilers. *Science and Technology in Canadian History: A Bibliography of Primary Sources to 1914.* Thornhill: HSTC Publications, 1987. This is a microfiche catalogue of over 58,000 entries by author, title and subject.

Jarrell, Richard A. *The Cold Light of Dawn: A History of Canadian Astronomy.* Toronto: University of Toronto Press, 1988.

Hodgson, J.H. *The Heavens Above and the Earth Beneath: A History of the Dominion Observatories Part 1 To 1946.* Ottawa: Geological Survey of Canada, 1989.

Thomas, Morley. *The Beginnings of Canadian Meteorology.* Toronto: ECW Press, 1991.

Ladell, John L. *They Left Their Mark: Surveyors and Their Role in the Settlement of Ontario.* Toronto and Oxford: Dundurn Press, 1993.

Canada Year Book is published annually by Statistics Canada, Ottawa, (formerly the Dominion Bureau of Statistics). Sections on science and technology include information on astronomical research, particularly at government institutions. A nine-page section "Astronomy in Canada" was prepared for the 1965 edition by Ian Halliday.

NOTES

Specific references for most of the facts in the text of *Looking Up* are not given since they can generally be traced by their date either to articles in the Society's publications or else to minutes of Council meetings. Therefore, references to factual information are provided only in two situations — (a) where the statements in the book are based on the Society's own publications or minutes but the approximate date of the source is not clear from the context, or (b) where sources other than the the Society's publications or minutes have been used.

Biographical information comes from many sources — especially the *Journal of the* RASC and *Canadian Who's Who*. (A very useful *Canadian Who's Who Index 1898–1984*, compiled by Evelyn McMann, was published by the University of Toronto Press in 1985.) The two General Indices to the *Journal*, the first covering the period 1890–1931 and the second from 1932–1966 can be used to find references to most people profiled in the book, to papers published by them and obituaries of them. Also reference can be made to "About our Authors," a regular feature of the *Journal* since 1964, and to lists of RASC Officers published in connection with annual reports. In addition, all people featured in the book who are still living have been given the opportunity to comment, revise, and add to their write-ups. As well, universities and other institutions with which people have been associated often have archives with helpful information. Therefore, except for C.A. Chant, H.S. Hogg, and L.W. Smith, for whom book-length biographies exist, biographical information has not been referenced here. Instead, the extensive notes and files made in the course of preparing the personal profiles for *Looking Up* have been placed in the RASC Archives.

Reports from RASC Centres are found throughout the *Journal* up to 1959 and subsequently in the annual *Supplement to the Journal* up to 1989. Since then, the Annual Reports of the Society and its Centres have been published as the April issue of the *Bulletin*. In addition, of course, are the newsletters of the Centres themselves which record the local activities. Except for retrospective articles, none of these sources are cited in the notes which follow.

In the following Notes *TAPST* is an abbreviation for *Transactions of the Astronomical and Physical Society of Toronto* (1890–99); *TTAS* is an abbreviation for *Transactions of the Toronto Astronomical Society* (1900–01); *JRASC* stands for the *Journal of the Royal Astronomical Society of Canada* and *NNL* for *National Newsletter* (1970–89).

PREFACE

Histories of the RASC include:

A.D. Watson, "Astronomy in Canada," *JRASC* 11, 47–78.

C.A. Chant, "The Fiftieth Anniversary of the Royal Astronomical Society of Canada," *JRASC* 34, 273–307.

J. Low, "The Early History of Amateur Astronomy in Canada and of the RASC," *JRASC* 59, 265–8.

R.J. Northcott, "The Growth of the RASC and its Guiding Mentor C.A. Chant," *JRASC* 61, 218–25.

H.S. Hogg, "The First Hundred Years of the Royal Astronomical Society of Canada," *Observer's Handbook 1990*, (Toronto: Royal Astronomical Society of Canada, 1990):1–3.

Histories of other societies include:

Ralph S. Bates *Scientific Societies in the United States* (3rd ed.), (Cambridge, Mass.: MIT Press,1965), includes information on the AAS, AAAS, ASP.

M. Rothenberg, "Organization and Control: Professionals and Amateurs in American Astronomy, 1899–1918," *Social Studies of Science* 11, 305–25 provides insightful analysis especially concerning the creation of the AAS.

The Journal of the AAVSO 15, no.2 (1986) contains a number of papers marking the seventy-fifth anniversary of the Association.

J. Vallières, "Mise au point sur l'A.G.A.A.," *Magnitude Zéro*, 1, nos. 1 and 2 (fev. & mai, 1977).

K. Bracher, "The Stars for All: A Centennial History of the Astronomical Society of the Pacific," *Mercury*, Sept./Oct. 1989.

The British Astronomical Association — The First Fifty Years (BAA *Memoirs* 42 part 1, 1989) and The Second Fifty Years (BAA *Memoirs* 42 part 2, 1990)

L.E. Dreyer and H.H. Turner, eds. *History of the Royal Astronomical Society*, vol. 1 (1820–1920); vol. 2 (1920–80) R.J. Tayler ed. (Oxford: Blackwell Scientific, 1987).

W. Stewart Wallace, ed. *The Royal Canadian Institute Centennial Volume 1849–1949*, (Toronto: RCI, 1949).

The Royal Society of Canada, *Fifty Years Retrospect Canada, 1882–1932*, (Toronto: The Ryerson Press, 1932).

The Royal Society of Canada, 1882–1957, (Ottawa: RSC, 1958).

H.S. Hogg, "Two Centenaries: The Royal Society of Canada and the Last Transit of Venus, 1882," *JRASC* 76, 362–70.

C.A. Chant, "The Golden Jubilee of the French Astronomical Society", *JRASC* 32, 321–7.

S.J. O'Meara, "Amateurs Triumph in Paris," *Sky & Telescope* 74, 480–3, contains information on SAF and IUAA.

CHAPTER 1

R.A. Stebbins, "Amateur and Professional Astronomers: A Study of their Interrelationships," *Urban Life* 10, 433–53. This study, carried out in the late '70s, was based on the Calgary Centre of the RASC and professional astronomers in western Canada.

R.A. Stebbins, *Amateurs: On the Margin Between Work and Leisure*, (Beverly Hills: Sage Publications, 1979).

R.A. Stebbins, *Amateurs, Professionals, and Serious Leisure*, (Montreal: McGill-Queen's University Press, 1992).

P.M. Kelly, ed., "Survey Results," *Bulletin* 3, no. 6. (December 1993). This is a summary of all the numerical data from 447 responses to a survey sent to all RASC members.

Information on the early names of the Society in TAPST for 1892, 91; TTAS for 1900, 11, 13; RASC *Selected Papers and Proceedings* (1902 and 1903), xi–xiv; *JRASC*, 11, 57-8; 25, 235-6; 28, 432.

Larratt Smith's diaries are in the Baldwin Room of the Metropolitan Toronto Library. The earlier years have been published as *Young Mr. Smith in Upper Canada*, ed. by Mary Larratt Smith. (Toronto: University of Toronto Press, 1980). Summer Hill is described in *Aristocratic Toronto* by L.B Martyn (Toronto: Gage, 1980).

CHAPTER 2

Information on the Society prior to incorporation in 1890 can be found in the general histories of the Society cited above and in TAPST for 1893, 123; for 1896, 90; for 1897, 107; TTAS for 1901, 83–4; *JRASC* 25, 233–4; 37, 382–3; 69, 249; 75, 281–8; 76, 26–34, 149–56, 235–44, and *The Centennial of the Toronto Centre*, by James Kemp, published in Toronto Centre's newsletter, 'Scope, 1968. (Portions of this were published in *JRASC* 75, 326–8.)

A photograph of the Draft Constitution written by R. Ridgeway in 1868 was published as the frontispiece (Plate VI) to *JRASC*, 35, no. 3 with an explanatory note on p. 92.

L. Herzberg, *A Pocketful of Galls* (not yet published), contains valuable information on the Recreative Science Association and the Natural History Society of Toronto. Some of the members in the 1870s were also pioneers in the Astronomical Society.

C.A. Chant *Astronomy in the University of Toronto: The David Dunlap Observatory*, (Toronto: University of Toronto Press, 1954) is part of Chant's unpublished autobiography. It and several boxes of correspondence are in the U. of T. Archives. J.F. Heard's oral history, also at the U. of T. Archives, is the source of the quote about Chant and also the source of some information about Heard himself.

CHAPTER 3

D.H. Armstrong, *Profile of Amateur Astronomy in Ontario* is a thesis submitted in fulfilment of the requirements of Geography 490 at The University of Western Ontario, 1985.

J.K. Beatty, "Who the Heck *Are* You, Anyway?" *Sky & Telescope* 74, 572, in which the main characteristics of a "typical" amateur astronomer are confirmed.

The quotation by Dan Brunton was published in Ottawa Centre's newsletter, *Astronotes* 26, (Dec. 1987): 9.

The Calgary Centre's Long Term Planning Meeting and the recommendations which came out of it were published in the Centre's newsletter, *Starseeker*, in January and April, 1985.

The quotation by Becca Stone is from *Astronotes* 19, no. 4 (April 1980): 1.

The quotation by Mary Lou Whitehorne is from "Where are the Women," *Bulletin*, 3, no. 3, (1993): 1. See also Couper, H. "Where Are All the Women Amateur Astronomers?" *Sky & Telescope* 75, p. 4. *Mercury* (the Journal of the ASP) 21, no. 1, is a special issue on women in Astronomy, primarily in the profession.

Michael Webb, *Helen Sawyer Hogg: A Lifetime of Stargazing* (Toronto: Copp Clark Pitman, 1991). This 28-page book is aimed at young readers.

For a broader discussion of amateur-professional co-operation in Astronomy see the special issue of *Sky & Telescope*, 74 (November, 1987). Also *The Contribution of Amateurs to Astronomy* (Proceedings of Colloquium 98 of the IAU, June 20–24, 1987, [Berlin: Springer-Verlag, 1988]) and the paper by Rothenberg cited above in the notes for the Preface.

CHAPTER 5

The story of Chant's visit to the Provincial Treasurer comes from *JRASC* 34, 296–7.

CHAPTER 7

W.J. Calnen, "Astronomy at King's College, Windsor, Nova Scotia," *JRASC* 74, 57 and G. Herzberg, "Astronomy and Basic Science," 74, 70.

Journal Citation Reports are published annually by ISI, Philadelphia.

The description of the *Handbook* as "the single most useful book ..." is from a review by Dave Bruning in *Astronomy*, (March, 1993): 90.

CHAPTER 8

D.H. Levy, *The Universe for Children: How Astronomy-Minded Adults Can Help Children to Love the Sky*, (Oakland, CA: Everything in the Universe, 1985).

D.H. Levy, *Astronomy Day: Bringing Astronomy to the People*, (Belmont: Sky Publishing Corporation). This 62-page handbook covers everything a group could need to make astronomy day a success. Published and distributed as a public service by the publisher.

E.R. Paterson, "Report on Observation of Total Eclipse by Boy Scouts, August 31, 1932," *JRASC* 27, 201–10.

J.M. Pasachoff ed. and J.R. Percy ed. *The Teaching of Astronomy*, (Cambridge: Cambridge University Press, 1989). Proceedings of the 105th colloquium of the IAU, 1988.

CHAPTER 9

J. Lankford, "Amateurs and Astrophysics: A Neglected Aspect in the Development of a Scientific Specialty," *Social Studies of Science* 11, 275–303 outlines the careers of seven amateurs in the last half of the nineteenth century who made important contributions to science.

Regarding the survey in 1984, see *Profile ...* , cited in the notes for Chapter 3.

J.R. Percy, ed., *The Study of Variable Stars Using Small Telescopes*, (Cambridge: Cambridge University Press, 1986). Proceedings of a symposium held at the University of Toronto, July 11–14, 1985 sponsored by AAVSO, IAPPP and RASC.

Some references documenting the long struggle for a RASC observatory include TAPST for 1890, 30; for 1892, 85; for 1893, 144; for 1894, 72, 79; for 1895, 62, 146; for 1896, 24– 5, 104, 119; for 1897, 120; for 1898, 44–60, 60-1; RASC *Selected Papers and Proceedings* (1902–3): 57–60; *JRASC* **34**, 304–6; **20**, 157; **8**, 197–8; **11**, 69; **13**, 283, 299–302, 469–70; **14**, 126; **22**, 199. A prospectus "The People's Observatory" is filed with the 1898 Council minutes, f433.

References to the establishment of the DDO include *JRASC* **22**, 199; **25**, 44–5; **29**, 265–313; **73**, 130–1; *Sky* **4**, 12–14; *Telescope* **1**, 58–65; *Popular Astronomy* **44**, 349–53, and especially *Astronomy in the University of Toronto* cited in Chapter 2. RASC Council minutes of December 26, 1923 refer to a meeting of the university Board of Governors on December 13, 1923, appointing a deputation to meet D.A. Dunlap to see if he would be willing to contribute a part or whole of the cost of an observatory. The quote by Chant at the end of the chapter comes from *JRASC* **29**, 273. Though Chant did not specifically say so, the lecture he referred to was apparently a special public lecture held on Saturday, May 28, 1921. (See RASC Council minutes of May 17, 1921.)

CHAPTER 10

A. Dyer, "A Question of Balance," *National Newsletter* (1979), L70.

CHAPTER 11

For reports of the Meaford Astronomical Society see Appendices to TAPST and TTAS (1895–1901) and TAPST for 1894, 8, 53; for 1895, 69; for 1896, 55; for 1899, 27. Orillia: TAPST for 1897, 90–2, 153; for 1898, 15. Owen Sound: RASC *Selected Papers and Proceedings* (1904), Appendix I. Seaforth: TAPST for 1898, 13, 15. Port Dalhousie: TAPST for 1898, 43. Simcoe: TAPST for 1897, 109; TTAS for 1900, 34 and Appendix; TTAS for 1901, 146–7. Tavistock: TAPST for 1896, 73. Uxbridge: TAPST for 1895, 35. Wiarton: TAPST for 1897, 30.

For Ottawa: F. Lossing, *Astronotes* article abstracted for "A Short History of the Ottawa Centre 0.4-m Telescope (1968–1980)," *JRASC* 75, 325–6. R. Lavery, "Astronotes, Etc. — A Very Short History", *Astronotes* 11, no.3, 3–5. R. Meier, "The History of a Centre Observatory," *Astronotes* 20, no.8, 6–9. (This latter reference is the source of the quotations on p. 183). F. Roy and K. Tapping, "A Brief History of the Indian River Radio Interferometer," *JRASC,* **84**, 260.

For Hamilton: See a special edition of the Centre newsletter: K.E Chilton, ed. *Sixty Years in Orbit* (1969); and R. Lang, "Astronomy in the Hamilton Area," *Papers and Records of The Head-of-the-Lake Historical Society* (Hamilton), no. 10 (1973): 53–8. Many relevant clippings and photographs are in the Archives of the Hamilton Public Library. References relevant to the early Hamilton Centre in TAPST for 1895, 141–2; for 1896, 24; TTAS for 1901, 148; *JRASC*, **3**, 3, 56, 61–4, 149, 235–6, 460. For information about the Planetarium see *JRASC* **53**, 149 and **86**, 286, and A. Vowles, "New projector brightens planetarium's future," *The McMaster Courier,* Sept. 29, 1992.

CHAPTER 12

For Winnipeg: See Phyllis Belfield, ed. *A History of the Winnipeg Centre 1911–1977*, (1977), and a very brief account in *NNL* (1978), L20–21. References to events prior to 1910 are TAPST **6**, 71; *JRASC* **3**, 326; **12**, 5; **65**, 206.

J. Hodges, "The Regina Astronomical Society, June 1955." *JRASC* **49**, 231–4 describes some of the local history from 1910–55. Roger Nelson prepared a further history in 1990 which is now in the RASC Archives. The Centre's newsletter, *The Regina Astronomer* for January, 1990, summarizes the use of the club's observatory 1955–89.

E.S. Keeping, *The Earlier Years of the Edmonton Centre RASC*, (typescript), was reprinted in five parts beginning in *Stardust* **25**, no. 2 (November 1979): 10, and is complemented by *A Short History of the Mathematics Department*, University of Alberta, by E.S. Keeping, (1971) which provides some biographical information particularly on Campbell and Keeping. Also see an illustrated article, "Edmonton's Observatories" in the Centre's newsletter, *Stardust*, (May, 1980): 12–3.

The formative years of the Calgary Centre were described in the Centre newsletter by W. Stilwell, "R.A.S.C. Calgary Centre History," *Starseeker*, **31**,(September 1990): 4.

For Victoria: Events prior to 1914, see *JRASC* **1**, 128; **2**, 13– 4; **3**, 3; **8**, 75–6; **11**, 77; **17**, 45; **20**, 159; **31**, 25; **79**, 98. R.M. Peters, "Summer Evenings with the Stars", *JRASC* **52**, 158–60 gives a historical summary of the Centre's very successful summer program and M.J. Enock, "Sixty-seven Years of R.A.S.C. Victoria," *JRASC* **75**, 329 provides a very brief history of the Centre.

CHAPTER 13

For Montreal: I.K. Williamson, ed. *Fifty Times Around the Sun: A History of the Montreal Centre RASC 1918 to 1968*, and a special fortieth anniversary issue of the Centre's newsletter, *Skyward,* (February, 1988). A short article is by D. Levy, "Times Past: A Look at Earlier Days of the Montreal Centre's Observatory," *NNL* (1985) L13. For matters prior to the formation of the Centre in 1918, see especially S.B. Frost, *Histoire des Sciences au Quebec*, cited at the beginning of the notes, *McGill University*, Vol. I, (Montreal: McGill University Press, 1980) and TAPST for 1892 **3**, 23, 63; **5**, 89; **9**, 56; *JRASC* **8**, 119; **10**, 145. Records of the Astro-Meteorological Association are in Montreal's McCord Museum (M19597 and M19599).

For Centre Francophone: F. Laforest, "Le Centre français de Montréal a dix ans," *JRASC* **52**, 95; J. Lebrun, "25 Joyeux Anniversaire," *Le Québec Astronomique* (mai, 1972) 2–3, and a special issue of *Le Québec Astronomique* (juin, 1978) commemorating the tenth anniversary of SAM. For information on the role of J. Edgar Guimont, see C.A. Chant, "The Ville Marie Observatory at Montreal", *JRASC* **36**, 137–9 and R. Marion, "Naturaliste honoré par la fondation Marie-Victorin," *JRASC* **47**, 203–6.

P-H. Nadeau, "The Quebec Centre of the RASC," *JRASC*, **38**, 313–9, gives a brief sketch of the origin and growth of the Centre. For further background information, see *JRASC* **35**, 43, 421–7, 433; **36**, 218; **37**, 111, 237. Newspaper articles on the formation of the La Fédération des Groupements astronomiques appeared in *L'Action Catholique* and *Le Soleil*, 5 juin, 1961.

Space precludes a proper bibliography of references to early observatories in the Maritimes (and elsewhere). See *The Cold Light of Dawn*, cited in the bibliography above and Paul A Bogaard, ed. *Profiles of Science and Society in the Maritimes prior to 1914*, (Victoria: Morriss, 1990).

A very brief history of the Halifax Centre is W.J. Calnen, "Halifax Centre," *JRASC* 75, 323–4. Also there is a condensed version of a talk by M. Cunningham, "Astronomy for Everyone," *Nova Notes* 16, no. 2 (1985): 85–90, reprinted in *Nova Notes* 24, no.3 (1993): 4–5. Reference to early attempts at starting a Centre are found in *JRASC* 31, 116; 46, 59–60; 49, 91 and 50, 80. More details on the planetarium are given by W.L. Orr, "The Halifax Planetarium," *JRASC* 58, 226–8.

CHAPTER 14

For London: H.R. Kingston, "The Twenty-Fifth Anniversary of the London Centre," *JRASC* 41, 45–8 and E. Clinton, *Astronomy London* (Oct. '90): 1(?)–12. References to the various examples given of early astronomical interest in south-western Ontario include P. Mozel, "The Woodstock College Observatory," *JRASC* 76, 168–180. For Barker's "observation" of T Coronae Borealis, see *The Observatory* 2, 224; *MNRAS* 27, 57 and *JRASC* 4, 256. Edw. Flood's telescope is described in correspondence of J.C. Watson at the University of Michigan. Some of Petit's correspondence is in the RASC Archives. For early, unsuccessful attempts to form a London Centre, see *TAPST* for 1896, 2; for 1898, 1–2, 12, 111.

For Niagara: See the Centre newsletter: *Whirlpool* (Sept. '88).

I.D. Howarth, "The Barr Effect: A Statistical Study," *The Observatory*, 113, 75–78 is the most recent paper on this subject and it contains several other references.

For Kingston: A.V. Douglas, "Astronomy at Queen's University," *JRASC* 52, 82–6 and L. Enright, "A Resurgent Kingston Centre," *JRASC* 75, 324–5.

For events described in the Thunder Bay area prior to the formation of the Centre, see *TAPST* for 1892, 11 and Minutes for 1892, f. 179, *JRASC* 48, 238 and 56, 127 and some newspaper clippings from 1958–59 in the RASC Archives.

James Kemp *The Centennial of the Toronto Centre*, (Toronto: Toronto Centre of the Royal Astronomical Society of Canada, 1968).

INDEX

Proper names of individuals, institutions, organizations or places to which only passing reference is made may not be indexed. If husband and wife or other family members are referred to in the text, only the surname may appear in the index.

Abbe, Cleveland, 20
Acadia University, 107
Adams, Walter S., 37, 101
Adamson, W., 16
"Advances in Astronomy", 100, 145
Advertising in RASC publications, 107, 110, 121
Affiliated societies, 10, 166, 177, 179
Aikman, Christopher, 220
Ainslie, Donald S., 66, 68, 253
Aitken, Robert G., 94
Alberta Science Centre, see Calgary Centennial Planetarium
Alberta Star Party, 114, 134, 213
Algonquin Radio Observatory, 55
Aller, Lawrence H., 255
Almanach graphique, 236, 240, 242-3
Altman, Morris, 36
Amateurs - interaction with professionals, 1, 50-55, 93-7 *passim*, 101, 105, 143, 165, 180, 277, 279, 280
American Association for the Advancement of Science (AAAS), ix-xi, 52, 71, 168, 269, 279
American Association of Variable Star Observers (AAVSO), ix, xiii, 4, 9, 49, 96-7, 99, 132, 139, 147-9, 151, 153 173-4, 200, 229, 232, 259, 262, 279
American Astronomical Society (AAS), ix, 52, 99, 104, 279
Andreae, Peter, 254
Annuaire Astronomique de l'Amateur, 236, 238
Annual Meeting, see Meeting, annual
Antoniadi, E. M., 101
Arbogast, E. C., x, xi
Archives (RASC), viii, 279
Argyle, P., 166
Armstrong, Dale, 35
Armstrong, William S., 56

Arndt, Carl, 264
Asbury, H. E. S., 194-5, 227-8
Ashenhurst, Peter, 66, 192
Asselin, Jean, 235
Associate membership, 12, 40
Association des Groupes d'Astronomes Amateurs (AGAA) - see also *Magnitude Zero,* ix, 126, 174, 238, 242-3, 279
Association of Lunar and Planetary Observers (ALPO), 139, 231-2, 262
Asteroids, xiii, 97, 110, 117, 142, 262, 264
Astro-Meteorological Association, 227
Astronomical and Physical Society of Toronto, see Toronto Astronomical Society
Astronomical Society of the Pacific (ASP), ix, 50, 83, 87, 98, 128, 141, 174, 262, 279
L'Astronomie, xiv
Astronomie Québec, 238
Astronomy Day, vii, 55, 121, 123-5, 184, 193, 216, 233, 245, 249, 258, 260, 263, 265, 267, 275-6
Astrophotography, 60-1, 125, 127, 142, 158, 164-5, 182, 185, 201-2, 205, 207, 214-5, 219, 232-3, 243, 264, 267
At-Home, see Meeting, Annual
Atkinson, (Reverend) Robert, 65
Atmospheric phenomena, see Meteorological observations
Attwood, Randy, 118, 120, 129, 276
Auclair, 40-1, 174, 238
Audio-visual equipment, 166
Aurora, 3, 20-2, 49, 103, 114, 132, 142, 146, 148, 153, 164, 174, 204, 207, 219, 228-9, 232, 240, 263
Awards - see also specific names, vii, 17, 100, 172, 207-8, 213, 215, 237-8, 243, 263, 274, 276

Bakeef, Alex, 211
Baker, John F., 261
Bakos, Gus, 211
Balam, David, 97
Ball, George, 219
Ball, (Sir) Robert, 37, 143
Balmer, H., 65
Barbados Astronomical Society, 247
Barker, H., 27-8, 65-6, 71
Barker, William, 251, 282
Barnard, Edward E., 37, 100, 141, 162, 269
Barr, J. Miller, 108, 148, 258-9, 282
Barters, 21
Bastien, Pierre, 35
Batten, Alan, 66, 91-2, 96, 98, 105, 114
Bawtenheimer, Daniel, 255-6, 258
Beals, xiii, 66, 103-4, 151, 182, 217, 244
Beattie, Brian, 66
Beatty, Samuel, 71
Beaudry, Gerard, 235
Beginner's Observing Guide, vii, 80, 112, 133, 263
Beginners - programs for, 184-5, 202, 211, 221, 231, 237, 247, 267
Belfield, 201, 281
Bell, Charles, 256
Bequests, see Donations
Bernier, (Captain) J, x, 186
Berry, Richard, 35, 136
Best, D., 66
Bilingualism, see French language
Binary stars, see Double stars
Binoculars, 149, 153, 158-9, 177, 188
Bishop, Mr and Mrs Robert C., 266-7
Bishop, Roy L., 52, 66, 98, 107, 111, 174, 246
Black holes, 111, 166
Blake, F., 108, 151, 177
Boivin, Albéric, 103, 160, 239
Bok, Bart, 37, 254, 262, 274
Bolduc, (Reverend) J., 240
Bonin, Roger, 235
Book reviews, 92, 98, 114
Borra, Ermanno F., 105
Bouchard, Joseph, 239
Boudreau, Jean-Paul, 239
Bouteiller, Robert, 257
Bowen, (Lt-Gov) J., 206
Boy Scouts, see Scouts and Guides
Boyer, Marc, 63, 66
Bradley, Robert, 256
Brashear, John A, 37, 179, 186, 188, 203, 217
Brassard, Pierre, 57
Bridgen, 153, 228
British Association for the Advancement of Science (BAAS), xii, 197, 269
British Astronomical Association (BAA), xii, 83, 87, 98, 109, 114, 132, 136, 139, 152, 279
Broadfoot, Raymond, 66, 270
Broadhead, Norman, 189-90

Brook, Eva, 45, 138
Brooks, Randall, 66, 245-6
Broughton, Peter, 66, 89
Brown, Ernest W., 151, 228
Brown, (Hon Justice) J., 203
Brown, Peter, 126
Bruce, William, 65, 187-8
Brunton, Dan, 39, 280
Bryce, G., 65
Brydon, 14, 60, 66 101, 106, 218-9
Buchanan, Daniel, 221
Budd, Eva, 27, 71, 87, 270
Buffalo, 260
Building, see Headquarters
Building fund, 27-8, 76-7, 80, 153, 268
Bulletin (formerly *National Newsletter*), 92, 96, 112-5, 133, 248, 272
Burke-Gaffney, 16, 244-5
Burland, Miriam, x, xi, 56, 143, 180-1
Buscombe, William, 56
Buteau, Amédée, 235
Butler, R., 166
By-laws, 6, 7, 10, 15, 17, 39, 108, 168-9, 178-9, 231, 236, 268, 280

Calendar (RASC), 224
Calendar reform, 101, 106, 181
Calgary, 30, 43, 50, 121-3, 125-7, 129, 134, 136, 142, 150, 157, 173-6, 203, 207, 210-3, 218, 280-1
Calgary Centennial Planetarium, 55, 126, 210-212
Calnen, William J., 59, 280, 282
Campbell, G., 56
Campbell, John W., 66, 101, 204, 206, 281
Campbellford High School, 56
Camps and campers, 122, 191, 194, 230, 237, 246, 249, 254, 263, 265, 267
Canada-France-Hawaii Telescope, 51, 54, 117, 242
Canada-Wide Science Fair, see Science fairs
Canadian Almanac, 108
Canadian Astronomical Society (CASCA), xii, 8, 52, 56, 96, 99, 100, 104, 113, 117, 277
Canadian Coast Guard, 40 174
Canadian Geological Survey, 22, 67
Canadian Institute, see Royal Canadian Institute
Canadian Institute of Theoretical Astrophysics (CITA), 56, 105

Canadian National Exhibition, see Exhibitions
Canadian Pacific Railways, 126
Canadian Planetarium Association, 54
"Canadian Scientists Report", 38, 100
Cannon, Annie, 100, 228
Carnegie Institute, 99
Carpmael, Charles, 2, 23, 25, 64-5, 160-1
Carrier, M-L., 160
Caswell, (Reverend) D. J., 177
Cayrel, Roger, 117
Centennial (RASC), vi, vii, 80
Centre francophone de Montréal, 50, 112, 121, 126, 135, 140, 230, 234-8, 281
Centres - see also specific names of Centres, 10, 11, 76, 78, 80, 89, 100, 112-3, 123, 163-5, 168, 175, 177-8 - representation 15, 33, 73
Ceravolo, Peter, 185
Chambers, Robert, 56
Chant, Clarence Augustus, 11, 12, 24, 25, 28, 54, 65, 75, 77, 86, 91-5, 98, 101-3, 106, 108-11, 119-20, 128-9, 135, 146, 152-3, 156, 162, 164, 178, 186, 188, 194, 202, 204, 240-1, 268, 279-81
Chant Medal, vii, 55, 59, 77
Chaos, 166
Chapman, David, 59
Charbonneau, Paul, 57
Charge-coupled devices (CCDs), 61, 140, 148, 167, 185, 202, 265
Chau, W. Y., 166
Chester, Len, 66, 131
Chilton, Kenneth, xiii, 29, 113, 129, 157, 192-3, 248, 260, 281
Chilton Prize, 55, 59
China, 190
Choquette, (Monseigneur) C-P, 64-5, 227
Chou, Ralph, 9, 66, 115, 118, 272-3
Church connections, 2, 27, 110, 115, 120, 153, 190, 244, 252, 258
Citation indices, 105
Clare, 21-3, 101
Clark, B., 137
Clark, Charles, 251
Clark, Cyril, 32, 66
Clark, Robert J., 222
Clarke Institute of Psychiatry, 30
Clement, Christine, 273
Clinton, Eric, 282

Clipsham, Kenneth, 65, 71
Clusters, star, 47, 99, 109, 150, 163, 165, 180, 205
Coallier, 237
Coats, Darby, 129, 143, 198, 228
Cobalt-Haileybury High School, 128
Cobban, James, 273
Coleman, A. P., 228
Colgrove, (Reverend) W., 60, 252
Colleges, 59, 130, 242-4, 247-8, 254-8 passim, 265-7, 272
Collins, John R., x, xi, 5, 15, 65, 71, 152-3, 188, 210, 213, 268
Collins, Zoro M., 15, 65, 71, 105, 134, 159
Columbia University, 110
Combs, R., 65
Comets - 3, 13, 36, 60, 94, 97, 109-114 passim, 120, 123, 141-2, 148, 162, 174, 180, 190, 194, 204-5, 208, 220, 224, 228, 232, 252, 262, 270 - Halley, 43, 97, 108, 118-9, 121, 123, 141, 158, 202, 204, 208, 228, 238, 246, 249, 260, 275
Comision, Paul, 147
Committee on the Co-ordination of Centre Activities (COCOCA), 133
Committees - see also specific names of Committees, vii, 15-7, 268
Compagnie Internationale Papier (CIP), 135
Computers, 4, 8, 125, 166-7, 202, 216, 233, 242, 276-7
Comtois, (Honourable) Paul, 241
Concours Annuel des Fabricants de Télescopes Amateur (CAFTA), 237-8
Connon, J. R., 140-1, 146
Conservation areas, see Parks
Constantinescu, Clinton, 165
Constitution, see By-laws
Constitution Committee, 68
Cooper, Grant A., 16
Copernicus Nicolaus, 96, 119, 201
Copland, J. A., 164
Cornell University, 146, 229
Cosmic Background Radiation, 36
Cosmic Ray Observatory, 210, 212
Cosmology, 117
Council - see also specific names of Centres, 11, 13-6, 95, 158, 267
Courses, 119, 128, 184, 206, 211, 213, 218-21, 231, 234-5, 237, 243, 249, 262, 267, 269
Covington, 63-4, 66, 103, 182, 184, 203
Craig, Steve, 136
Craters - see also Moon, 104, 111, 169, 262-3
Crease, A., 65
Creighton, Harlan, 66, 115
Creighton, Warren, 266
Crombie, W. T. B., 227, 233
Cronyn, Hume Observatory 252-4
Crooker, Arthur M., 222
Crowdis, Donald, 244
Cunningham, Clifford, 142
Curator, 72

Dalhousie University, 245-6
Dalton, F. Keith, 60, 66
Dance Hill Observatory, 264
Darwin, (Sir) George, 37, 102
David Dunlap Observatory, see Dunlap, D.
Debates, 217, 221
Deep sky, 57, 111, 194, 208, 254
Defence Research, 79, 184, 214
DeKinder, Frank, ix, 60, 139, 143, 229-230
Delorier, Guy, 239
Delury, Alfred and family, 65, 102, 119, 152, 188, 197
Delury, Ralph, x, xi, 65, 100-2, 128
Demers, Serge, 98
Denison, Napier, 216, 218, 221
Denning, William, 37, 101, 108, 143
Dent, 48, 65
Depression, see Economic conditions
DeRobertis, Michael, 56
Desjardins, Jacques, 235
Detroit, 38, 255
Devon Observatory, 208
Dewar, R., 164
Dick, Robert, 44, 185
Dickinson, Terence, 35, 111, 129, 258, 267
Dinosaurs, 117
Displays, 55, 120-3, 125, 171-2, 184, 191, 194, 199-201, 207, 211, 216, 219, 220, 230, 238, 243, 246, 249, 254, 258, 260, 263, 265-7, 270, 273, 276
Dobson, John, 224
Dodson, Steven, 125, 183
Dominion Astrophysical Observatory (DAO), ix, xii, 11, 33, 36, 38, 47, 53, 55-6, 77, 92, 94, 99, 104-5, 119, 121, 149, 166, 169, 206, 217-221 passim
Dominion Observatory (DO), x-xii, 7-11 passim, 44, 51, 56, 63, 67, 76- 7, 92, 94-5, 99-104 passim, 121, 127, 132, 141, 145, 153, 156, 164, 166, 169, 179-82, 185, 227, 278
Dominion Radio Astrophysical Observatory (DRAO), 8, 51, 56
Donaldson, James C., 140, 147, 151
Donations, 7, 17, 25, 33, 37, 52-3, 73, 78, 80, 87, 120, 230-1
Donnelly, Alfred M., 230
Double stars, x, 11, 21, 92, 105, 109, 113, 132, 147, 150, 205, 220-1, 259
Douglas, Allie Vibert, 29, 46, 66, 128, 181, 227, 229, 261, 282
Dow Planetarium, 233, 238
Draper, John W., 59
Drew, (Honorable) George, 63, 66
Drew, Robert, 208
Drolet, Maurice, 60
Drukier, G. A., 56
Duchow, Louis, 46
Duff, James, 203, 214
Duffie, John, 229
Dufour, Yvon, 242
Dumont, Alfred, 241
Duncan, Martin, 166
Duncan, Robert, 72, 84
Dunlap, David, and Observatory (DDO), viii, xii, 24, 28-30, 47, 53, 56, 77-9, 86, 93-4, 99, 113-4, 118, 121, 136, 159, 162, 260, 269-275 passim, 281
Dunlop, (Honorable) William J., 63, 66
Dupuy, David, 98
Duric, Nebosa, 56
Dustheimer, Oscar Lee, 129
Duty, import, 15
Dyer, Alan, 35, 163, 281
Dyson, (Sir) Frank, 37, 77, 199

Eclipses - see also lunar and solar eclipses, 111, 123
Economic conditions, 42, 48-9, 75, 154
Eddington, (Sir) Arthur, 46, 100, 166, 181, 209, 269
Edison, Thomas, 99
Editors, 8, 9, 13, 15, 72-79, 91-3, 111, 115
Edmonton, 15, 16, 33, 54, 56, 116, 121, 132, 134, 142, 145-6, 155-8, 163, 165, 173-5, 178, 190, 204-210, 212, 218, 281
Edmonton Space Sciences Centre, 33, 174, 208-9
"Education Notes", 9, 98-9, 107, 128
Edward VII, 5
Edwards, B., 244
Einstein, Albert, 100, 152-3
Elections, 169
Electronics - see also Computers, 182, 184, 193
Ellis, David, 154
Ellis, John, 12, 55, 65
Elmwood Observatory, 187-8
Elvins, Andrew, xiv, 8, 19-23, 65, 134-141, 151, 164
Endowment Fund, 44, 74, 80
English Mechanic, 100, 141, 159, 177
Enright, Leo, 66, 112, 125, 146, 263, 282
Evans, Robert, 53, 196, 219, 220
Executive Committee, 14, 28
Executive Secretary, 15, 27, 29, 30, 68, 70, 72, 168
Exhibitions, 121, 124, 159, 222, 243, 272-3
Expulsion, 35
Extra-terrestrial life, 13, 221

Fahrner, J. Campbell, 213
Fédération des Astronomes Amateurs du Québec (FAAQ), ix, 238
Fédération des Groupements Astronomique du Québec, 242, 281
Fédération Québecoise du Loisir Scientifique (FQLS), 238
Fees, membership, 11, 45, 73-6, 178, 232
Ferguson, (Honorable) G. Howard, 63, 65
Fernie, J. Donald, 66, 114
Fidler, Marie, 29, 30, 36, 66, 92, 210
Films, 77, 88, 96, 125, 167, 199, 207, 211, 217, 222, 247, 254, 255, 257, 259, 269, 271
Finances, 14, 17, 68-70, 73-81, 112, 168, 225, 267
Findlay, William, 65, 110, 143, 190, 261
Fireballs, see Meteors
Fitzgerald, Diarmid, 30, 66, 271
Flammarion, Camille, xiv, 37, 101, 126
Flood, Edw., 251, 282
Floods, 202
Flynn, (Honorable) Jacques, 241
Forsyth, H. C. B., 210, 221
Fortier, George, ix, 148
Fortin, Raymond, 240
Fox, Herbert, 16
Freeman, Rosemary, 29, 72
French, C. A., x-xi
French language, 6, 15, 35, 40, 78, 92, 95, 103, 112, 114, 234, 239, 242, 246
Fuller, Edna, 28
Fund-raising, 192, 201-2, 207, 209-212 passim, 216, 221, 224, 233, 240, 247, 261
Funds - see also specific names, 76-81

Gaherty, Geoffrey, 141
Galaxies - see also Milky Way, 103, 111, 147, 150
Galichan, Lionel, 241
Gamba, (Maestro) Piero, 202
Garneau, 60, 66, 141, 226-230, 234
Garrison, Robert, 174
Gartlein, C. W., 49, 146, 148, 229
Gaspar, Michael, 56
Gelinas, Marc, 114, 140
Gemini Project, 55
General Assemblies (GA), vi, xii, 7, 9, 12, 15, 40, 55, 78, 81, 100, 114, 130, 163, 169-175, 183, 203, 210-1, 213-4, 220, 231, 245, 249, 260, 266, 271
Geodetic Survey of Canada, 127
Geophysics - see also Magnetism, Solar-Terrestrial, Seismology, 6, 67, 104, 180, 253
George, Douglas, 44, 60, 142, 183
Georgia State University, 56
Gies, Douglas, 35, 56
Gifts - see Donations
Gilchrist, Lachlan, 65, 67, 168
Gingerich, Owen, 37, 117
Gingrich, Curvin, 197
Girl Guides - see Scouts and Guides
Giscard d'Estaing, Valery, 234
Glenlea Observatory, 196, 200-2
Goddard, Walter T., 188-9
Gold Medal, 53, 55-7, 77
Goldberg, Leo, 37, 255
Gonzales Observatory, 216, 218
Good, Charles M., ix, 139, 151, 230, 233
Goodacre, Walter, 14
Gordon Southam Observatory, see Southam Observatory
Goth Hill Radio Observatory, 64
Government of Canada - see also Grants and specific institutions, 2, 24, 63, 75-6, 99, 123, 127, 152, 154, 179, 228, 241, 254

Government of Ontario - see also Grants, 3, 63, 74-5, 128
Government of Quebec - see also Grants, 64, 156, 236, 239-42
Gower, J. F. R., 166
Grand Valley Astronomers, 264
Grant, Jack, 210
Grant, (Honorable) Robert Henry, 63, 65
Grants - see also Special Projects, xiii, 63, 73-6, 128, 154, 174, 178, 197, 202, 212-3, 236, 239-240, 268, 274
Gravitational lensing, 166
Gray, Annie, 45, 48, 143
Gray, Richard O., 57
Gray, Robert A., 65, 69, 86, 268
Green, 16, 66, 115, 165, 244, 259
Gretna, Manitoba, 201
Grey, Mary, 66, 79, 127
Groundwater, Randy, 258, 265
Guelph, 178, 194-5
Guimont, J. Edgar, 234-5, 281
Gupta, Rajiv, 224

Haas, Walter, 101, 139
Hale, George E., 37, 136, 137
Haley, Donna, 36
Haliburton, Robert G., 106
Halifax, 15, 54, 59, 121, 128-9, 134, 151, 160, 163, 166, 174-5, 243-7, 282
Hall, Cathy, 74
Hall, G. Harper, 16
Hallam, Cyril, 255-6
Halliday, Ian, 56, 66, 91, 100, 103, 145, 171, 264, 278
Hamilton, xiii, 50, 54, 59, 110, 115, 121, 123, 125, 140, 156, 165, 168, 172-8 passim, 186-195, 253, 259, 281
Handbook, see *Observer's Handbook*
Harcourt, (Honorable) Richard, 63, 65
Hare, W. Almon, 256
Harper, William Edmund, x, xi, 56, 77, 94, 98, 109, 129, 166, 186, 197, 217, 253, 268, 278
Harrington, Mary Anne, 39
Harrower, George A., 181, 261
Hartley, Charles, 14
Harvey, Arthur, 3, 65, 75, 91, 165, 177
Hassard, Albert, 140, 159, 268
Haurwitz, B., 106
Headquarters - see also Property, 10, 25-33, 68, 72, 75-8, 80, 84-9, 235, 237-8, 270

Heard, John F., 16, 28-9, 43, 66, 78-9, 94, 156, 181, 244, 269, 280
Heaton, Mona, 16
Helm, Walter, 53, 78-9
Henderson, J. P., x-xi, 180
Henroteau, F. C., 166
Henry, Richard, 56
Herzberg, Gerhard, 64, 66, 280
Hibben, J. P., 14
Hicks, John, 136
Higgs, Lloyd A., 8, 9, 57, 66, 79, 91, 93, 96, 103, 204
Hildebrand, Alan R., 117
Hincks, William, 19
Hinks, Arthur, 98
Historical Committee, 17, 64, 88
History, vii, 52, 67, 79, 89, 98, 100, 107, 114, 127, 169, 200-1, 209, 231, 233, 240, 244-5, 278-82
Hladiuk, Don, 43, 129, 212
Hobby shows, 121, 195
Hodges, 203
Hodgson, Ernest, 56, 66, 128
Hogg, Frank, 44, 47, 53, 65-6, 77, 93-4, 109, 111, 120, 129, 195, 269
Hogg, Helen Sawyer, ix, xiv, 14, 16, 33, 37, 41, 47, 52, 56, 58, 66, 94, 98, 127, 129, 173, 244, 255, 264, 269, 279, 284
Hogg Lecture, 55, 100, 117
Holden, Edward S., ix, 37
Honorary Members, 37, 100
Honorary President, 13, 63-6
Horning, Harry, 16, 28, 65, 69, 76
Horton, A., 134
Hossack, William R., 28, 66, 170, 173
Howard, John G., 160
Howell, David J., 21-2, 65, 72, 134, 152, 166
Howell, Eva C., 134
Howell, John, 113, 211
Hubble Space Telescope, 117
Hube, 66, 208-9
Hughes, Mrs Patrick, 257
Hull, Gordon Ferrie, 103
Hume Cronyn Observatory, see Cronyn
Huneault, Herbert, 256-7
Hunter, Andrew F., 22, 65, 67, 146, 268

Incorporation, 23, 25, 220, 231, 236, 261, 270, 274
Indian River Observatory, 44, 142, 183-4
Ingalls, Albert G., 159

Innanen, Kimmo A., 117
Institut Astronomique, 234
Institut Canadien, 227
Institute of Astronomy, Cambridge, 56
Institutio Solar, 3
Insurance, 125, 178
International Astronomical Union (IAU), xii-xiii, 9, 52, 92, 154, 182, 198, 262, 263 - National Committee for Canada, xii, 52, 95, 137, 181, 241
International Geophysical Year (IGY), 144, 200, 219
International Polar Year, 204
International Union for Co-operation in Solar Research, xii, 202
International Union of Amateur Astronomers (IUAA), xiii, 113, 192, 279
Ireton, H. J. C., 71, 103
Irish Astronomical Society, 270
Irwin, John B., x-xi
Ishii, Shigeo, 127
Iwanowska, Wilhelmina, 37, 117

Jackson, Walter E. W., x-xi, 65, 141, 143
Jacobsen, 69
Janisse, Margaret, 257
Jedicke, Peter, 59, 129, 130, 255
Jenkins, G. Parry, 187
Jesuits, 244
Johns, 66, 190, 214
Johns Hopkins University, 56
Johnson, Gus, 149
Johnson, Hugh, 139
Joint meetings (of Centres), 203, 207, 212, 220, 224, 231, 234, 253, 255, 260
Jones, F. Shirley, 56, 143, 269
Jones, V. C., 199
Journal of the RASC, ix, 8, 27-8, 30, 38, 72, 76, 79, 86, 91-107, 131, 133, 143-6, 151, 168, 178, 181, 236, 239
Junior members, see Youth
Jupiter, xiii, 20, 109, 139-141, 151, 190, 193, 232, 275

Kaitting, Murray, 148, 264
Kalbfleisch, Moody, 133
Kalium Observatory, 204
Keeping, 204, 206, 209, 281
Kelly, Patrick, 115
Kelsch, Rudy, 275
Kelvin, Lord, xii
Kemble, Lucien, 60

Kemp, James, 282
Kennedy, E. J. Arthur, 57, 65, 71, 268
Kennedy, Gerald, 264
Kennedy, J. Edward, 16, 66, 79, 114, 210, 214, 259
Kenney-Wallace, Geraldine, 191
Ketchum, Jesse, 4, 140, 159, 270
Kinahan, Blake, 56
Kindy, Al, 259, 261
King, Harold, 210
King, Henry C, 66
King, Mary, 245
King, William Frederick, 8, 10, 63, 65, 152, 169, 179, 186
King's College, Windsor, NS, 59, 244
Kingston, 46, 125, 166, 171, 175, 261-4, 282
Kingston, George Templeman, 2, 19
Kingston, Harold R., 29, 50, 65, 71, 188, 198, 251-5, 282
Kirkaldy, June, 224
Kirkwood, Daniel, 37
Kitchener-Waterloo, 123, 142, 175, 264-5
Klotz, Otto, 63, 65, 152, 179-80, 197, 227
Knight, Gerry, 129
Knowledge, 164, 177
Koenig, Ray, 264-5
Kormendy, John, 56
Krosney, William, 43
Kubrick, Stanley, 167

Lafontaine, Gérard, 239
Laforest, Fleurange, 236
Lakehead Astronomical Society, 266
Lakehead University, 266
Laval University, 15, 156, 238-240, 242
Lectures, public - see also Hogg, Northcott, Speakers, 117-121, 125, 184
Leduc, Joseph, 235
Lee, Henry and Mamie, 66, 257
Lee, Man Hoi, 56
Leeds Astronomical Society, 202
Leger, (Governor General) Jules, 47
Lemaitre, Georges, 228
Lemay, Damien, 60, 61, 66, 141, 146, 158, 174
Lemieux, Pierre, 236
Leonard, Frederick C., 12, 103
Leonard, Peter, 57
Leonard, Reuben Wells, 86
Lethbridge Astronomical Society, 134

Levy, David, 49, 57, 60, 61, 97, 114, 130, 142, 174, 233, 263, 280-1
Librarian (for Asst. Librarian, see Executive Secretary), 13, 25, 68, 72
Libraries, Centre, 11, 129, 139, 184, 205, 209, 230, 233, 235, 237-9, 248, 264
Library - RASC, 6-7, 10, 14, 17, 25, 27-9, 31, 73, 76, 78, 83-89, 270
Lick Observatory, ix, 83, 117, 152
Life Membership, 12, 36, 76
Light pollution, 4, 123, 125, 134-6, 183, 207, 213, 230, 232, 233, 242, 258, 274-5
Lindsay, Thomas, 27, 65, 91-2, 152, 177
Ling, G. H., x, xi
Link, Theo. A., 211
Litchinsky, Sam - see also Fidler, 29-30, 210, 213
Loblaw, Robert, 139
Locke, Jack L., 16, 51, 66, 156
Lockhart, Robert, 16, 66, 68, 96
Lockyer, Norman, 20
Lodge, Oliver, 198
Loehde, 33, 66, 205, 209
Loewy, Maurice, 37, 83
London, xiii, 43, 50, 126, 129, 130, 140, 156, 174-5, 204, 250-5, 265, 282
Longworth, Gerald, 28
Lossing, Frederick P., 133, 182, 281
Low, James, 139, 141, 172, 279
Lumsden, 3, 4, 8, 23, 65, 74, 91, 119, 123, 135, 139-141, 147, 151-2, 159, 161
Lunar eclipses, 45, 151, 207, 224, 235, 243, 260, 267, 270
Lunar observations, see Moon

Maclean, M. (Hugh), 259-260
MacLean, Neil Bruce, 197-8
MacLennan, A. R., 16
MacLennan, Dan, 146
MacMillan Planetarium, 174, 214, 222, 225
MacRae, Donald A., 56, 143, 181
Madore, Barry, 117, 174
Magee, Gordon, 66
Magnetic Observatory, see Meteorological Observatory
Magnetism (terrestrial), x, xii, 3, 15, 16, 21, 99, 152-3
Magnitude Zéro, 238, 243
Mainguy, Roland, 241
Maktomkus Observatory, 107

INDEX • 285

Mall Displays, see Displays
Mallory, Wilfred S., 192
Mammoth House, 20
Manitoba Museum (and Planetarium), 96, 158, 199, 202
Manitoba Star Convention (MASCON), 134, 200
Marconi Company, 228
Marks and Spencer, 61
Marmet, Paul, 242-3
Mars, xiii, 44, 103, 117, 123, 133, 137, 139, 140, 151, 179, 180, 193, 212, 217, 228, 232, 239, 275
Marsh, 108, 152, 154, 156, 185-190, 193
Marshall, Roy K., 54
Martin, Peter G., 56
Maryland Academy of Sciences, 240
Massey, Hart, 161
Matte, Joseph, 150, 239
Matthews, Barry, 245
Max Planck Institut, 56
Maybee, J. Edward, 65, 71, 134, 152
McArthur, (Honorable) D., 63, 66
McBain, Miss, 143
McCallion, William J., 191, 194, 260
McCaw, Craig, 60
McClenahan, William S., x-xi
McClung, H. S., 202
McCrae, Linda, 142
McCullough, Ernest, 211
McCurdy, (Honorable) Arthur W., 216
McCutcheon, Mark, 56
McDiarmid, R. J., x-xi, 56
McDonald, Jean, 166
McEachern, Miss, 134
McFarlane, Alfred, 65
McGill University, 46, 197, 227, 230-1, 233, 281
McGregor, Ian, 100, 113, 115, 118, 273, 275
McKellar, Andrew, 36, 66, 120, 206, 211
McKone, E H, 71
McLaughlin Planetarium, 17, 81, 115, 126, 173, 254, 260, 271-4
McLennan, Ian, 54, 205-6, 273
McLennan, John C., 228
McLeod, Clement H., 227
McMaster University and Planetarium, 14, 54, 110, 115, 190-1
McNabb, Dale, 223-4
Meaford, 177, 281

Meanook, Alberta, x
Mechanics' Institutes, 19, 20, 227
Media, 119, 123, 128-9, 210, 230, 235, 241, 245, 249, 258
Meetings, Annual, 12, 13, 27-8, 30, 68, 70, 100, 167-9, 216, 243
- Centre - see also individual Centres, 10, 163-7
Meier, Rolf, 60-1, 142, 183, 185, 281
Meiklejohn, 210
Membership - see also Associates, Honorary, Life, New, Senior, Unattached, Women, Youth - in general, 12, 17, 35-61, 76 - Certificates, 55, 57
Memorial University, 247, 249
Mercury, 60, 71, 113, 125, 140, 259
Mercury, ix
Meredith, Edmund, 12, 68
Messier - Certificates, 49, 55, 57, 150, 246
- Club, 230, 232, 235
Meteor(s), 7, 44-5, 49, 79, 103, 107-8, 110, 126-7, 131-2, 139, 143-5, 148, 153, 180-2, 184-5, 199-200, 203, 208, 214, 219, 224, 228-9, 232, 235, 240-1, 243, 253-4, 263, 270
"Meteor News", 44, 98
Meteorite(s), 44, 60, 72, 97, 104, 126, 132, 142, 145, 208, 215, 243, 252-3
Meteorite Craters, see Craters
Meteorological - observations, 22, 106-7, 146, 204-5
- Observatory, x, 2, 19-23, 25, 151, 161, 185, 227
- Office, 15, 108-9
- Service, x, 2, 3, 87, 92, 99, 164, 177, 216, 221, 278
Milky Way, 103, 117, 123
Miller, 14, 21-3, 65, 101, 136-7, 140-1, 147, 164, 168, 177, 219, 251, 268
Millman, Peter M., - see also Endowment Fund, x, xi, 43-4, 49, 56, 65- 6, 79-80, 87, 93-4, 98, 103, 107, 128-9, 132, 143-5, 148, 156, 159, 171, 180-2, 199, 203, 211, 213, 229, 231, 239, 253, 261, 269
Milton, Earl, 60, 132, 146, 205
Misericordia Hospital (Winnipeg), 200
Mitchell, George, 166
Mitchell, Samuel Alfred, 37, 143, 188, 253

Moffat, Anthony, 56
Molczan, Ted, 145
Mont Mégantic Observatory, see Observatoire
Montreal - see also Centre francophone, ix, xii, xiii, 43, 46, 49, 64, 121-7 *passim*, 132, 139-146, 151, 153, 156-7, 166-178 *passim*, 194-5, 206, 226-234, 281
Montrose, Carmen, 256
Moon - see also Lunar eclipses and Occultations, 4, 20-1, 49, 59, 104, 108-113 *passim,* 132-3, 136-9, 150, 167, 185, 205, 208, 232, 235, 237, 270
Moonwatch program, see Satellites
Moore, Patrick, xiii, 37
Moore, W. J., 21
Morgan, Francis, 140
Morrison, Warren, 60, 149
Morton, Donald C., 56, 69
Motherwell, Robert M., 56, 141
Mott, Stanley, 184
Mount Kobau - see also Queen Elizabeth II Telescope, 114, 134, 208, 213, 224
Mount Wilson Observatory, xii, 202
Mozel, Philip, 59, 66, 282
Musson, 12, 65, 134, 148, 188

Nadeau, Paul-H., 60, 149, 158, 160, 239- 242, 281
Name (of the Society), 4-5, 280
National Museum of Science and Technology, 127, 161, 169, 185, 245
National Newsletter (NNL) - see *Bulletin*
National Research Council (NRC), xiii, 7, 44, 56, 64, 76, 132, 145, 158, 182, 215, 253 - Associate Committee, xii, xiii, 98, 242
National Science Foundation, 105
Natural History Societies, 3, 23, 216, 227, 247
Natural Sciences and Engineering Research Council (NSERC), 76
Nature, 3, 100
Naubert, Jean, 60, 234-5
Nautical Almanac, 107-8, 151, 183
Navigation, 40, 110, 199
Nebulae - see also Clusters, Galaxies, 18, 109, 117, 150, 166, 230, 270
Nelson, Robert, 210, 259
Neptune, 209
Neutrinos, 59, 166
New member programs, 231, 275-6

New York Central Railroad, 134
Newcomb, Simon and Newcomb Award, 37, 55, 59
"News and Comments", 100, 102
Newsletters, see also *Bulletin* or Specific Centre, 11, 178
Newspapers - see also Media, 3, 20, 44, 47, 71, 77, 94, 121, 127-9, 180-1, 206, 211-2, 214, 217-8, 227-8, 230, 240, 246, 248, 255, 260, 267, 270
Newton, Jack B., 60-1, 96, 133, 147, 149, 201, 273
Niagara, 80, 121, 126, 129, 150, 173-5, 192, 258-61, 282
Niagara Frontier Council of Amateur Astronomical Associations (NFCAAA), xiii, 192, 260
Noctilucent Clouds, see Meteorological Observations
Noël de Tilly, Roland, 126, 237
Norris-Elye, L. T. S., 199
North Mountain Observatory, 182-3
North York Astronomical Association, 133
Northcott, Ruth J., x, xi, 16, 57, 66, 79, 91-100 *passim,* 107, 111-3, 143, 153, 160, 214, 248, 255, 259, 261, 269, 278-9 - Fund, 78, 243 - Lecture, 78, 100
Northwestern University, 56
Noseworthy, Tom, 230
"Notes" and "Notes and Queries", 98-9
Nova East, 134, 246
Nova Scotia Museum of Science, 54, 244-6
Nova Scotian Institute of Science, 101
Novae, 49, 98, 132, 137, 148-9, 152-3, 158, 229, 232, 251
Nugent, D. Bert, x-xi
Nyberg, Gus, 258

O Canada, 101
Oakwood Collegiate, Toronto, 69
Objectives of Society, 6
O'Brien, Kennedy, xiii
Observatoire de Mont Mégantic, 233, 238
Observatories - see also specific names, 4, 11, 13, 80, 111, 125, 160- 2, 167, 178, 185-6, 192-3, 199-205, 208, 219, 223, 228-233, 237, 239-242, 246, 257, 261, 270, 275, 280

Observers Groups, 4, 64, 132-3, 142, 182, 205, 207-8, 222-3, 229, 254, 260, 269-271, 275
Observer's Handbook, 29, 44, 76, 79, 92-3, 106-112, 151, 168, 178, 181-2, 204, 211, 228, 236, 247, 253
Observing programs, xiii, 112, 132-3, 139, 141, 201, 205, 207, 220
Occultations, 45, 96, 108-11, 113, 131, 139, 140, 150-1, 179, 183, 202-3, 207, 212, 228, 232, 240, 248, 254, 263
O'Connor, John L., x-xi
Office, see Headquarters
Officers - see also specific positions, 25, 63-72, 100
Okanagan Astronomical Society, 133
Oke, J. B., 16, 66, 87
O'Neill, Gerald, 174
Ontario Field Naturalists, 253
Ontario Historical Society, 22
Ontario Hydro, 60, 67
Ontario Science Centre, 194, 254, 275
Open Houses at Observatories, 121, 123, 180, 194, 200, 204, 215-220 *passim,* 230, 232, 240, 264
Optics, 71, 107, 111, 202
Order of Canada, 47, 242
Organizations (outside RASC) - see also specific names, 120, 181, 198, 218, 220, 227, 230, 234, 235, 260, 265
Orillia, Ontario, 177, 281
Orr, 192
Ottawa, ix, xiii, 7, 8, 10, 44, 51, 63-4, 67, 102, 112, 121, 125, 127, 133, 142-3, 145, 151, 156-7, 166-7, 172-5, 178-185, 224, 245, 281
Owen Sound, Ontario, 281

Page Charges, 106
Parks, 122-3, 133-4, 162, 194, 200, 204, 208, 212-3, 223-4, 248-9, 254, 257-8, 260, 267, 270, 273, 275
Parsons, J. C., 53
Paterson, E. Russell, 78, 80, 127, 143, 229, 280
Paterson, John A., 2, 8, 12, 45, 65, 71, 120, 134, 139, 186, 188
Patterson, Gordon, 214-5
Patterson, John, 99
Patterson, Shirley, see Jones, S
Pawling, J., x-xi
Pearce Joseph A., 11, 14, 65-6, 119-120, 151, 190, 206, 218, 255

Pearce, Raymond, 270
Pelletier, Arthur, 273
Penchuk, Dale, 133
Penitentiary, 262-3
Pepin, (Honorable) Jean-Luc, 54
Percy, John R., ix, 9, 50, 56, 66, 98, 107, 11, 113, 115, 128, 280
Petellier, Eugène, 50
Peterborough, 178, 185-7
Peters, Robert and Mrs, 218-9, 281
Peters, William and Celeste, 201, 273
Petit, H., 251, 282
Petrie, xii, 35-6, 38, 53, 66, 100, 103, 148, 166, 173, 214, 217, 231
Photographic Zenith Telescope (PZT), 7, 173
Photography of celestial objects, see Astrophotography or name of specific object
Photometry, 4, 53, 114, 125, 142, 147, 166, 181-2, 197, 202, 264
Pickering, Edward C., 37, 83, 148
Pickering, William H., 37, 101
Pike, Robert, 136, 273, 275
Pitcairn, Douglas, 246
Planetarium Association, 174
Planetarium Committee, 15, 55
Planetariums - see also specific names, 53-5, 96, 111, 113, 115, 121, 125, 130, 190-1, 194-5, 199, 206, 211, 234, 244-6, 248, 257, 259, 261, 271-2
Planets and their satellites - see also specific names, xiii, 4, 20, 49, 111, 123, 132, 136-7, 139-141, 200, 202, 205, 208, 263, 270
Plaskett, 10, 11, 29, 36-7, 47, 53, 64-5, 100-1, 103, 117, 119, 169, 179-80, 198, 210, 216, 228
Plaskett Medal, 55, 57
Pluto, 141, 145, 258
Poisson, Eric, 57
Pope, (Sir) Joseph, 5
Popular Astronomy, 3, 87, 99, 108, 177, 197
Port Dalhousie, Ontario, 281
Porter, (Honorable) Dana, 63, 66
Povenmire, Harold, 248
Powis, Leslie, 66, 192-4
President, 2, 3, 13, 64-68, 167
Printers, 105-6
Prizes - see also Awards, 120, 126, 183
Proctor, 117, 152, 198, 228
Project Gemini, 55
Project Z[ubenelgenubi], 130

Promotional items, 74, 183, 238
Property Committee, 17, 30-2, 80
Proulx, Robert, 239
Public Awareness - see also Astronomy Day, Displays, Lectures, Media, Open Houses, Planetariums, Star Nights, 1, 9, 52, 117, 127, 184, 191, 245, 248, 254, 263, 274-5
Publications - see also specific names, 6-7, 14, 17, 28, 33, 52, 73-5, 90-115, 237, 243
Publicity, 119, 120, 123-4, 128, 187, 189, 211, 224, 230, 238, 243, 268, 273
Pulsars, 166
Pursey, 21-2, 25, 48, 65, 84, 120, 134-6, 164, 177
Purton, Christopher, 56
Pyke, A. J., 214
Pyne, (Honorable) Robert Allan, 63

Quasars, 51, 166, 183
Quebec, ix, 7, 123, 128-9, 142-3, 150, 156-8, 174-5, 281
Québec Astronomique, ix, 40, 236-8
Québec Téléphone, 158
Queen Elizabeth II Planetarium, 33, 54, 116, 145, 173, 206-7
Queen Elizabeth II Telescope, 38, 54
Queen's University, 46, 145, 166, 261-3

Racine, René, 117
Radio - see also Media, 14, 44, 60, 64, 77, 119, 129, 153, 164, 194, 198, 200, 213, 217-8, 220, 228, 230, 251
Radio Astronomy, 8, 51, 64, 80, 103, 111, 183-4, 194, 232, 265, 281
Ramsay, Vernon, 112, 132, 270-1
RASCANA - 142
Reber, Grote, 37, 79-81
Recession, see Economic conditions
Recorder and Recording Secretary, 13, 70, 72
Recreative Science Association 23, 280
Reed, E. Baynes, 216
Reeves, Hubert, 117, 235
Regina, 123, 143, 175, 202-4, 214, 281
Reinhardt, Carl, x, xi, 15, 16, 28-9, 53, 78, 87
Relativity, 67, 102, 152, 166, 204
Rent, 25, 27, 73-4, 80

Reports of RASC Meetings, 70, 100
Reprints, 106
Reynolds, Peter, 56
Ridout, John G., 65
Rivard, Father, 200
Riverside Telescope Makers, 133, 208
Robert, Frère, 234
Roberts, S., 22
Robertson, Elizabeth, 24
Roeder, Robert, 166
Roger, David, 214
Rogers, Christopher, 56
Rosebrugh, David, ix, 149, 258
Ross, (Honorable) George W., 63, 65, 74
Ross, Hugh, 149
Rothney Observatory, 212
Roy, Frank, 281
Roy, Jean René, 103, 128
Royal Astronomical Society (RAS), xii-xiii, 2, 45, 83, 152, 279
Royal Canadian Institute (RCI), xiv, 9, 19, 20, 23, 25-7, 84, 161, 165, 167, 269, 274
Royal Commissions, 15
Royal Society of Canada, xiv, 3, 52, 67, 83, 105, 279
Russell, Dora, 248-9
Russell, Henry Norris, 37, 100
Rutherford, Ernest, 117, 228
Rutkowski, 50, 59
Ryback, Peter, 128
Rystrom Observatory, 214-5

Sabatini, Denise, 146, 263
Sagan, Carl, 274
Saint John Astronomical Club, 246-7
Saint John's, 88, 125, 129, 157, 160, 175, 234, 247-9
Saint Mary's University, 166, 244-5
Sampson, Russ, 137
San Francisco Sidewalk Astronomers, 224
Sarnia, 175, 265-6
Saroch, Zdenko, 265
Saskatchewan Astronomical Society, 202
Saskatchewan Science Centre, 204
Saskatoon, 79, 80, 121, 123, 128, 157, 174-5, 213-6
Satellites, artificial, 96, 117, 144, 165, 176, 201, 211, 232, 270
Satterly, John, 65
Saturn, 45, 110, 139-141, 193, 232
Saunders, 253
Savigny, Annie, 45, 48

Scatliff, John, 200-1
Schlesinger, Frank, 37, 186, 194
Schneider, Frank, 193
Schools, 25, 109, 115, 123-8 *passim*, 130, 140, 159, 181, 191, 199-202, 213, 216, 229-237 *passim*, 240, 246, 248, 254-5, 258, 260, 264, 266
Science fairs, 55, 125-6, 130, 211, 213
Science North, 125
Scientific American, 3, 83, 151, 159, 177, 205
SCITEC, xii
Scott, Alfred, 272-3
Scouts and Guides, 60, 115, 120, 122, 125-7, 143, 153, 159, 191, 194, 198-200, 204, 211-3, 229-232, 235, 240, 245-9 *passim*, 254, 258-266 *passim*, 274
Seaforth, Ontario, 194, 281
Seal, 12, 55
Seattle Astronomical Society, 223-4
Secrétan, 240
Secretary, 14, 70
Seismology, 99, 181, 216, 227
Seminary of St Hyacinthe, 64
Senior membership, 39
Service Award, vii, 55, 57-8
Serviss, Garrett P, 153
Shapley, Harlow, 37, 47, 77, 100, 117, 166, 180, 228, 230, 269
Sharpe, Steven, 149
Shaw, Norman, 230
Shearmen, T. S. H., 136, 152
Shelton, Ian, 36, 114, 150, 174
Sheppard, Ross Stanley, 56
Sherwood, William, 56
Shinn, 61, 96-7, 115, 199
Shrum, Gordon, 217, 221
Sidereal Messenger, 83, 164
Simard, Henri, 66, 135
Simcoe, Ontario - see also Wadsworth, 45, 138, 177, 281
Simon Newcomb Award, see Newcomb
Simpson, (Honorable) Leonard J., 63, 65
Sinclair, Angus, 65, 84
Sisman, Frank, 59-60
Sky & Telescope, vii, 87, 97, 125, 157, 205
Slides, xii, 84, 88, 120, 128, 130, 166-7, 172, 177, 202
Slocum, Frederick, 141
Smallwood, Charles, 227
Smith, C. C. and Mrs, 180
Smith, Larratt, 2, 12, 64, 83, 160,

280
Smith, Lewis, 141
Smith, Lyons, 33
Social activities, 167-8, 171-3, 185-6, 192, 213, 217, 221, 222, 230, 235, 247, 253-4, 270-1
Société Astronomique de France (SAF), xiv, 86-7, 126, 237, 239, 279
Société d'Astronomie de Montréal (SAM) - see also Centre francophone, ix, xiii, xiv, 126, 135, 236-8
Solar Eclipses, 2, 22, 53, 71, 79, 99, 113, 151-8, 188, 198, 200, 202, 209, 211, 216-7, 223, 226-7, 241-2, 251, 253, 256, 265-6, 272-5 *passim*
Solar-terrestrial relationships, 3, 20-1, 102, 146, 227
Sorensen, 50, 57
Southam Observatory, 174, 223, 225
Southam Press Fund, 77
Space exploration - see also Satellites, xiii, 103, 105, 111, 113, 117, 133, 144, 151, 158, 165, 167, 210, 238, 254, 277
Sparling, Charles, 25, 65, 105
Speakers, exchange, 178, 192, 207, 211-4, 254, 258, 265
Speakers, visiting, 100, 165, 177-8, 180-1, 186, 188, 190, 194, 197-9, 202, 204, 211, 217, 219, 228, 237, 248, 253-5, 257, 260, 262, 267, 269, 271, 274
Special Projects, 80, 178, 192, 202, 257-8
Speck, Robert, xiii
Spectroscopic work, 11, 21, 38, 77, 79, 94, 103-5, 108, 136-7, 141, 145, 168, 180, 206, 220, 259
Spencer Jones, (Sir) Harold, 37, 271
Spratt, Christopher, 60-1, 97, 105, 142, 149
Stackhouse, 266
Stanford, F. T., 65, 160
Star - maps, 108-9, 120, 143, 240, 251
- tables, 109
Starfest, 133, 258, 265-6, 275
Starkman, Glenn, 56
Starnights and Star parties, 79, 121-5 *passim*, 133-4, 139, 182, 184, 188-190, 194, 199-201, 204, 207, 211-3, 216, 218, 220, 223-4, 230-7 *passim*, 239-40, 243, 245, 253-5, 257-8, 260, 263-5, 267, 271-5 *passim*

Stebbins, Robert, 1, 279
Steel Company of Canada, 16, 193
Stellafane, 133, 135, 160, 182-3, 231, 237, 254, 258, 260, 275
Stetson, Harlan T., x-xi
Stewart, Louis B., 65, 100, 161
Stewart, R. Meldrum, x-xi, 64-7 *passim*, 166, 179-180
Stilwell, Walter, 210-212, 281
Stoddart, D. G., 53
Stone, Becca, 44, 280
Storch, R. A., 165
Strasenberg Planetarium, 271
Stratton, C. and J., 185-6
Strong, J. J., 247
Struve, 37, 92, 101
Student membership, see Youth
Study groups, 8, 195, 233, 270-1, 275
Stupart, (Sir) Frederic, 3, 5, 15, 65, 161, 185-6, 268
Subscriptions, 106
Summerside Astronomy Club, PEI, 247
Sun - see also Solar ..., 4, 20, 59, 96, 101-3, 111, 125-6, 131-2, 136-7, 139, 146, 161, 164, 186-7, 194, 200-2, 205, 208, 228-9, 232, 240, 263, 272
Sun Life Company, 49, 197, 228-9
Supernovae, 114, 150, 174
Supplement to the *Journal*, 100
Surcharges, 11, 178
Surveying - see also Canadian Geological Survey, Geodetic Survey and Yukon Survey, 179-80, 197, 278
Surveys of members, 35-6, 40, 207, 280
Sydney, 174-5
Sykes, Paul, 128

Tanner, Richard, 56, 181
Tapping, Ken, 281
Tatum, Jeremy, 91, 96, 103
Tavistock, Ontario, 281
Taylor, Dan, 258
Taylor, Mrs E. L., 50, 198
Taylor, Sarah, 48, 120, 143
Teaching of Astronomy - see also Courses, ix, xiii, 9, 15, 24, 79, 107, 128, 190, 199-201, 206, 208, 213, 252, 261, 280
Teece, Philip, 60
Tekatch, 225
Telescope making, 4, 14, 21, 28, 59, 64, 94, 96, 106-7, 125, 133, 135, 142, 159-160, 180, 182-3, 191, 193-4, 198-205 *passim*, 207, 210, 215-6, 219, 221-3, 229-245 *passim*, 254, 260, 266, 270, 275
Telescopes, 11, 15, 25, 60, 72, 80, 105, 111, 114, 125, 132, 147, 159-160, 177, 185-6, 189, 193, 201, 203, 205, 208, 210, 214, 218-221, 223, 228-9, 233, 238-248 *passim*, 251, 254, 257-8, 263-5, 269, 274-5
Television - see also Media, xiii, 129-130, 167, 224, 233, 254-5, 267, 276
Theft, 203, 208, 232, 238
Thompson, 4, 60, 148
Thomson, Andrew, 66, 71, 99
Thomson, (Sir) Joseph J., xii, 197
Thomson, Malcolm, 7, 66, 143, 180-1, 242, 266, 278
Thunder Bay, 175, 266-7, 282
Time-reckoning, 7, 15, 67, 179, 227, 241
Tindall, David, 66, 267
Todd, Mrs E., 27
Todhunter, James, 65, 160
Tokyo Astronomical Observatory, 127
Tombaugh, Clyde, 258, 262
Topham, Bertram, 60, 148, 269
Toronto, ix, xii, 5, 19-21, 25, 75, 135, 160
Toronto Astronomical Club and Society, ix, 4, 5, 20, 22-5, 74, 101, 177, 179, 186, 216, 251, 266, 280
Toronto Centre, vii, xiv, 3, 4, 28-9, 32, 39, 53, 57, 60, 64, 67-71, 76, 79, 87-8, 121-9 *passim*, 136, 140, 148, 157-162 *passim*, 170-4 *passim*, 267-76, 280, 282
Toronto General Hospital, 137
Toronto Observatory - see also Meteorological Observatory, 2, 54, 160-1
Tothill, Tom, 133
Townsend, G. H., 231, 233-4
Transactions (of the Society), 6, 25, 27, 83, 91-3, 179
TransAlta Utilities, 213
Trant, William, 202
Travel assistance and expenses, 15, 33, 73, 165, 167, 178, 197, 202, 268
Treasurer, 14, 15, 68-70, 168
Trombino, Donald, 59
Troyer, Frederic, 66, 159, 272-3
Turnbull, Mungo, 21, 22, 165
Turner, Herbert H., 37, 198, 279
Tutte, William, 66
Tyrell, John B., 98

Unattached members, 12, 40-42, 178-9, 268
United States Naval Observatory, 83, 207
"Universe", 167
Université de Montréal, 15, 233-4, 237-8, 241
Université Laval - see Laval
Universities - see also specific names, 55, 99, 109, 119, 172
University Lowbrow Astronomers, 258
University of Alberta, 53, 145, 157, 204-210 *passim*
British Columbia, 56, 166
Calgary, 150, 174, 210-2, 221-2
Lethbridge, 132
Manitoba, 53, 196-9, 206, 252
Michigan, 255, 257
New Brunswick, 79, 244
New Mexico, 56
Ottawa, 242
Regina, 204
Saskatchewan, 79, 132, 214-5, 222
Toronto - see also Dunlap Observatory, 2, 9, 11, 24-9 *passim*, 44, 47, 55-7, 67, 86, 93, 102-3, 109, 114, 118, 150, 153, 161-2, 168, 273
Toronto Press, 33, 105-6
Victoria, 97, 119, 217-9
Waterloo, 53
Western Ontario, 251-4
Winnipeg, 201, 206
Urania, 12
Uranus, 141, 151
Uxbridge, Ontario, 281

Van de Hulst, H C, 37, 271
van den Bergh, Sidney, 103, 142
van Steenburgh, W. E., 63, 66
Vancouver, 53, 125, 128-9, 157, 166, 173-5, 218, 220-5
Vanderbyl, Leo, 220-1
Variable stars - see also AAVSO, ix, 4, 9, 38, 98, 102, 108, 111, 114, 142, 147-9, 181, 183, 202, 224, 228-230, 232, 240, 254, 264, 275
Vatican Observatory, 165
Veeder, W. A., 132, 146
Venus, 2, 59, 60, 71, 114, 137, 140, 161
Veverka, Joseph, 117
Vice-President, 13-14, 64-68
Victoria, ix, xii, 11, 14, 19, 29, 36-8, 50, 53-4, 61, 77, 92, 94, 104, 119-121, 125, 129, 137, 140, 151, 153, 166, 173, 175, 214, 216-221, 239, 281
Video, 121, 167, 224, 247
Ville-Marie Observatory, 229-230, 234, 281

Wadsworth, J. J., 45, 138-9, 159, 164, 177
Wales, William, 98
Warren, William, 232
Waterfield, W. H. F., 149
Wates, Cyril, 53, 60, 205-6
Watson, Albert, 65, 100-1, 216, 279
Watson, Michael, 33, 118, 158, 273, 275
Wayne University, 255
Webb, (Canon) T. W., 15, 187
Webber, 65, 134
Wehlau, William, 16, 255
Welch, Douglas, 35, 142, 191
West Shore Railroad, 134
Westcott, Guy, 129
Western Amateur Astronomers, 174
Western Development Museum, 204, 216
Westinghouse Company, 193
Westoby, H., 194-5
White dwarfs, 166
Whitehorne, Mary Lou, 50, 60-1, 105, 280
Whyte, Anthony, 208
Wiarton, 282
Wilfrid Laurier University, 264-5
Williams, Frederick, 53
Williams, K. Miss, 14
Williamson, Dorothy and Ralph, 28-9, 79
Williamson, Isabel, 49, 57, 60, 132, 139, 156-7, 229-231, 281
Willistead Library, 255
Wilson, (Sir) Adam and Lady, 25, 72, 83, 160, 165
Wilson, Christine, 142
Wilson Coulee Observatory, 127, 136, 212-3
Winder, Daniel, 21, 23
Windsor, 43, 126, 166, 175, 253, 255-8, 265-6
Winger, James, 193
Wingham, T. H., 143, 190
Winnearls, Bert, 9
Winnipeg, ix, xii, 11, 15, 50, 61, 96, 121, 123, 125, 128-9, 133-4, 155-8, 165, 174-5, 196-202, 251-2, 281
Witton, (Honorable) H. B., 188
Women - see also specific names, 12, 44-50, 103, 111, 143, 197, 280
Wonnacutt, Thomas, 253
Woods, Jack, 248
Woodstock, 59, 152, 251, 282
World War I, 42, 46, 67, 69, 75, 95, 109, 119, 148, 153, 160, 162, 180, 195, 197, 203, 206, 209, 210, 217-8
World War II, 29, 36, 38, 42-44, 50, 54, 68, 94, 110, 139, 156, 181-2, 195, 198-9, 203-4, 218, 222, 229, 239, 240, 271
Wright, James, 210
Wright, Kenneth O., 14, 54, 56, 66, 93, 105, 203
Wunder, W., 65

X-ray, 67, 168
Xerox Corporation, 106

Year Book and Almanac, 3
York, Duke and Duchess, 5
York University, 56, 274
"You and the Universe" Symposium, vii
Young Men's/Young Women's Christian Association (YM/YWCA), 25, 120, 185, 217
Young, Reynold K., 56, 65, 79, 93-4, 114, 119, 128, 143, 151-2, 156, 159, 188
Youth, 7, 12, 38-9, 49, 123, 125-8, 130, 132, 160, 180-2, 191, 193-4, 199-200, 207, 211, 218, 222, 246, 254, 275
Yukon Survey, 167

Zalcik, Mark, 146,
Zarins, John, 30
Zodiacal light, 20, 135